QGIS FOR AGRICULTURAL STATISTICS

A PRACTICAL GUIDE

DECEMBER 2025

ASIAN DEVELOPMENT BANK

ADB

6 ADB Avenue, Mandaluyong City, 1550 Metro Manila, Philippines
Tel +63 2 8632 4444; Fax +63 2 8636 2444
www.adb.org

ISBN 978-92-9277-615-2 (print); 978-92-9277-616-9 (electronic); 978-92-9277-617-6 (ebook)
Publication Stock No. TIM250542-2
DOI: http://dx.doi.org/10.22617/TIM250542-2

Notes:
In this publication, "$" refers to United States dollars. ADB recognizes "Korea" as the Republic of Korea and "Vietnam" as Viet Nam.

Cover design by Cleone Flores Baradas.

Featured on the cover, starting from left to right, are images capturing the remote sensing training session and fieldwork in Viet Nam (photo by JFPR TA 6856 Project Team).

Contents

Tables and Figures

Tables

Figures

Preface and Acknowledgments

Geospatial data, or data disaggregated by geographical locations, are essential to provide governments and companies with a more detailed context to nationally aggregated information. They enable users to understand the quantitative variations across physical space, which are often unavailable from official statistics. Such data open the way for more textured analyses and better targeted government policies and market decisions.

Geospatially granular data are, however, very costly to compile through surveys and administrative data. These data sources also lack timeliness and require frequent updating. One solution is the use of satellite images, trained on a representative sample of ground truth data, with machine learning and other artificial intelligence methods. This approach economizes on the time and effort in data collection, while achieving high accuracy. It is thus of significant potential benefit to national statistics offices working under tight constraints.

This handbook, *QGIS for Agricultural Statistics: A Practical Guide*, was prepared by the Data Division of the Economic Research and Development Impact Department at the Asian Development Bank. It aims to assist national statistics offices and agricultural researchers in harnessing open-source geographic information system (GIS) tools for spatial data analysis and survey design in the context of agricultural statistics. Although the applications of GIS in this handbook feature agricultural data, this guide is also useful for data practitioners in national statistical systems, as well as in the private sector, who wish to learn GIS methods more generally.

The preparation of this handbook was supported by the Japan Fund for Prosperous and Resilient Asia and the Pacific (JFPR) as part of Technical Assistance (TA) 6856: Development of New Statistical Resources and Building Capacity in New Data Sources and Technologies. We are grateful to the People and Government of Japan for supporting the research and development activities and the statistical capacity building in the TA. We also extend our sincere appreciation especially to Manabu Takami and Ella Cecilia R. Arienda, of the Climate Change and Sustainable Development Department, for their valuable support in the implementation of JFPR TA 6856.

The publication team was led by Takaaki Masaki, with key contributions from Anthony Burgard, Anna Christine Durante, and Pamela Lapitan. The guidance and support from Mahinthan Joseph Mariasingham, Yating Ru, and Macario Cordel II were instrumental in advancing the team's crop-mapping initiatives, which yielded several knowledge products—including this handbook—that demonstrate the practical applications of geospatial tools in agricultural statistics.

We extend special thanks to Arturo Y. Pacificador Jr., whose expertise in sampling design and area frame sampling greatly enriched the methodological foundation of this handbook. His insights were instrumental in developing the modules on spatial sampling and survey design, ensuring their technical rigor and operational relevance.

We also appreciate the National Statistics Office of Viet Nam for its collaboration and provision of data, which served as the foundation for the case study featured in this handbook. Their support was essential for demonstrating the applied use of QGIS in agricultural mapping and analysis.

The copyediting, typesetting, and design of this handbook were made possible through the expertise of Joanne Gerber (editing); Cleone Flores Baradas (cover and graphic design); and Jonathan Yamongan (typesetting). Their contributions ensured the clarity, visual appeal, and professional presentation of the publication.

Acknowledgment is also due to the open-source communities and platforms whose datasets and tools were used throughout this handbook. These resources enabled the development of practical exercises and demonstrations that reflect real-world agricultural challenges.

Elaine S. Tan
Director, Data Division
Economic Research and Development Impact Department
Asian Development Bank

Abbreviations

AOI	Area of interest
CRS	Coordinate reference system
CSV	Comma-separated values
EORC	Earth Observation Research Center
EPSG	European Petroleum Survey Group
FAO	Food and Agriculture Organization of the United Nations
GDAL	Geospatial Data Abstraction Library
GIS	Geographic information system
GPS	Global Positioning System
GPX	GPS Exchange Format
JAXA	Japan Aerospace Exploration Agency
km	Kilometer
km^2	Square kilometer
LULC	Land use and land cover
m	Meter
m^2	Square meter
NVDI	Normalized Difference Vegetation Index
QGIS	Quantum Geographic Information System
UN DESA	United Nations Department of Economic and Social Affairs
URL	Uniform Resource Locator
UTM	Universal Transverse Mercator
WGS 84	World Geodetic System 1984

Introduction

The integration of geographic information systems (GIS) into agricultural statistics has become increasingly vital, as data becomes central to decision-making. With the increasing availability of open-source platforms such as QGIS and of satellite-based remote sensing data,[1] national statistics offices, agricultural researchers, and planners now have access to advanced tools that support visualization, analysis, and interpretation of spatial data with greater precision and accessibility. This handbook offers a comprehensive guide for applying QGIS to the production of agricultural statistics. It is structured across eight modules that progressively build users' capabilities—from foundational concepts to advanced spatial analysis. This handbook targets professionals in agricultural statistics with no prior GIS background, including statisticians and agricultural economists seeking practical, work-ready GIS skills. The only requirement is access to a computer on which the free QGIS software can be installed.

Upon completing this handbook, learners will be able to:

- work with and map spatial data in QGIS,
- build and manage geodatabases,
- perform spatial analyses and calculations with vector and raster data, and
- design and construct area frame samples for agricultural surveys.

The handbook begins by establishing a conceptual foundation based on GIS through **Module 1**, which introduces users to the QGIS environment. Users are guided through the structure of a GIS database; the installation and setup of QGIS; and the navigation of its interface, including plugin management. These skills will enable users to begin working with spatial data and prepare them for more advanced applications in agricultural mapping and analysis.

Module 2 focuses on geospatial data models, coordinate reference systems (CRSs), and geodatabase management. Users learn to work with both vector and raster data formats, configure CRS settings, and organize spatial layers within a structured file system. These skills are essential for ensuring spatial consistency and data integrity when integrating multiple datasets. The ability to manage and style spatial layers establishes a foundational framework for more advanced tasks in later modules, such as raster analysis, land cover classification, spatial sampling design, and the creation of map outputs for agricultural surveys.

Module 3 introduces key concepts and practical tools for handling raster data within the QGIS environment. It focuses on preparing and managing raster datasets, including importing, clipping, and converting raster layers into vector formats. The module highlights the value of remote sensing in capturing land use and environmental patterns that inform agricultural monitoring and sampling

1 The original name of the system was "Quantum GIS" (for "Quantum Geographic Information System"). However, it was later renamed simply "QGIS."

design. These skills support subsequent modules that apply land cover data to spatial analysis and survey planning.

A central theme throughout the handbook is the effective use of attribute tables, which link geographic features to statistical data. This is explored in detail in **Module 4**, where users are guided through operations such as editing attributes, calculating new variables, and merging tabular data with spatial boundaries. These techniques are vital for producing analytical outputs such as area estimates and thematic visualizations, which support agricultural monitoring and policy development.

Mapping and cartographic visualization are emphasized as key competencies. **Module 5** guides users in creating thematic maps, incorporating essential map elements such as legends and scale bars, and in preparing outputs suitable for reporting and presentation. These skills ensure that spatial data is not only analyzed but also communicated clearly and effectively to stakeholders.

Module 6 introduces the integration of GPS data into spatial analysis to improve agricultural data collection and analysis. It covers the basics of GPS technology and guides users through the importation of GPS coordinates; digitization of geographic features; and the construction of spatial representations such as points, lines, and polygons. The module also includes techniques for creating polygons from GPS points and measuring distances between features, which are essential for precise mapping and crop monitoring.

Module 7 equips users with essential GIS vector-analysis tools for managing spatial boundaries in agricultural surveys. It introduces operations such as splitting and merging polygons, correcting geometry errors, calculating areas, and editing attribute tables. The module emphasizes working with spatial units commonly used in survey planning, showing users how to calculate areas and count features within these zones to support accurate mapping, resource allocation, and data analysis.

Building on these foundations, **Module 8** explores applications of GIS in agricultural survey design, with a focus on spatial sampling and the integration of remote sensing data. Users learn how to construct stratified sampling frames using raster-derived land cover information, using such tools as reclassification, zonal statistics, and zonal histograms to extract pixel-level indicators and identify dominant land cover types. These techniques are particularly valuable in agricultural contexts where field data collection is limited, costly, and logistically challenging.

The module further shows users how to leverage open-source raster datasets to extract agricultural indicators and generate maps that can be validated against ground truth data. These techniques enhance the accuracy of crop area estimation and land cover classification, contributing to more reliable and evidence-based agricultural statistics. By combining spatial sampling with remote sensing analysis, users gain practical tools for scaling GIS applications across diverse agricultural landscapes, while maintaining consistency and analytical robustness—culminating in a multistage sampling design that supports the broader goals of the handbook.

Throughout the handbook, the province of An Giang, in Viet Nam, serves as a study area, providing the geographic context for exercises and demonstrations.[2] It is also the focus of a study published by the

[2] The maps of An Giang, Viet Nam, reflect administrative boundaries prior to the province's merger with Kien Giang Province in June 2025, and so do not represent the newly configured provincial boundaries.

Asian Development Bank (ADB) in 2025 titled *Mapping Rice in Viet Nam: Estimating Rice Area Using Remote Sensing and Area Frame Survey Methods in An Giang Province*, which demonstrates a scalable approach to mapping rice cultivation using satellite remote sensing and probability-based ground truth sampling. Shapefiles, raster data, and other datasets are sourced from open-access platforms, which are listed in the handbook; some can also be accessed through the shared folder[3] linked in the relevant modules. These datasets are used as frameworks for illustrating GIS techniques and outputs.

While the level of analysis may vary depending on the user's context, the overall GIS process remains consistent. For example, in Viet Nam, the typical level of analysis is based on specific administrative units. However, the steps demonstrated in this handbook can be applied to other levels of administrative or geographic boundaries, depending on the needs of the user.

By the end of the handbook, users will have developed a robust skill set in QGIS—from basic mapping and data management to advanced spatial analysis. These competencies will enable them to produce high-quality spatial outputs, support evidence-based agricultural decision-making, and contribute to the development of agricultural statistical systems.

3 For access to shapefiles and other open-access datasets referenced in the handbook, go to the GitHub repository: https://github.com/ADB-Data-Division/QGIS-Handbook-2025

Module 1: Introduction to Geographic Information Systems

In recent years, the availability of open-source geospatial tools and satellite-based remote sensing data has expanded significantly, making these technologies more accessible to a wider range of users. Geographic information systems (GISs) provide powerful capabilities for visualizing, analyzing, and interpreting spatial data—capabilities that are increasingly essential for understanding agricultural patterns, monitoring crop areas, and supporting evidence-based decision-making in the sector.

This module introduces users to the foundational concepts of GIS and the practical use of QGIS, a free and open-source GIS software solution. By equipping users with the skills to access and utilize geospatial data, the module aims to strengthen the integration of spatial analysis into agricultural research and statistical systems, ultimately contributing to more informed policy formulation.

The topics covered in this module include:

1.1 What is a Geographic Information System?
1.2 The Concept of the Database Structure
1.3 Introduction to QGIS
- 1.3.1 Why QGIS?
- 1.3.2 Software Installation Procedure
- 1.3.3 Getting Started with QGIS
- 1.3.4 QGIS Graphical User Interface
- 1.3.5 Plugin Manager

By the end of this module, users will be familiar with the core functionalities of QGIS and its plugins. These foundational skills will serve as a starting point for more advanced applications in agricultural mapping and spatial data analysis.

1.1 What is a Geographic Information System?

A GIS is a computer-based tool that creates, analyzes, manages, stores, and visualizes descriptive data related to any position on the earth's surface. These functions are bundled into a single software system that integrates various types of data from multiple data sources. GIS can synthesize a vast amount of data through thematic data layers to represent real surfaces at any scale, from local to global.

The application of GIS to connect "what" is happening and "where" things are happening in different sectors—such as business, infrastructure, agriculture, and the environment—is critical for understanding social, economic, and environmental issues. Integrating statistics and geospatial information will provide better insights than visualizing them separately (UN DESA 2019). In essence, GIS provides a means of linking tabular data (on attributes such as crop type, yield) to known spatial features such as point locations (e.g., farmer households), lines (e.g., rivers, irrigation canals), or areas represented by polygons (e.g., parcels and plot boundaries).

The data intelligence compiled by GIS is helpful in the public and private sectors for data-driven and evidence-based policymaking and decision-making. In agriculture, GIS enables the integration of diverse datasets such as land use, soil characteristics, and climate variables into a unified spatial framework. This supports data-driven policy formulation by revealing patterns and relationships that are not easily captured through traditional tabular data. When combined with remote sensing technologies, GIS allows for the timely monitoring of agricultural landscapes, offering insights into crop health, seasonal changes, and land productivity at various scales.

GIS and remote sensing have been increasingly adopted in agricultural statistics to improve data quality and efficiency. For example, satellite imagery and vegetation indices such as the Normalized Difference Vegetation Index (NDVI) are used to estimate crop areas and yields, while geospatial tools assist in designing sampling frames and guiding field data collection.

These technologies have been used in agricultural censuses, surveys, and monitoring systems to enhance coverage, reduce costs, and support policy formulation. Organizations such as the Food and Agriculture Organization of the United Nations (FAO) and the United Nations Department of Economic and Social Affairs (UN DESA) have documented the value of GIS in strengthening agricultural data systems and promoting sustainable development (FAO, n.d.; UN DESA, n.d.). Examples include sampling frame construction, through which GIS divides administrative areas into land segments, enabling the geocoding of farm holding locations. For stratification, land use and crop maps help in the allocation of samples, improving the efficiency and precision of estimates. During field operations, Global Positioning System (GPS) point locations can be used to verify survey locations and monitor progress in data collection. And GIS-calculated areas can provide objective measurements for validating owner-reported farm sizes (ADB 2024). These are some practical applications of GIS used by national statistics offices of countries in Southeast Asia and the Pacific for improving the agricultural survey life cycle.

1.2 The Concept of the Database Structure

A database is a structured collection of data that models the real world. Conceptually, a geographical database ("geodatabase") is a collection of structured thematic data layers, vertically organized as stacks within a common reference framework based on location to create an integrated data layer (Figure 1.1). The multi-data layers can be created outright, drawn from different data sources online, derived or extracted from an existing data layer, sourced from scanned and georeferenced hard-copy maps, or obtained from ground-truth datasets.

Figure 1.1: Map Layers in a Geodatabase

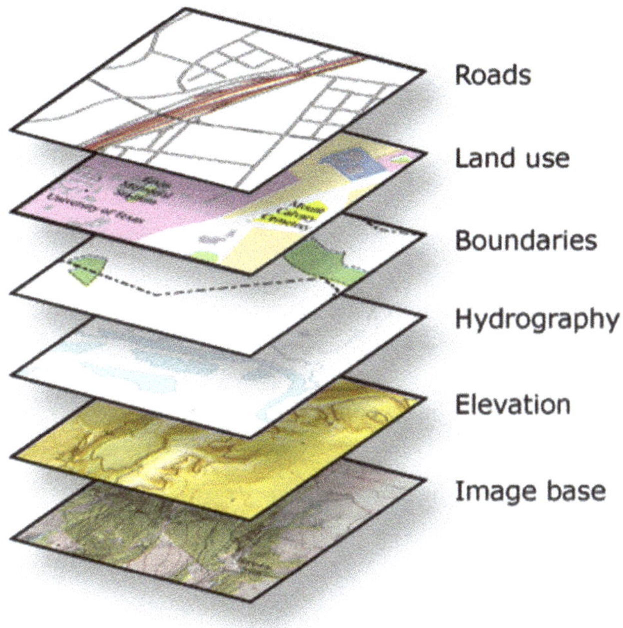

Source: Esri. Topics of: Fundamentals of GIS Data; Elements of Geographic Information. http://webhelp.esri.com/arcgisdesktop/9.2/printBooks_topics.cfm?pid=22.

The design and content of a geodatabase depend on the user's application. A common digital enumeration area database for census operations may contain these data layers:

(i) satellite imagery,
(ii) regional boundaries,
(iii) district boundaries,
(iv) statistical area boundaries (e.g., enumeration areas, census districts),
(v) dwelling frame,
(vi) landmarks or places of interest points (e.g., schools, health facilities),
(vii) roads,
(viii) rivers, and
(ix) cadastral or parcel boundaries.

A more detailed discussion on the types of GIS data, such as points, lines, and polygons (vector data), and on raster formats such as satellite imagery, is included in Module 2 under the overview of geospatial data models. This module elaborates on the distinction between vector and raster data formats, helping users understand how different data types are represented and used in spatial analysis.

A geodatabase integrates spatial data with attribute (tabular) data. Spatial data are typically represented by feature classes such as point (e.g., location of weather stations), multipoint, line (e.g., irrigation canals), multiline, polygon, multi-polygon (e.g., farm boundaries), and continuous surfaces and imagery (raster data). Attribute data provides descriptive information about spatial features, and can be either quantitative (e.g., crop yield, area size) or qualitative (e.g., crop type, soil classification). Common attribute data types include integer, float, real number, decimal, text, and date. For example, attribute data might describe the crop type grown in a field or the yield recorded at a specific location. A geodatabase can be stored as a collection of data files or as a single spatially enabled relational database. A relational database consists of linked tables, allowing access to and querying of large volumes of information.

Each table is made up of rows and columns, with the rows representing individual records (e.g., a farm plot), and columns containing specific attributes (e.g., crop type, area, yield). Tables can be linked using fields with common values. In Figure 1.2, Table A contains geographic data for provinces in the Mekong Delta region of Viet Nam, while Table B presents agricultural productivity data. Fields such as **ADM1_PCODE** (province identification number) and **ADM1_EN** (province name) appear in both tables and can be used to link them.

Figure 1.2: An Example of a Relational Database

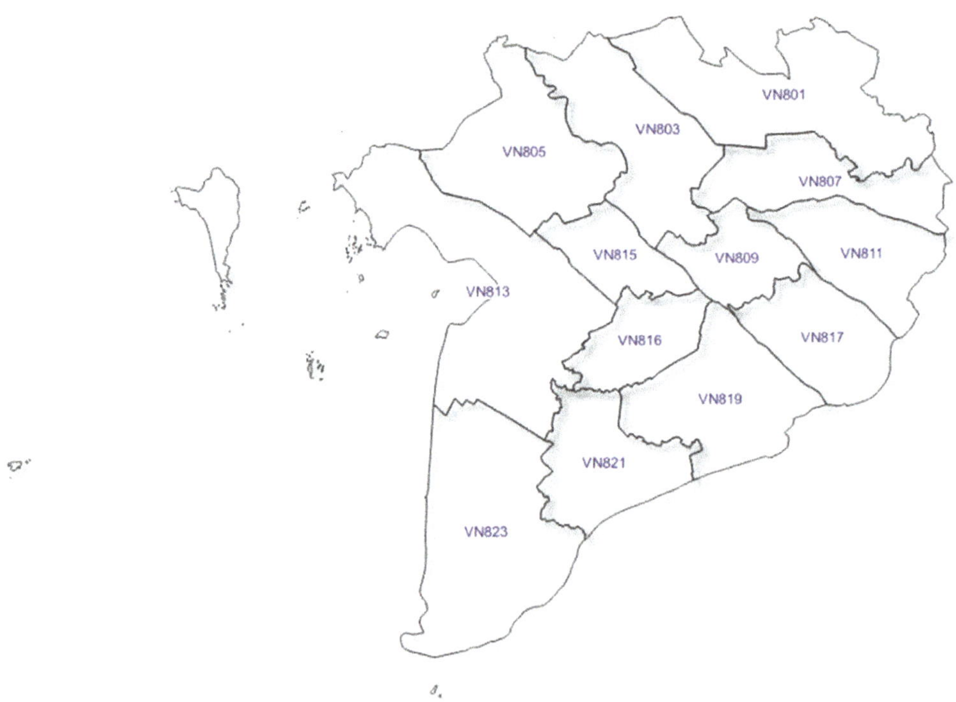

Sources: Asian Development Bank screenshot; Government of Viet Nam, National Statistics Office. n.d. *Statistical Yearbook of Viet Nam 2023*. https://www.gso.gov.vn/wp-content/uploads/2024/06/NIEN-GIAM-THONG-KE-2023_Ban-quyen.pdf; Humanitarian Data Exchange. Viet Nam—Subnational Administrative Boundaries. https://data.humdata.org/dataset/cod-ab-vnm.

Table A: Number of Administrative Units

ADM1_PCODE	ADM1_EN	Town/Districts	Communes
VN801	Long An	15	161
VN807	Tiền Giang	8	142
VN811	Bến Tre	10	139
VN817	Trà Vinh	10	85
VN809	Vĩnh Long	6	87
VN803	Đồng Tháp	9	115
VN805	An Giang	18	110
VN813	Kiên Giang	10	116
VN815	Cần Thơ	5	36
VN816	Hậu Giang	11	51
VN819	Sóc Trăng	12	80
VN821	Bạc Liêu	5	49
VN823	Cà Mau	9	82

Table B: Production of Main Agricultural Crops
('000 tons)

ADM1_PCODE	ADM1_EN	Paddy Rice	Maize	Sugarcane
VN801	Long An	3081.0	1.2	1.7
VN807	Tiền Giang	791.0	8.4	5.5
VN811	Bến Tre	109.7	0.6	
VN817	Trà Vinh	1110.3	23.3	147.1
VN809	Vĩnh Long	676.2	4.6	15.8
VN803	Đồng Tháp	3325.3	41.8	0.1
VN805	An Giang	4067.7	42.1	0.7
VN813	Kiên Giang	4561.9	2.4	62.8
VN815	Cần Thơ	1362.2	5.9	
VN816	Hậu Giang	1186.7	15.7	327.3
VN819	Sóc Trăng	2077.2	8.9	351.3
VN821	Bạc Liêu	1239.0	1.4	17.0
VN823	Cà Mau	568.4	3.3	13.1

Notes:

1. Table A contains the geographic data for the Mekong Delta region in Viet Nam. Table B contains the agricultural productivity data for each province in that region.
2. "ADM1_PCODE" indicates the field for the province identification number and "ADM1_EN" indicates the field for the province name.
3. The administrative boundaries shown reflect those of provinces in the Mekong Delta region prior to the merger implemented in June 2025. As such, they do not represent the newly configured administrative boundaries across the region.

Sources: Asian Development Bank screenshot; Government of Viet Nam, National Statistics Office. n.d. *Statistical Yearbook of Viet Nam 2023*. https://www.gso.gov.vn/wp-content/uploads/2024/06/NIEN-GIAM-THONG-KE-2023_Ban-quyen.pdf; Humanitarian Data Exchange. Viet Nam—Subnational Administrative Boundaries. https://data.humdata.org/dataset/cod-ab-vnm.

In a relational database, tables do not need to have the same number of rows to be linked. As long as a common or shared field exists—such as province ID or farm code—relationships between tables can be established (Esri, n.d.). This approach enables the integration of data from different sources while minimizing duplication and maintaining consistency. Relating tables in this way helps reduce data redundancy and ensures data integrity, which is important when managing large datasets in agricultural statistics.

The common field is useful for two reasons: (i) It can be set as a primary key of a table to build a relationship with another table; and (ii) it can be used to query records easily and quickly. In database terminologies, the **primary key** is the unique identifier within a table, while a **foreign key** refers to the matching field in another table used to establish a relationship.

For example, in an agricultural database, one table might contain spatial data showing the boundaries of provinces, while another table holds crop production statistics. The field **ADM1_PCODE** (province ID) can act as the primary key in the geographic table (Table A) and as a foreign key in the productivity table (Table B). Linking these tables allows the mapping of crop yields by province, assessment of productivity across regions, and the generation of insights without duplicating data. This relational structure supports complex queries and ensures consistency, which is essential for geospatial analysis and agricultural planning (FAO, n.d.).

1.3 Introduction to QGIS

1.3.1 Why QGIS?

QGIS is free and open-source software for geographic information systems and remote sensing applications. It is known for its intuitive interface, cross-platform compatibility, and extensive feature set, including support for plugins and a strong global user community. The software is continuously evolving through community-driven development (QGIS Development Team, n.d.).

As QGIS is open-source software, it has major advantages over commercial software. For instance, aside from being freely accessible, QGIS allows users to run, copy, distribute, study, modify, contribute to, and improve the software (QGIS Development Team, n.d.). QGIS is licensed under GNU General Public License (GPL) version 2 (Free Software Foundation 1991). It can be downloaded freely from the internet and has similar functionalities to those of commercial software.

QGIS supports a wide range of operating systems, including Linux, Unix, macOS, Microsoft Windows, and Android. It supports diverse vector and raster data formats, including database formats and functionalities. Beyond desktop use, QGIS includes server and web-based components that allow data publication through services offered by the Open Geospatial Consortium (OGC), such as the Web Map Service (WMS), Web Map Tile Service (WMTS), Web Feature Service (WFS), and Web Coverage Service (WCS). QGIS can be integrated with other open-source GIS platforms, including PostGIS, GRASS GIS, and Map Server (QGIS Development Team, n.d.).

The platform's functionality is extended through plugins, which are developed by individual users, developers, and organizations. These plugins, written in C++ or Python, can be installed directly from the QGIS Plugin Manager. While most plugins are freely available, some are available commercially (QGIS Development Team, n.d.)

Like other open-source software systems, QGIS has multiple versions, including the long-term release (LTR) version for everyday reliable use and the developer version (DEV) for testing new features. Although updates are frequent, the core functionality remains consistent. The LTR version is released once a year and is considered the most stable, making it ideal for corporate and academic uses. The latest release version, updated every four months, includes newly developed and tested features. The DEV version contains the most recent developments, but is usually unstable and intended primarily for testing and evaluating new features.

This handbook uses QGIS 3.34.14 Prizren (the LTR version for 2024), which at the time of writing, is the most stable released version maintained by OSGeo. The application can be downloaded from the official QGIS website. Additional references related to QGIS 3.34.14 and links to the online user community are also available on the website.

1.3.2 Software Installation Procedure

The following is a step-by-step guide on where to download QGIS and how to install it in Windows:

To download the software, an internet connection is required. Visit the following Uniform Resource Locator (URL) in a web browser: https://download.qgis.org/downloads/. The site provides access to both the latest release and LTR versions (Figure 1.3). For the Windows operating system, select the QGIS LTR version (QGIS 3.34.14 LTR). For other operating systems, such as macOS and Linux, compatible installers are available under the **Download for Windows** tab on the same website.

Figure 1.3: Downloading QGIS Long-Term Release

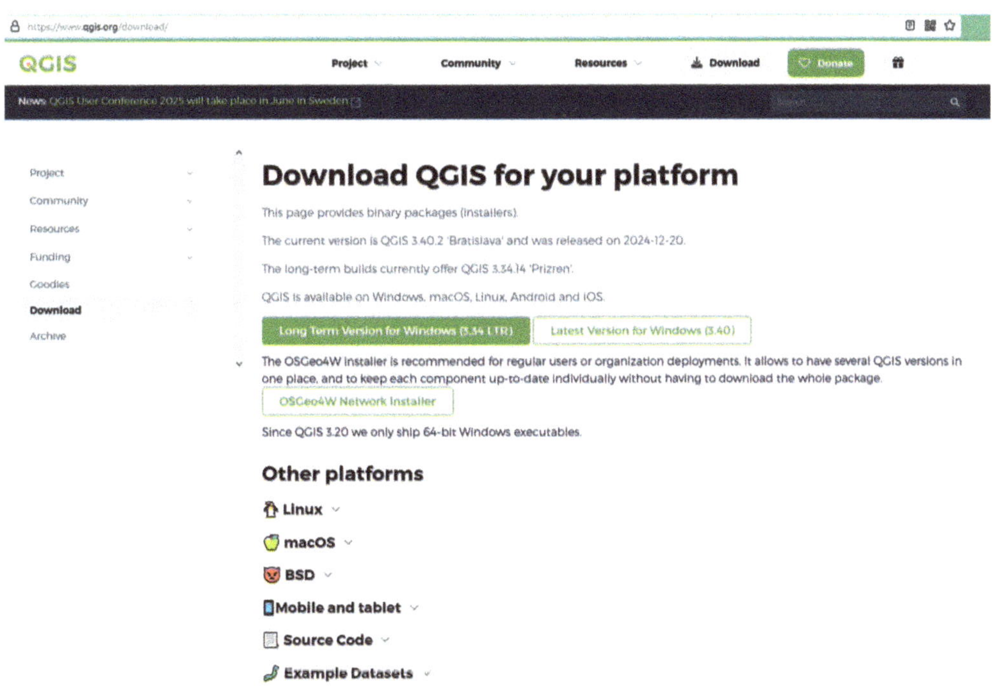

Source: QGIS. https://www.qgis.org/en/site/.

(i) Once the installer has been downloaded, navigate to the designated download folder on the computer. Double-click it to begin the installation, or right-click it and select Install (Figure 1.4).

Figure 1.4: QGIS Windows Stand-Alone Installer

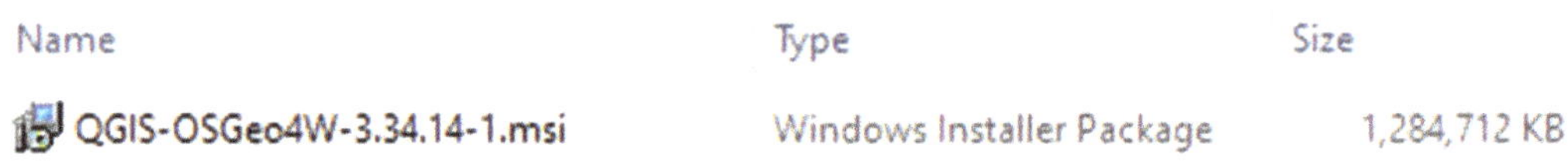

Source: QGIS. https://www.qgis.org/en/site/.

(ii) In the **Welcome to the QGIS 3.34.14 'Prizren' Setup Wizard** dialog box, click on Next (Figure 1.5).

Figure 1.5: Setup Wizard Dialog Box

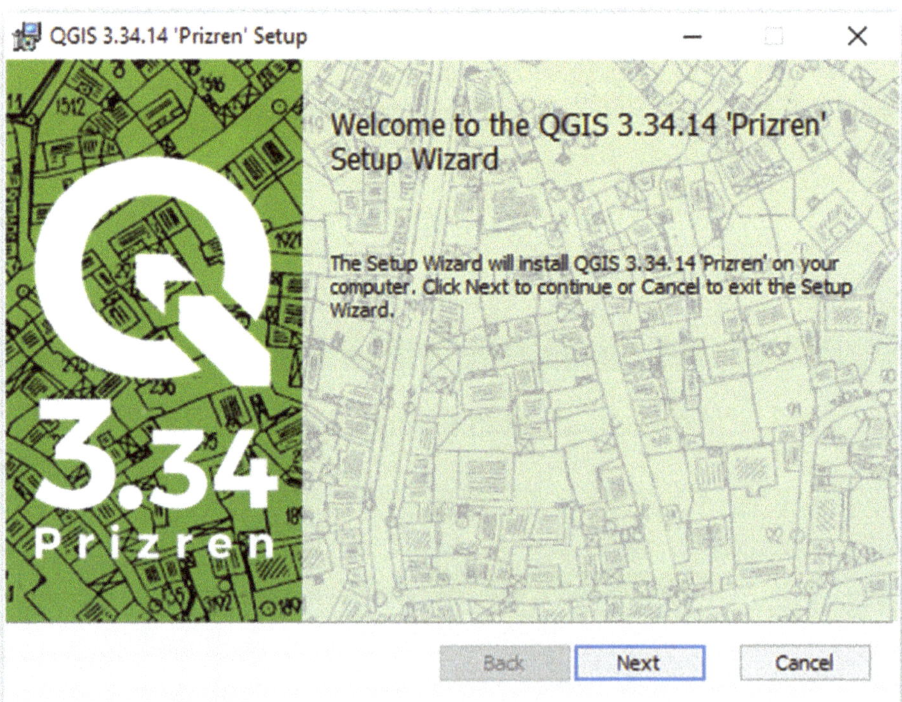

Source: QGIS. https://www.qgis.org/en/site/.

(iii) In the **End-User License Agreement** dialog box, click on the tiny box near the bottom to accept the terms in the License Agreement, then click the **Next** button (Figure 1.6).

Figure 1.6: End-User License Agreement Dialog Box

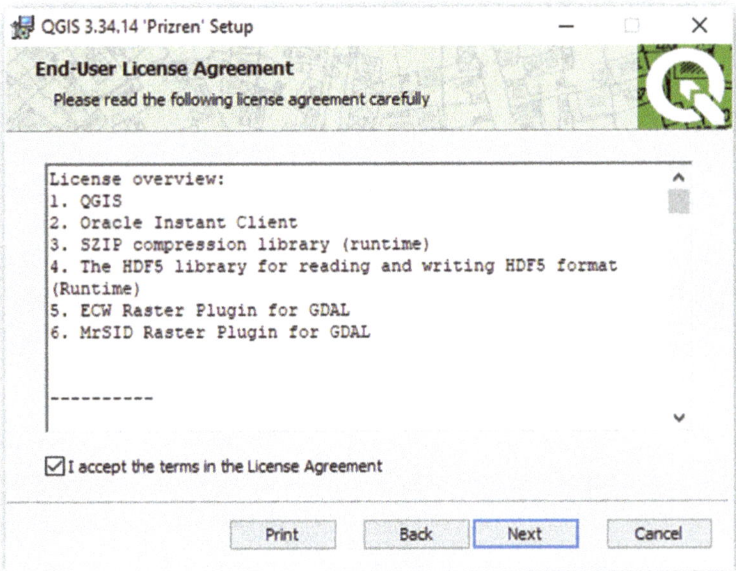

Source: QGIS. https://www.qgis.org/en/site/.

(iv) In the **Destination Folder** dialog box, click **Next** to install to the default folder or click **Change** to choose another folder, and then click **Next** (Figure 1.7).

Figure 1.7: Destination Folder Dialog Box

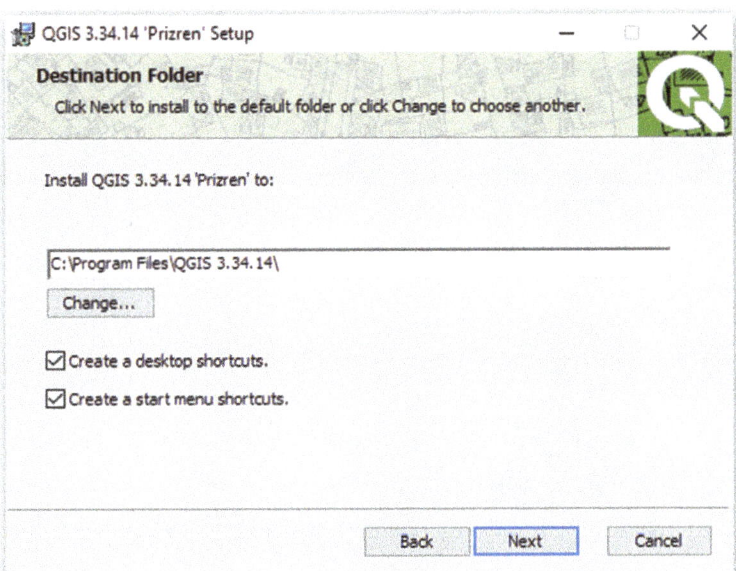

Source: QGIS. https://www.qgis.org/en/site/.

(v) Click **Install** to begin the installation. The installation normally takes 5–10 minutes to complete (Figure 1.8).

Figure 1.8: QGIS 3.34.14 Installation Dialog Box

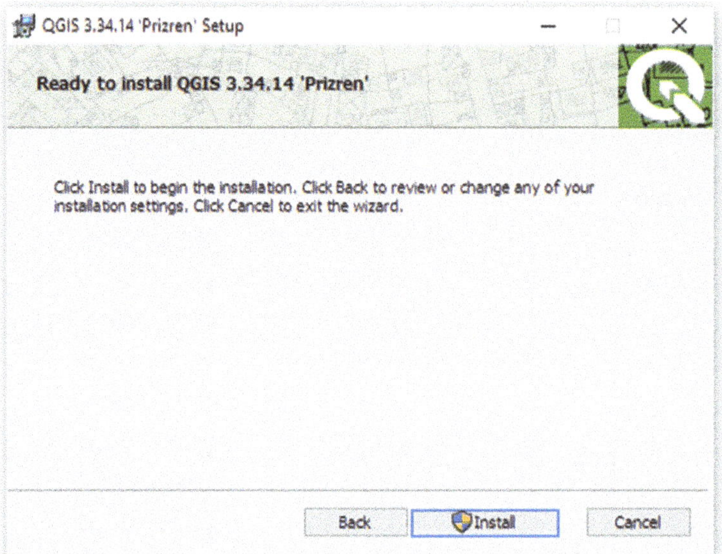

Source: QGIS. https://www.qgis.org/en/site/.

(vi) When the **Completed the QGIS 3.34.14 'Prizren' Setup Wizard** dialog box appears, the software has been successfully installed on the computer. Click **Finish** to exit. (Figure 1.9).

Figure 1.9: Installation Completed Dialog Box

Source: QGIS. https://www.qgis.org/en/site/.

(vii) Once the installation is completed, a new folder named "QGIS 3.34.14" will appear on the desktop. The folder contains shortcuts to the various programs included in the installation package (Figure 1.10). For the purposes of this handbook, QGIS Desktop 3.34.14 will be used.

Figure 1.10: Programs in the QGIS Installation Package

Name	Type	Size
GRASS GIS 8.4.0	Shortcut	2 KB
OSGeo4W Setup	Shortcut	2 KB
OSGeo4W Shell	Shortcut	2 KB
QGIS Desktop 3.34.14	Shortcut	1 KB
Qt Designer with QGIS 3.34.14 custom ...	Shortcut	2 KB
SAGA GIS 9.5.1	Shortcut	2 KB

Source: QGIS. https://www.qgis.org/en/site/.

1.3.3 Getting Started with QGIS

When QGIS is first launched, a welcome page appears (Figure 1.11). The **Recent Projects panel** is blank if no projects have been previously opened. To proceed, click **New Empty Project** to start exploring the interface.

Figure 1.11: Welcome Page of QGIS

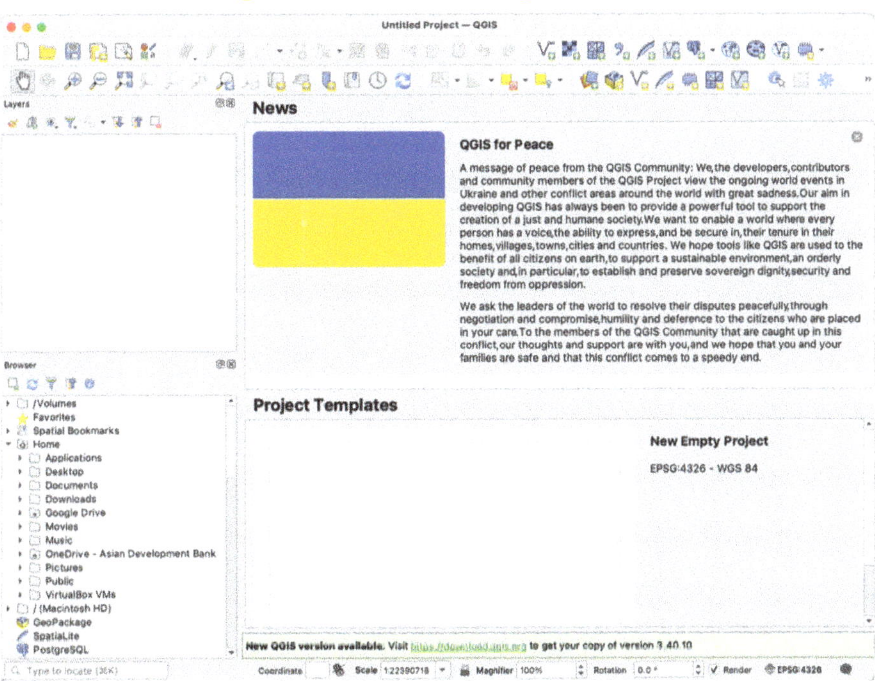

Source: Screenshot generated by Asian Development Bank using QGIS 3.34.14. 2025.

1.3.4 QGIS Graphical User Interface

The QGIS Graphical User Interface (GUI), also called the "main window," allows the user to interactively create, visualize, edit, store, and analyze data. It has five components: (i) menu bar, (ii) toolbars, (iii) panels, (iv) map display, and (v) status bar (Figure 1.12). With the exception of the map display, the components are equipped with graphical icons and text labels that facilitate map creation and data analysis within QGIS. When the mouse cursor hovers over icons, layers, or toolbars, their names are displayed—a feature that helps to quickly identify functions.

Figure 1.12: The QGIS Graphical User Interface

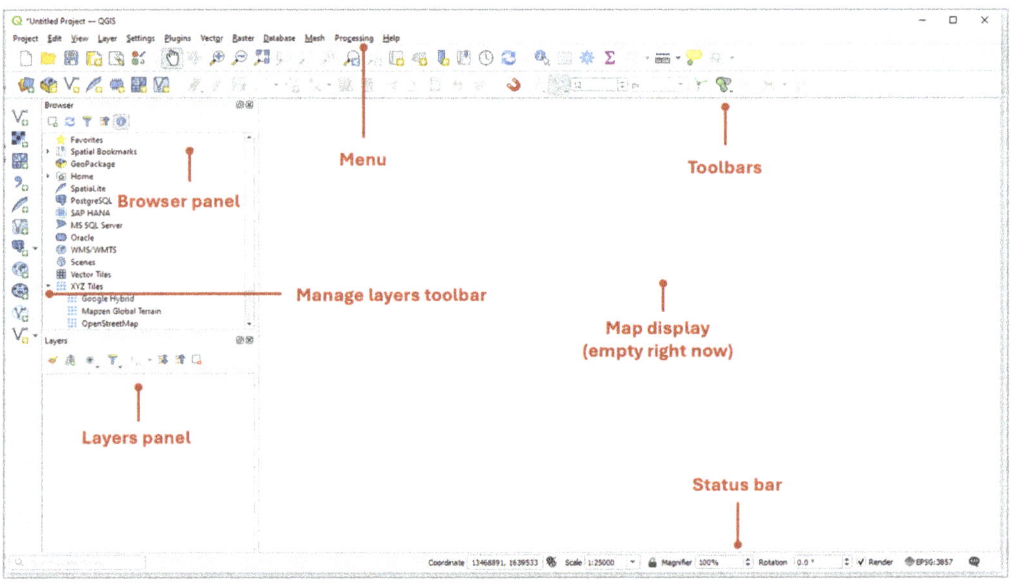

Source: Screenshot generated by Asian Development Bank using QGIS 3.34.14. 2025.

The main functions, composition, and commonly used features of the GUI components are described below:

Menu Bar

The menu bar, located at the topmost part of the QGIS interface, provides access to functions through hierarchical and cascading menus. There are 12 menus (Figure 1.13), each containing functions that are also accessible from the toolbars. Adding new plugins may add new options or a new menu to the menu bar.

Figure 1.13: The Menu Bar

Source: Screenshot generated by Asian Development Bank using QGIS 3.34.14. 2025.

Follow the instructions in the exercise box below to learn how to navigate menus and access different functions in QGIS:

- Click **Raster** in the **menu bar**. The Raster menu is opened. It contains options, functions, and operations specifically for raster data.
- The triangle ▸ to the right of each command indicates that there are cascading options related to the command. Hovering over the **Projections** command will reveal a submenu with additional options.
- Click the **Raster Calculator** command to open the Raster Calculator dialog box. This dialog box allows calculations to be performed on raster values based on pixel data. Click the close button (X) to close the dialog box.
- Click on other menus to explore. These menus will be used frequently in the next modules.

Toolbars

The toolbars each contain a set of tools that are represented by tool icons. The icons provide quick access to the functions and other interactive features. Related functions are grouped into each toolbar. The toolbars, not individual icons, are adjustable and movable around the interface by hovering the mouse over the dotted button, , on the left side of each toolbar, and holding a left-click.

Figure 1.14: Toolbars

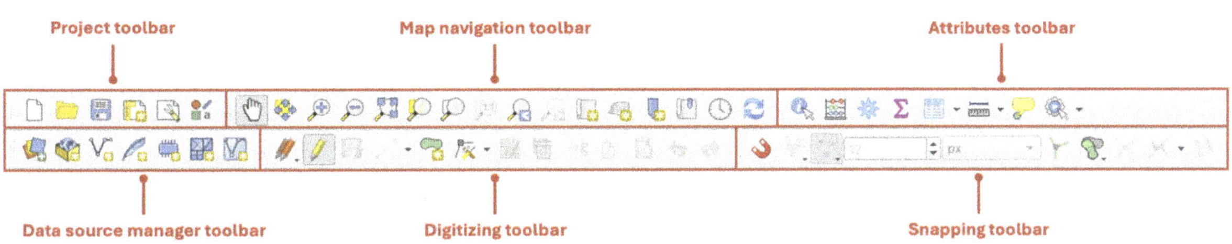

Source: Screenshot generated by Asian Development Bank using QGIS 3.34.14. 2025.

Some icons in certain toolbars may appear inactive or grayed out—for instance, those in the Digitizing toolbar (Figure 1.14). These icons are enabled when the relevant features or map layers are available or selected. Figure 1.14 shows the enabled toolbars in the interface, by default, when QGIS is launched.

To add or view other toolbars, go to the menu bar, click View | Toolbars or right-click on any empty space in the toolbars. A list of toolbars will be displayed (Figure 1.15). Click on the small box to check (or uncheck) and activate (or deactivate) any toolbar in the interface. This feature provides flexibility by allowing only frequently used toolbars to remain visible, while rarely used ones can be hidden. The same customization options apply to panels.

Figure 1.15: Activating the Toolbars

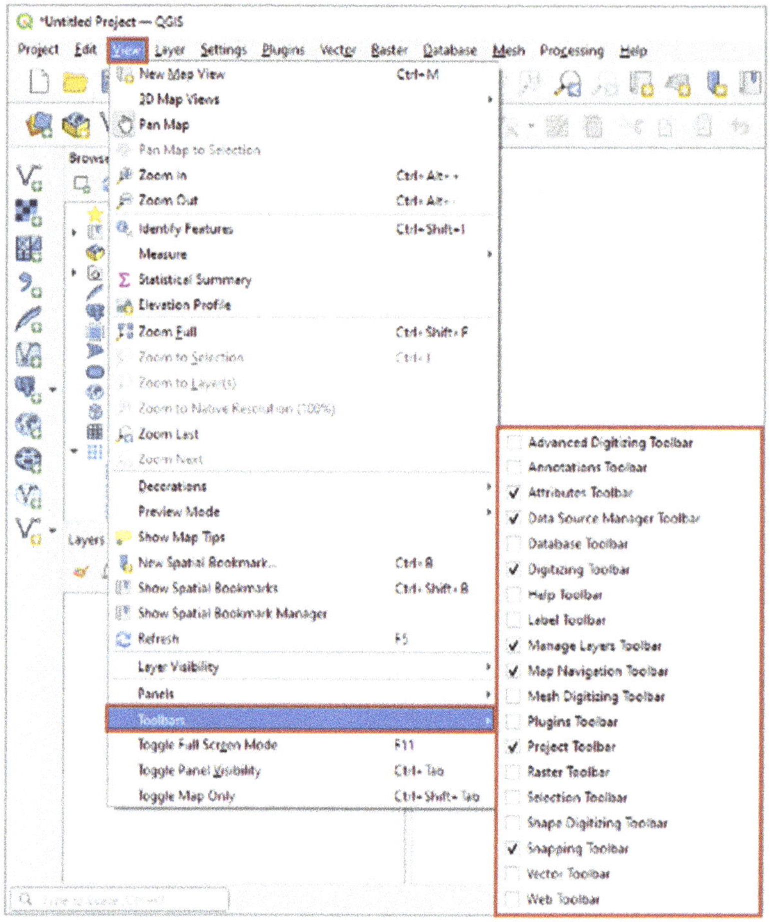

Source: Screenshot generated by Asian Development Bank using QGIS 3.34.14. 2025.

The functions of commonly used toolbars and icons are briefly discussed just below:

Project Toolbar

The Project toolbar contains tools used to manage QGIS projects. A QGIS project is a saved workspace that stores all the map layers, styles, layouts, and settings used in a mapping session, allowing work to be resumed or shared later. The paper icon is used to create a project. The icon allows a project to be opened from a directory. is used to save a project. opens the print layout, a widget used for laying out and adding elements to the map.

Figure 1.16: Project Toolbar

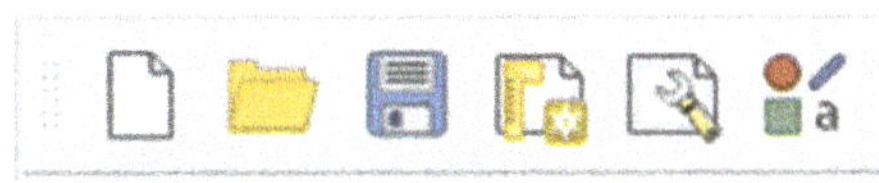

Source: Screenshot generated by Asian Development Bank using QGIS 3.34.14. 2025.

Map Navigation Toolbar

The Map Navigation toolbar contains tools that allow the user to interact with the maps. The icon is used to pan maps in the map view, which is found in the Map Navigation toolbar. is used to pan a map to selection. The icons are used to zoom in and zoom out of maps, respectively. is used to show the map in full. is used to zoom the map according to selection. updates the map view.

Figure 1.17: Map Navigation Toolbar

Source: Screenshot generated by Asian Development Bank using QGIS 3.34.14. 2025.

Attributes Toolbar

This Attributes toolbar is used to manage map attributes. "Attributes" refer to the information about the features of the map. The icon is used to identify features on the map. opens the attribute table. opens the field calculator. opens the Processing Toolbox. Σ displays a statistical summary. is used to measure a line or distance on a map. shows map tips, which appear beside the Search bar in the status bar, at the bottom of the interface. provides two options when the toggle editing mode is enabled: The "Duplicate Feature" tool creates an exact copy of selected features, while the "Duplicate Feature and Digitize" tool duplicates features and activates the digitizing mode for editing geometry.

Figure 1.18: Attributes Toolbar

Source: Screenshot generated by Asian Development Bank using QGIS 3.34.14. 2025.

Manage Layers Toolbar

The Manage Layers toolbar is used to add data layers to a project. By default, it appears docked on the left side of the interface. The icon is used to load or add a vector layer. is used to import or add a raster layer. The icon or the "Add Delimited Text Layer" option bring in tabular data with spatial components. These icons will be used in Module 2 for adding data layers.

Figure 1.19: Manage Layers Toolbar

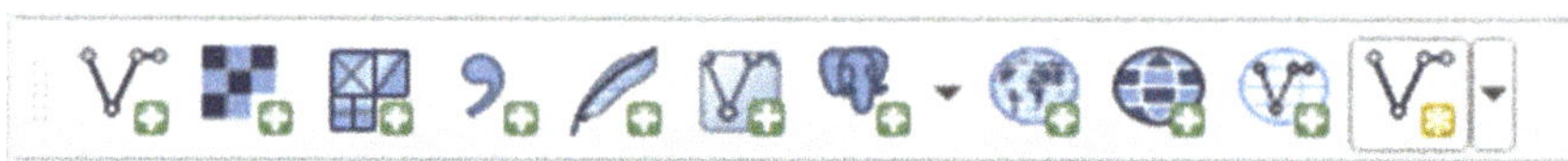

Source: Screenshot generated by Asian Development Bank using QGIS 3.34.14. 2025.

To explore the configurations of the toolbars, follow the exercise instructions below:

(i) Add the Selection toolbar to the interface. In the menu bar, click View | Toolbars | Selection Toolbar.
(ii) Position the mouse pointer on the dotted button of the Attributes toolbar and drag it to the left side. Notice that its neighboring toolbars are compressed, and a double arrow » is created, indicating that other available icons are hidden.
(iii) Try to drag and move other toolbars anywhere in the workspace.

Panels

Panels are movable and dockable widgets designed to perform complex tasks, including complex data operations. Among the most commonly used are the Layers and Browser panels, which are docked by default on the left side of the interface (Figure 1.20). Another important panel is the Processing Toolbox, which automatically docks to the left side of the map view when activated.

Layers Panel

The Layers panel is like the Table of Contents in QGIS (Figure 1.20). This is where the list of added data layers is displayed and managed. The toolbar at the top of the panel is used to adjust the visibility of the layers.

Figure 1.20: The Layers Panel

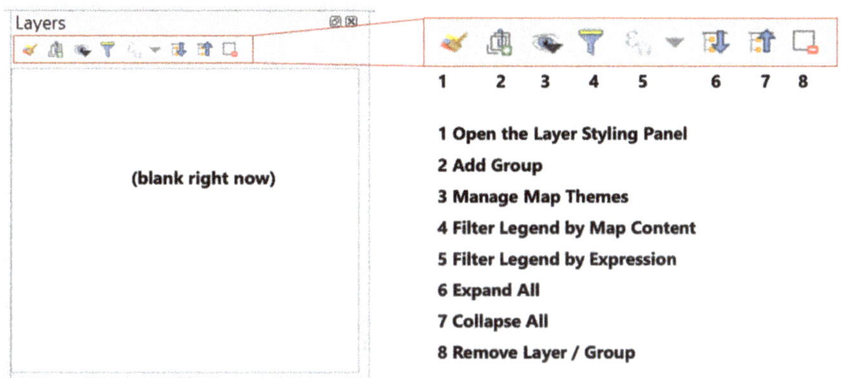

Source: Screenshot generated by Asian Development Bank using QGIS 3.34.14. 2025.

(i) The icon opens the Layer Styling panel. This panel appears by default at the left side of the interface when the paintbrush icon is enabled. The Layer Styling panel provides settings for styling and rendering map layers.

(ii) adds a group to the Layers panel, which is useful for organizing multiple map layers into groups.

(iii) displays a drop-down menu that offers options for showing or hiding layers, and for creating new map themes or replacing existing ones.

(iv) filters the legend by map content, displaying only relevant legend items based on the current map view.

(v) filters legends by expression, applying custom expressions to refine the legend display.

(vi) expands all map layers and sub-layers, enabling the full visibility of all layers without manual selection.

(vii) collapses all map sub-layers, minimizing all sub-layers simultaneously.

(viii) removes layers or groups, deleting the selected layers or groups from the map interface.

Browser Panel

The Browser panel is used to browse, inspect, copy, and add data to projects (Figure 1.21). It enables navigation through the file system and the management of QGIS-supported geodata, such as data layers (raster, vector, table) and other data-source formats (databases, web services, compressed files). Directories can be set up in the Favorites tab for easy access. Files can be added to the project by being dragged and dropped into the map view or Layers panel, and both layer and file properties can be previewed.

Figure 1.21: The Browser Panel

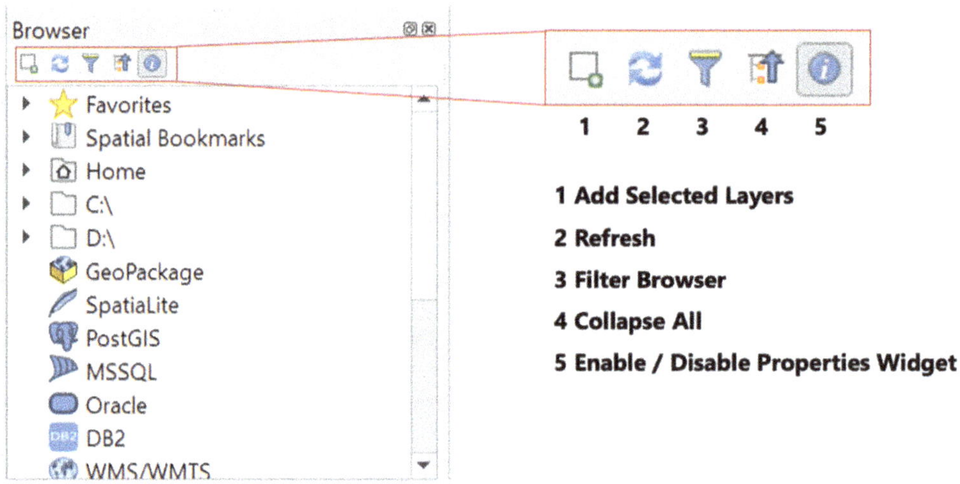

Source: Screenshot generated by Asian Development Bank using QGIS 3.34.14. 2025.

Like the Layers panel, the Browser panel is equipped with toolbars, located at the top of the panel.

(i) The icon adds selected files to the project, requiring prior selection of the file(s) before clicking the icon to add them.

(ii) refreshes the browser tree, updating the displayed directory and data structure.

(iii) searches files using entry names, thereby enabling the quick location of specific files.

(iv) collapses all top-level entries, minimizing the browser tree for a cleaner view.

(v) shows the properties widget, which contains useful information about the data files. The widget allows the inspection of metadata, preview maps, and attributes; and it can be enabled and disabled (Figure 1.22).

Figure 1.22: The Properties Widget in the Browser Panel

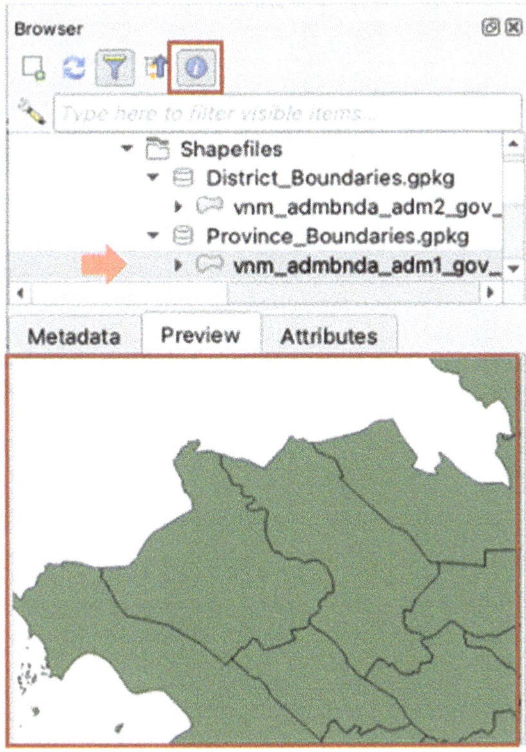

Source: Screenshot generated by Asian Development Bank using QGIS 3.34.14. 2025.

Map Display

The map display occupies the largest area in the interface, as it includes the loaded and enabled maps in 2D. This section allows interaction with the maps and the visualization of the applied rendering styles.

The **Map Navigation toolbar**, previously discussed, provides tools for interacting with the maps in the map view.

Right-clicking on the map display enables the copying of the coordinates of selected points in the map's coordinate reference system (CRS) (Figure 1.23).

Figure 1.23: Copying Coordinates in the Map Display

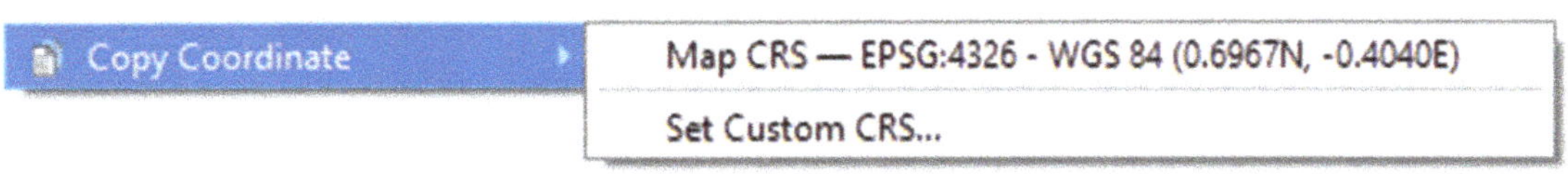

CRS = coordinate reference system, EPSG = European Petroleum Survey Group, WGS = World Geodetic System.
Source: Screenshot generated by Asian Development Bank using QGIS 3.34.14. 2025.

Status Bar

The status bar provides quick access to information about the current map overview and navigation options (Figure 1.24). Located at the bottom of the QGIS window, it displays the mouse pointer's position in terms of coordinates, the current scale, and the current CRS. The buttons could be used to quickly set a fixed scale, set a fixed zoom level for the map, and to quickly open the CRS dialog box to set the CRS of the project. The status bar also contains toggles for switching the extents and mouse-position display, and a button to turn the map rendering on and off. The messages button provides access to log messages, potential errors, and updates about plugins.

Figure 1.24: The Status Bar

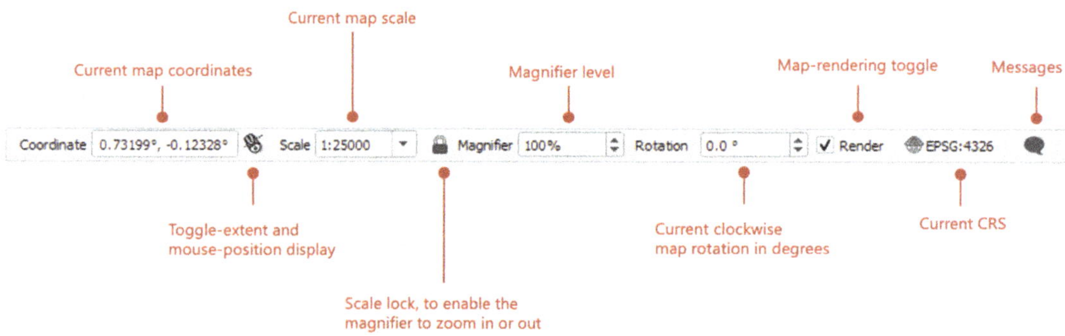

CRS = coordinate reference system

Source: Screenshot generated by Asian Development Bank using QGIS 3.34.14. 2025.

The locator bar, positioned at the left end of the status bar (Figure 1.25), enables quick searches for functions, datasets, and more. Key words can be entered directly into the search field; for instance, entering the key word "export" will generate a display of all the functions related to data export (Figure 1.25). Shortcuts for enhanced navigation and functionality are available, as shown in Figure 1.26.

Figure 1.25: Searching Functions in the Locator Bar

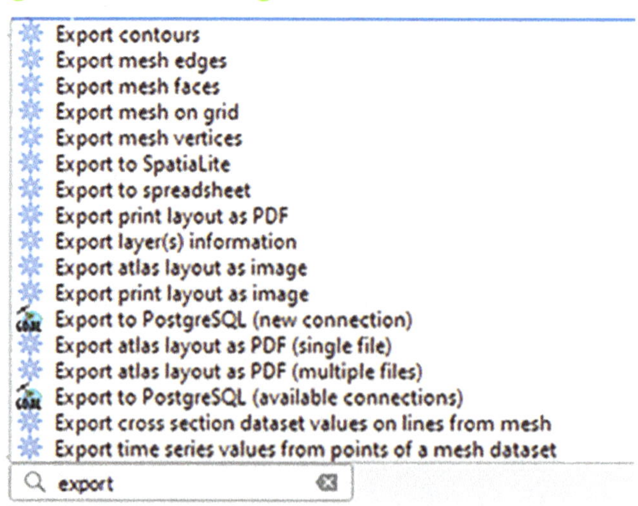

Source: Screenshot generated by Asian Development Bank using QGIS 3.34.14. 2025.

Figure 1.26: Shortcuts in the Locator Bar

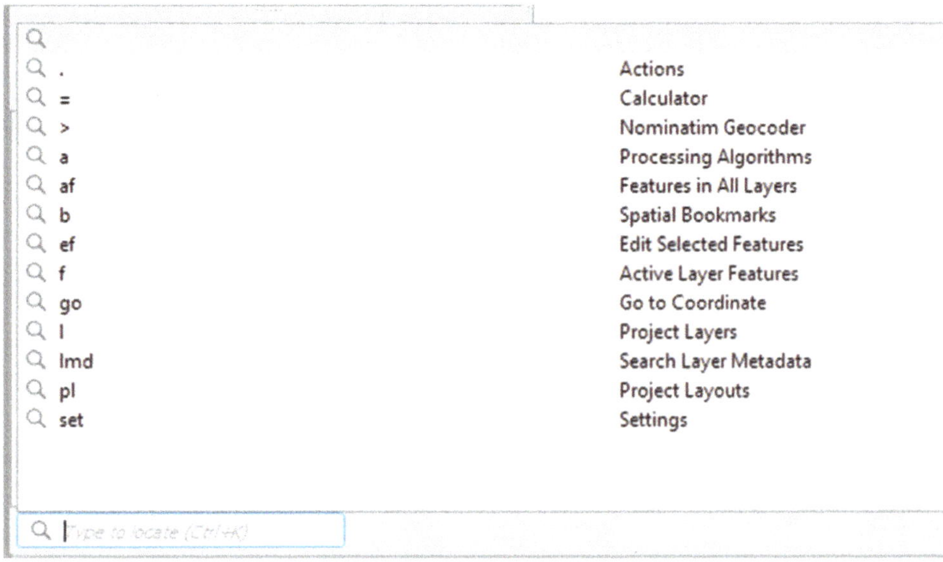

Source: Screenshot generated by Asian Development Bank using QGIS 3.34.14. 2025.

1.3.5 Plugin Manager

QGIS offers numerous functions that can be extended through plugins developed by independent organizations and developers.

There are two types of plugins: core plugins and external plugins. Core plugins are built into the QGIS application. These plugins can be enabled or disabled by the user, but cannot be uninstalled. The external plugins are in the external repository and are maintained by the authors. These can be installed and uninstalled using the Plugin Manager.

Several plugins are utilized in the exercises included in the next modules. The following are recommended for installation:

- **autoSaver.** This plugin automatically saves the file and the modified layers currently in edit mode.
- **Batch GPS Importer.** This plugin allows the conversion of multiple GPX files into a single layer with multiple features based on the features in each GPX file.
- **Calculate Geometry.** This plugin calculates the area of a polygon and the lengths of common units, without typing expressions.
- **Geometry Checker.** This is a core plugin that checks for errors in the geometry of a layer. As a core plugin, it can be enabled or disabled, but not uninstalled.
- **QuickMapServices.** This plugin is used to add base maps from such mapping services as OpenStreetMap, Bing Maps, and Google Maps.
- **Topology Checker.** This is a core plugin that is used to find topological errors in vector layers by applying topological rules.

To install the recommended plugins, take the following steps:

(i) In the menu bar, click **Plugins | Manage and Install Plugins...** (Figure 1.27).

Figure 1.27: Plugins in the Menu Bar

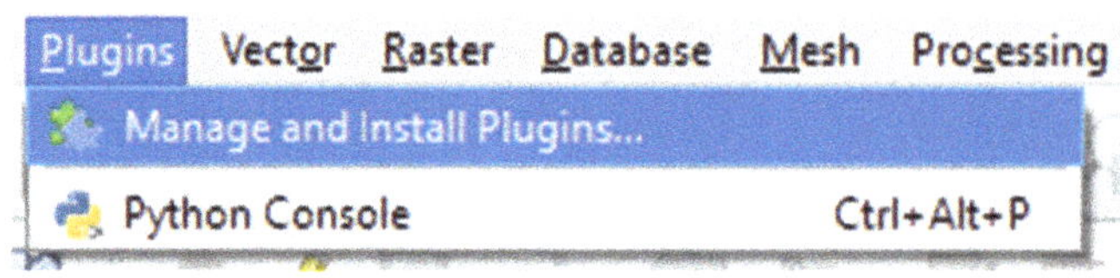

Source: Screenshot generated by Asian Development Bank using QGIS 3.34.14. 2025.

(ii) In the search tab, type the name of the plugin you wish to search for and install.

For example, type "Batch GPS Importer" to search for that plugin. Select Batch GPS Importer and click **Install Plugin** to install. Repeat similar steps to install other recommended plugins (Figure 1.28).

Figure 1.28: Plugin Dialog Box

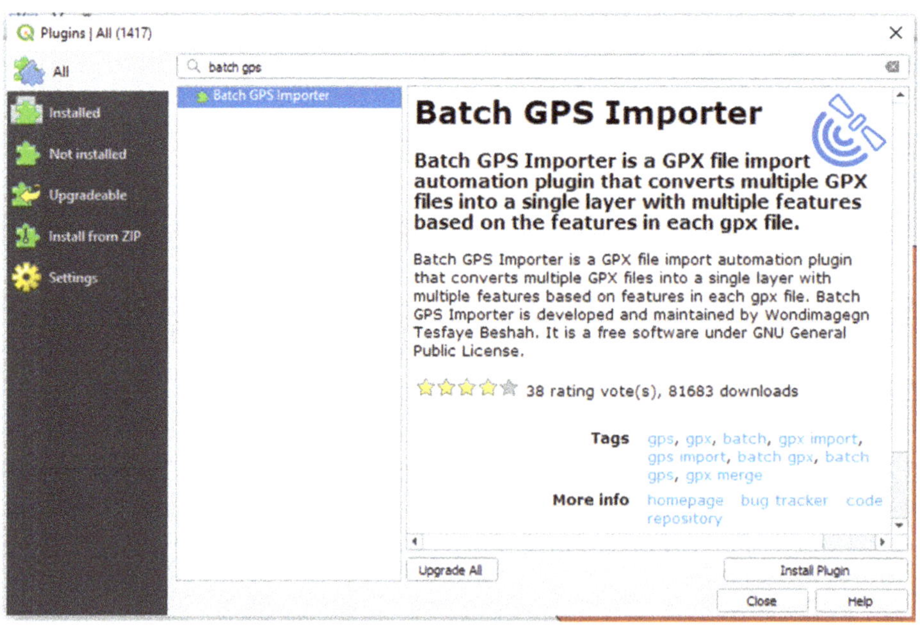

Source: Screenshot generated by Asian Development Bank using QGIS 3.34.14. 2025.

Module 2: Geographic Information System Data Models, Coordinate Reference Systems, and Geodatabase

This module introduces the foundational concepts of the geographic information system (GIS), focusing on geospatial data models, the coordinate reference system (CRS), and on the creation and management of a basic geodatabase using a structured file folder approach. It also covers techniques for importing, organizing, and styling various types of spatial data layers, including vector and raster formats.

The topics covered in this module include:

2.1 Overview of Geospatial Data Models
2.2 Coordinate Reference System
  2.2.1 Layer Properties
  2.2.2 Project Properties
  2.2.3 General Settings
2.3 Creating and Managing a Basic Geodatabase (Map Layers)
  2.3.1 Creating and Saving a Project
  2.3.2 Importing Base Maps (Satellite Images)
    1. QuickMapServices
    2. XYZ Tiles
  2.3.3 Importing Map Layers
  2.3.4 Exporting Map Layers as New Layers
  2.3.5 Interacting with Map Layers
    1. Reordering Map Layers
    2. Navigating the Map Area
    3. Identifying Data
    4. Checking Layer Properties
2.4 Visualizing and Styling Data Layers
  2.4.1 Styling Points—Landmarks
  2.4.2 Styling Lines—Roads
  2.4.3 Styling Polygons—Administrative Boundaries
  2.4.4 Styling Raster Layers—Land Cover Map
2.5 Where to Find Data to Download

For demonstration purposes, the province of An Giang, in Viet Nam, will be used throughout the succeeding parts of this handbook. The map of An Giang presented is based on the administrative boundaries prior to the province's merger with Kien Giang Province, in June 2025, and so does not reflect the newly configured provincial and commune boundaries.

By the end of this module, users will be able to create a styled GIS map layout that blends the An Giang land cover map with satellite imagery using different rendering modes for layered visualization.

2.1 Overview of Geospatial Data Models

GIS represents real-world features such as roads, rivers, and administrative boundaries, using geometries (e.g., points, lines, polylines, and polygons) or grid cells. These models are referred to as "data models." Each feature in a data model contains attributes (descriptions) that can be stored in different file formats and structures, according to the intended use.

The two primary categories of data models in GIS are the vector and raster data models (Figure 2.1):

The **vector data model** contains discrete data using geometries such as points, lines, and polygons, defined by x and y coordinates to represent a location on the earth's surface. Table 2.1 presents the different vector data types, along with their definitions and illustrative examples. Vector data represent topography better than raster data because they are more accurate and precise.

Table 2.1: Vector Data Types

Data Types	Definition	Example
Point	A point is a nondimensional feature that uses a pair of coordinates (x, y) to define its location. Points represent discrete data that can be spatially linked to form more complex features. Examples are buildings, tourist attractions, banks, markets, and schools.	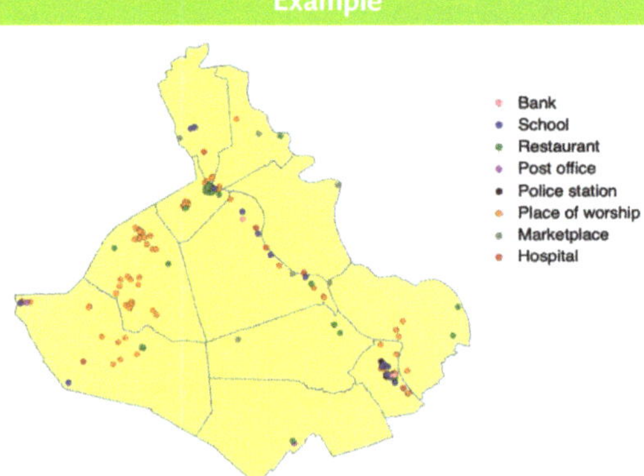
Line	Lines are one-dimensional, and are used to represent linear features. They are composed of multiple connected points, usually defined by a start and an end point, and can be used to measure length. Examples are roads and waterways.	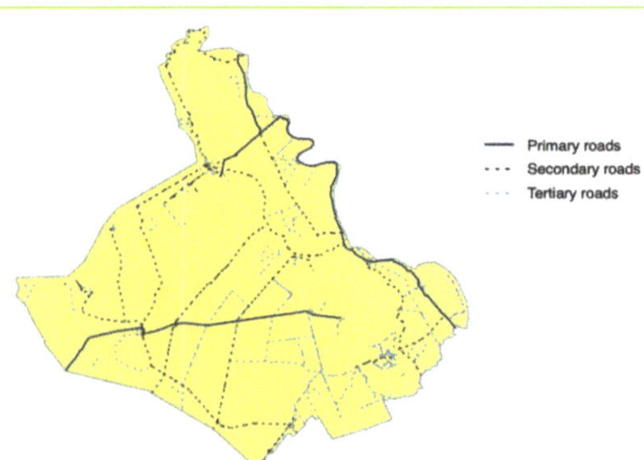

Data Types	Definition	Example
Polygon	Polygons are two-dimensional and can be used to measure the perimeter and area of a geographic feature. They are created when lines are connected to form an enclosed shape, with the start and end points having the same coordinates. Examples are administrative boundaries.	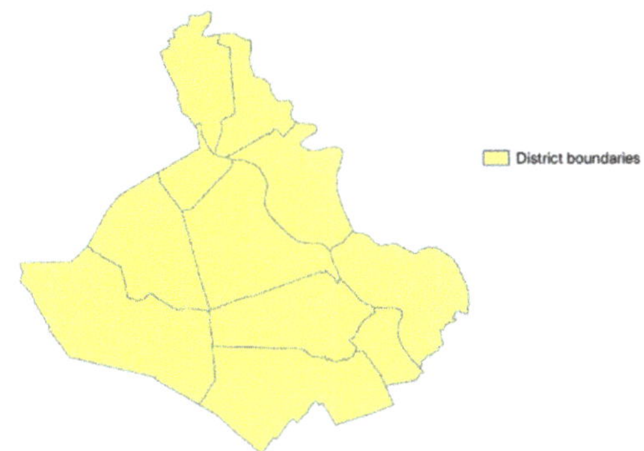

Sources: Asian Development Bank screenshots; Humanitarian Data Exchange. Viet Nam—Subnational Administrative Boundaries; https://data.humdata.org/dataset/cod-ab-vnm; OpenStreetMap. Roads and Points of Interest: https://data.humdata.org/dataset/hotosm_vnm_roads. https://data.humdata.org/dataset/hotosm_vnm_points_of_interest.

The **raster data model** represents continuous or discrete data across an area (e.g., on elevation or rainfall), which are stored in a matrix of grid cells called "pixels" and organized into rows and columns. The pixels are normally equal in size, contain specific information that describes the features, and may be georeferenced. The size of the pixels determines the resolution and the granularity of details the raster can contain. The most common example of raster data is a satellite image.

Figure 2.1: Examples of Vector and Raster Data

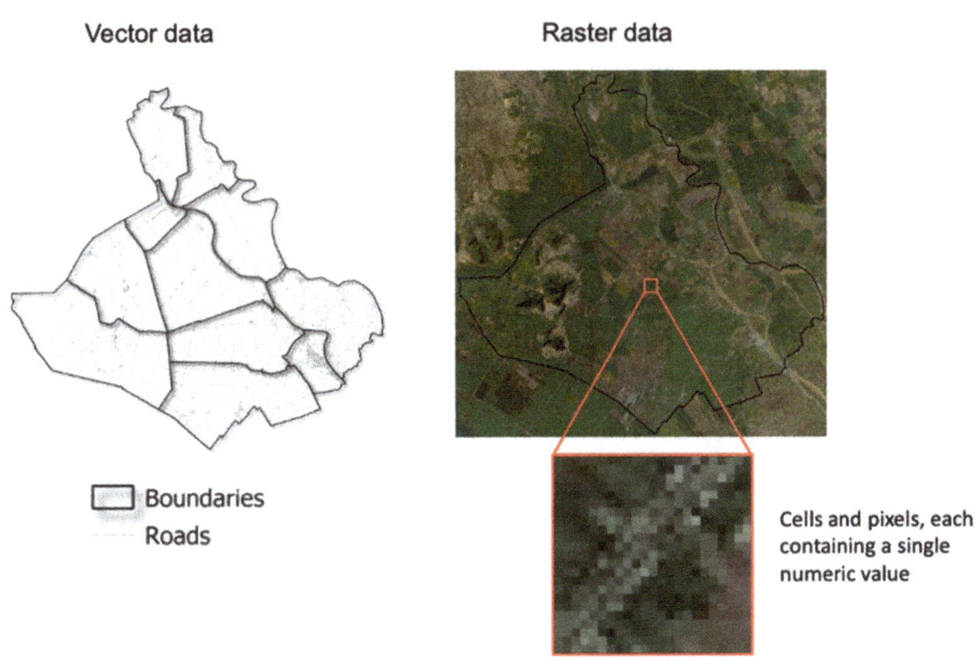

Note: The administrative boundaries are those of An Giang Province, Viet Nam, prior to its merger with Kien Giang Province in June 2025. As such, they do not reflect the newly configured administrative boundaries.

Source: Screenshot generated by Asian Development Bank using QGIS 3.34.14. 2025.; Humanitarian Data Exchange. Viet Nam—Subnational Administrative Boundaries. https://data.humdata.org/dataset/cod-ab-vnm; Esri. XYZ Tiles Satellite Map. https://server.arcgisonline.com/ArcGIS/rest/services/World_Imagery/MapServer/tile/{z}/{y}/{x}.

Vector and raster data are stored in different file formats. Shapefiles are the most common format for vector data. However, one shapefile can only contain one geometry type (point, line, or polygon), which means that multiple geometry types cannot be combined into one shapefile. To have a functioning shapefile, three file formats should be placed in the same file folder structure database (*.shp, .dbf, and .shx*). Each vector shapefile contains files in a geodatabase that thus share a common name, but have different extensions (Figure 2.2). The file extensions determine the file format and the type of data it contains. These include:

(i) ***.shp,*** for a file that stores a feature's geometry;
(ii) ***.dbf,*** for a database file that stores the feature's attributes; and
(iii) ***.shx,*** for a file that stores the indexed information regarding the feature's geometry, linking the ***.shp*** to the .dbf file, and indicating the shape's index position.

Other file extensions include:

(i) ***.prj,*** for a file that stores the coordinate system information,
(ii) ***.cpg,*** for a file that identifies the character set (to be) used, and
(iii) ***.sbn*** and ***.sbx,*** for files that store the spatial index of the features.

Figure 2.2: Example of Files in a File Folder Structure Database

vnm_admbnda_adm1_gov_20201027.CPG	CPG File	1 KB
vnm_admbnda_adm1_gov_20201027.dbf	DBF File	39 KB
vnm_admbnda_adm1_gov_20201027.prj	PRJ File	1 KB
vnm_admbnda_adm1_gov_20201027.sbn	SBN File	1 KB
vnm_admbnda_adm1_gov_20201027.sbx	SBX File	1 KB
vnm_admbnda_adm1_gov_20201027.shp	SHP File	6,466 KB
vnm_admbnda_adm1_gov_20201027.shx	SHX File	1 KB

Note: The files framed in red are mandatory for creating a functioning shapefile.
Source: Asian Development Bank screenshot.

Common image file formats such as .jpg (or .jpeg), .tiff (or .tif), and .png are often used to store raster datasets. By contrast, vector data formats tend to be more complex than those used for raster data.

2.2 Coordinate Reference System

In QGIS terminology, the spatial reference system is called the "coordinate reference system" (CRS). The CRS provides a frame of reference for locating features on the surface of the earth. The CRS in QGIS can refer to both a geographic coordinate system (a global positioning in latitude and longitude expressed in degrees, e.g., the World Geodetic System 1984, or WGS 84) and a projected coordinate system (x and y coordinates expressed in metric units, e.g., Universal Transverse Mercator or UTM) (Figure 2.3).

Figure 2.3: Spatial Coordinate Systems

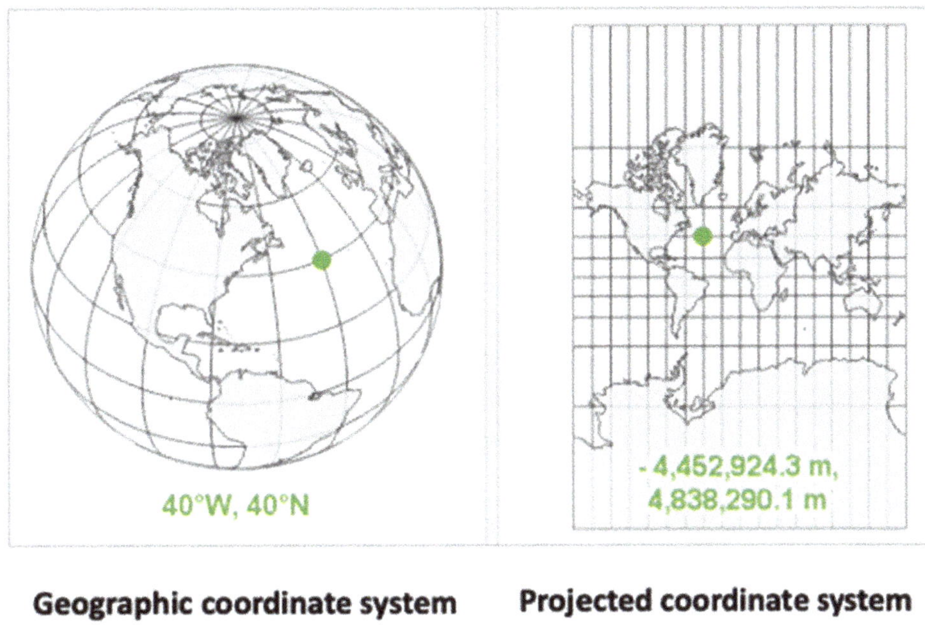

Source: Esri. ArcGIS Pro Resources. https://pro.arcgis.com.

Map projection is the transformation of the three-dimensional earth onto a two-dimensional plane using mathematical algorithms. Through map projections, geographic coordinates are converted into a metric coordinate system, measuring the x and y coordinates. It is useful for creating maps and spatial analyses (Esri 1998).

The earth is nearly spherical, but there is a small bulge at the equator that makes it slightly larger there than at the poles. An ellipse is a simple model that describes the basic shape of the earth. A way of explaining a map projection is to imagine the earth as a ping pong ball and dissecting it into different ways to lay it flat on a planar surface. Regardless of the method used, flattening a curved surface inevitably introduces distortions. Similarly, projecting the relatively ellipsoidal shape of the earth onto a planar view on the computer screen is quite a challenge for geospatial analysts, given that distortions are inevitable.

Different types of map projections have been created for different applications. The existing projections involve a trade-off among angularity, distance, and area. Some projections preserve equal distances, while others preserve the angular relationships or areas. The choice of CRS depends on the analysis being conducted, the geographic location and regional extent of the project, and often on the availability of data. There are certain map projections recommended for each country to reduce distortions. It is important to check which CRSs are standard for the area being mapped, as using an inappropriate CRS could poorly represent the position and dimensions of features (Esri, n.d.).

One commonly used projected CRS is the UTM system, which divides the earth into zones and uses metric units (meters) for coordinates. The system is particularly useful for area-based calculations, as it minimizes distortion within each zone and enables the accurate measurement of distances and surface areas. This is especially important in agricultural statistics, where precise area estimation is required for crop mapping and land use classification.

For example, when estimating rice-cultivation areas in An Giang Province, Viet Nam, using a UTM zone appropriate for the region ensures that the calculated field sizes from satellite imagery or drone surveys are accurate and consistent with ground measurements. By contrast, using a geographic CRS such as WGS 84 for such calculations could result in significant errors due to distortions in area representation.

QGIS predefines approximately 2,700 CRSs. The CRSs in QGIS are based on the registry of well-known CRSs maintained by the International Association of Oil & Gas Producers (IOGP), and they are used in geospatial analysis for energy exploration. IOGP was formerly known as the European Petroleum Survey Group (EPSG), so the CRSs are traditionally identified using EPSG codes (QGIS Project 2024).

In QGIS, the CRS can be set up at different levels: Layer Properties, Project Properties, and General Settings. Figure 2.4 illustrates the steps for setting up the CRS at these three levels. Defining the CRS at the Project Properties level will override the settings in General Settings. Similarly, setting up the CRS in Layer Properties will override the CRS in Project Properties. These different methods of setting up the CRS for a project or map layer are most often misunderstood by beginners (De la Rúa, Bright, and Le Gall-Queguineur 2020). The following sections explain the distinctive characteristics and significance of each level.

Figure 2.4: Levels for Setting the Coordinate Reference System in QGIS

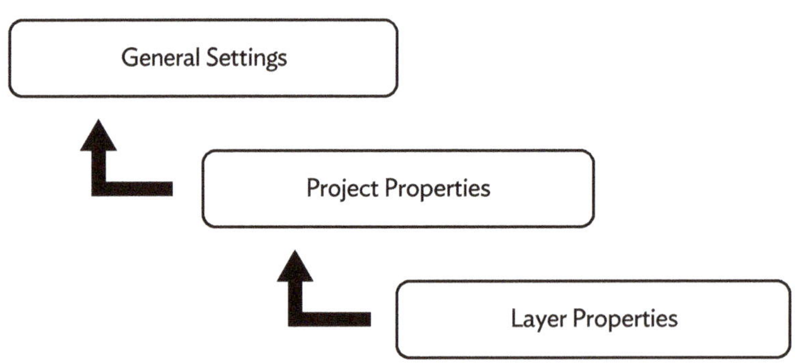

Source: Asian Development Bank interpretation.

2.2.1 Layer Properties

Once the data have been added to the project, the **Layer Properties** can be accessed by right-clicking the layer in the **Layers panel** and selecting **Layer Properties** in the menu (Figure 2.5).

In the Information section of the Layer Properties, the default CRS of the layer is displayed along with other information.

Figure 2.5: Information Provided in Layer Properties

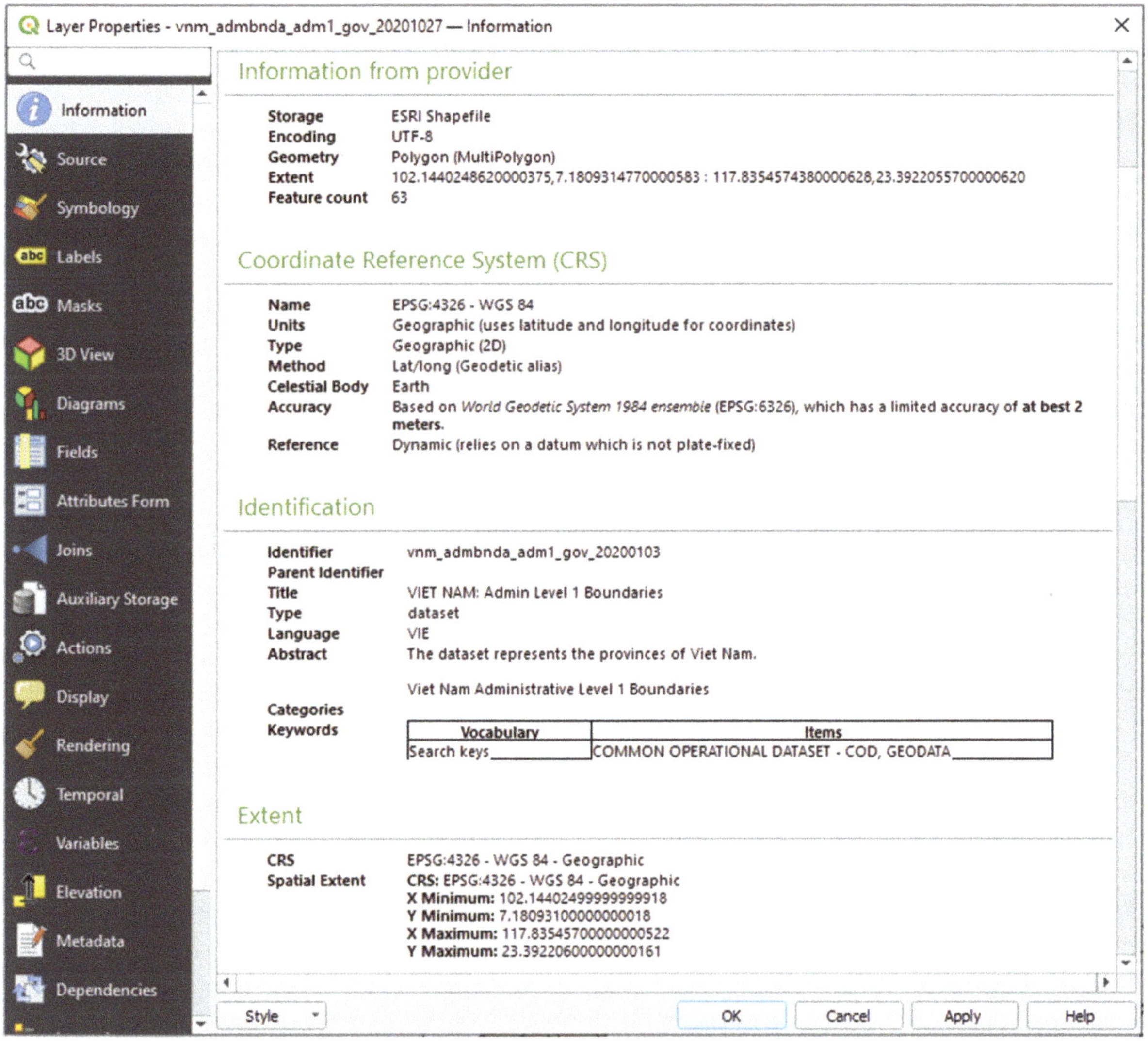

Source: Screenshot generated by Asian Development Bank using QGIS 3.34.14. 2025.

In the Source tab (Figure 2.6), the CRS can be assigned to the layer. It is important to note that the assigned CRS does not alter the underlying data source; instead, it only changes the way QGIS interprets the raw coordinates from the layer. Virtually, QGIS reprojects all layers into the project's CRS such that all the layers will be rendered in the correct position with respect to each other.

Figure 2.6: Assigning the Layer Coordinate Reference System in the Layer Properties Dialog Box

Source: Screenshot generated by Asian Development Bank using QGIS 3.34.14. 2025.

To reproject the CRS of a layer into a target CRS, different tools are used, depending on the data type. For a raster layer, the reprojection process can be accessed through **Menu Toolbar | Raster | Warp (Reproject). For a vector layer, go to the Main Menu | Vector | Data Management Tools | Reproject Layer**. An exercise demonstrating the reprojection of a vector layer is included in Section 4.4 of Module 4. It is recommended that all layers within a project share the same CRS to ensure spatial consistency and accurate rendering.

2.2.2 Project Properties

When a new project is created in QGIS, the default CRS is set to WGS 84, identified by EPSG code 4326. QGIS reprojects data "on-the-fly" (called "OTF" in previous versions). The on-the-fly feature allows QGIS to display multiple layers with different CRSs in alignment with the project CRS. The tab is displayed at the bottom-right corner of the QGIS interface (Figure 2.7). Clicking it will open the CRS settings in **Project Properties**.

Figure 2.7: On-the-Fly Projection

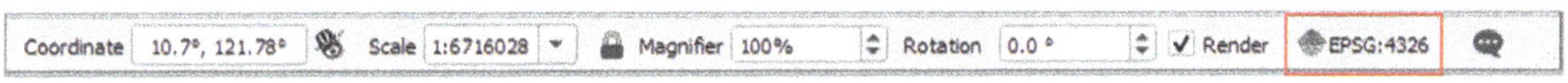

Source: Screenshot generated by Asian Development Bank using QGIS 3.34.14. 2025.

Another way to open the CRS in Project Properties is to click **Menu Toolbar | Project | Properties** (Figure 2.8). The Project Properties dialog box will pop up.

Figure 2.8: Opening Project Properties

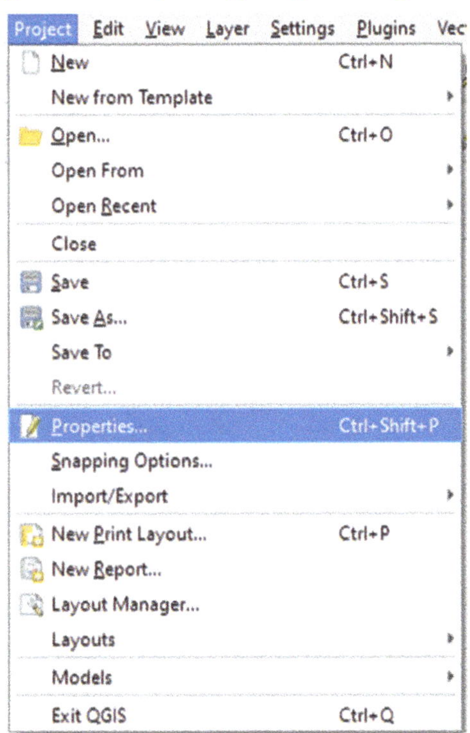

Source: Screenshot generated by Asian Development Bank using QGIS 3.34.14. 2025.

In the Project Properties dialog box (Figure 2.9), the CRS for the project can be configured. A suitable CRS may be located using the search bar. Once selected, clicking **Apply** will set the selected CRS for the project.

Figure 2.9: The Coordinate Reference System in Project Properties

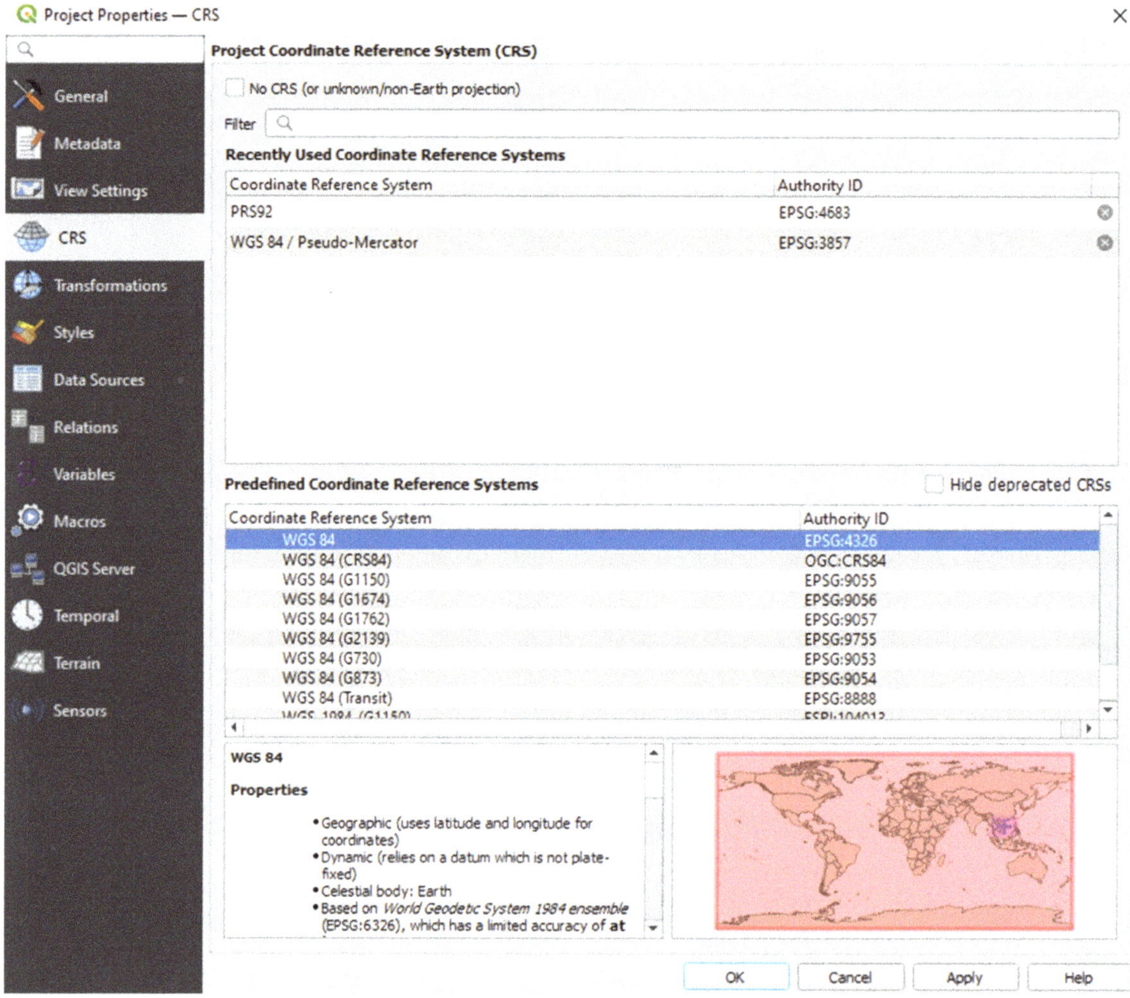

Source: Screenshot generated by Asian Development Bank using QGIS 3.34.14. 2025.

There is a significant difference between the CRS of the data layer and that of the QGIS project. The CRS of the data layer refers to the coordinate system in which the layer's spatial data are stored. The CRS of the project defines how all layers are displayed within the map canvas, enabling the visualization of layers with differing CRSs in a unified spatial context. QGIS supports on-the-fly reprojection, which functions as a dynamic "filter"—allowing spatial data to be displayed in the project without altering the original CRS of the data layer. This ensures consistent visualization while preserving the integrity of the source data.

2.2.3 General Settings

"General Settings" refers to the setup of the entire application, which is like an umbrella above different projects containing different data layers. In the CRS section of General Settings, a default CRS can be specified for new projects. When a new project is created, it will automatically adopt the CRS defined in these settings. By default, this CRS is set to WGS 84 (EPSG:4326).

To find General Settings, navigate to the Menu Toolbar and select **Settings | Options | CRS** (Figure 2.10).

Figure 2.10: Opening the Coordinate Reference System in General Settings

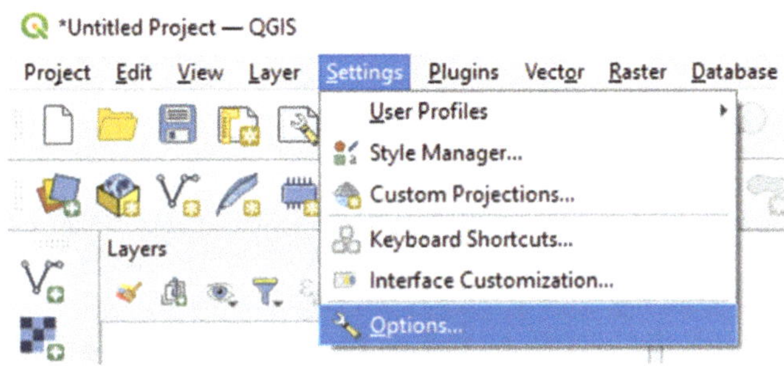

Source: Screenshot generated by Asian Development Bank using QGIS 3.34.14. 2025.

The CRS tab in the Options menu allows the configuration of default CRS settings and other related options for both projects and layers (Figure 2.11).

Figure 2.11: Setting the Coordinate Reference System in General Settings

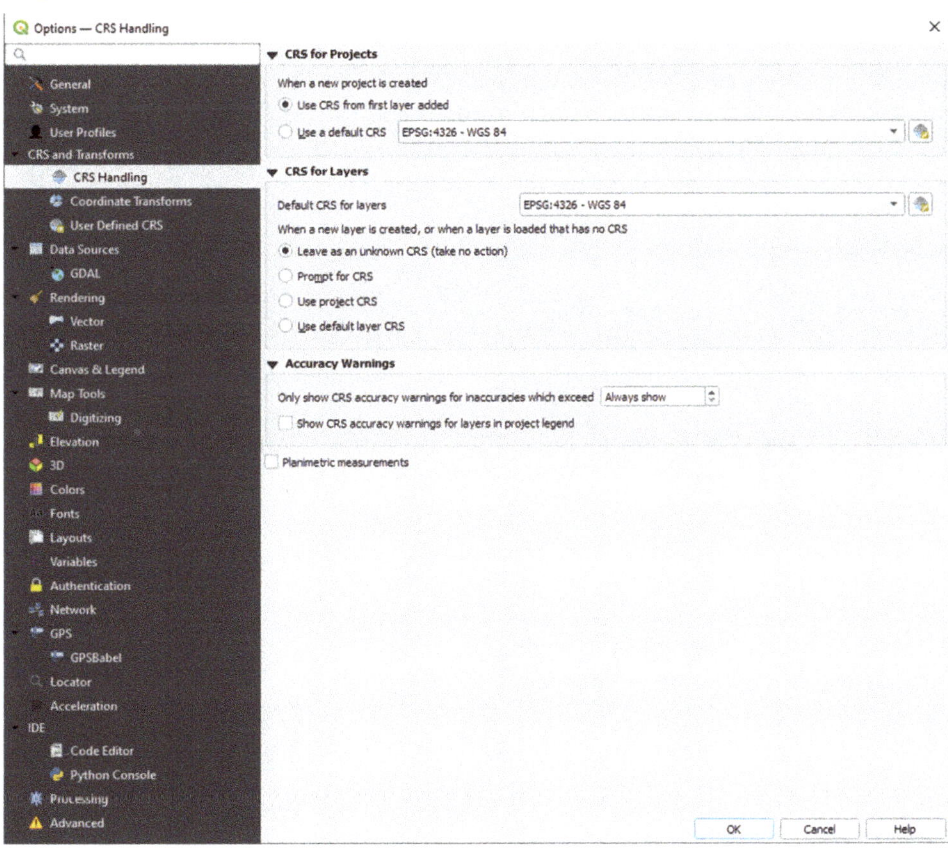

Source: Screenshot generated by Asian Development Bank using QGIS 3.34.14. 2025.

2.3 Creating and Managing a Basic Geodatabase (Map Layers)

This section outlines the process of designing a basic geodatabase using a structured file folder system and sample data from Viet Nam. The steps include importing spatial data, interacting with map layers, and applying the appropriate symbols when styling those layers for visualization and analysis.

How can the dataset be downloaded for the exercises?

The raster data should be downloaded from the Japan Aerospace Exploration Agency (JAXA), but user registration is required. Registration is free and involves confirming your email address. To begin, visit the **JAXA registration page** (https://www.eorc.jaxa.jp/ALOS/en/lulc/registration.htm) and follow the instructions for creating an account. Once registered, you will be able to access the Advanced Land Observing Satellite (ALOS) Annual Land Use and Land Cover Maps (1990–2020) for Viet Nam through the following link: https://www.eorc.jaxa.jp/ALOS/en/dataset/lulc/lulc_vnm_v2109_e.htm. The dataset covers the entire country, and is divided into multiple raster tiles. For this exercise, download and add the tiles **N10E104** and **N10E105** onto the map canvas, which covers An Giang Province, in the Mekong River Delta region. A comprehensive step-by-step guide to downloading raster data is available in Module 3, section 3.2.

Download the administrative boundaries of Viet Nam from the Humanitarian Data Exchange at: https://data.humdata.org/dataset/cod-ab-vnm.

Download the shapefiles for Buildings, Roads, and Points Of Interest from the Humanitarian Data Exchange, exported from **OpenStreetMap**, through the following links:

Buildings: https://data.humdata.org/dataset/hotosm_vnm_buildings

Roads: https://data.humdata.org/dataset/hotosm_vnm_roads

Points: https://data.humdata.org/dataset/hotosm_vnm_points_of_interest

2.3.1 Creating and Saving a Project

To create a new project, open an existing project, or save a project, use the Project toolbar (Figure 2.12).

Different options for creating, opening, and saving a project are also available in the **Menu Toolbar**, under the **Project** tab (Figure 2.13).

Figure 2.12: Project Toolbar

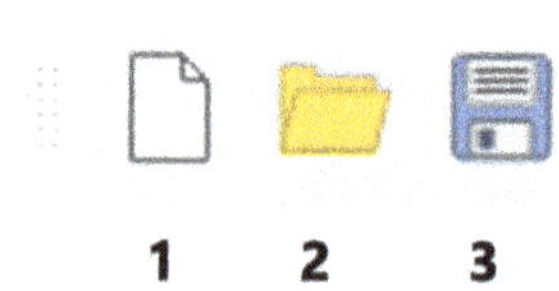

1. Create new project

2. Open existing project

3. Save a project

Source: Screenshot generated by Asian Development Bank using QGIS 3.34.14. 2025.

Figure 2.13: Project Tab in the Menu Toolbar

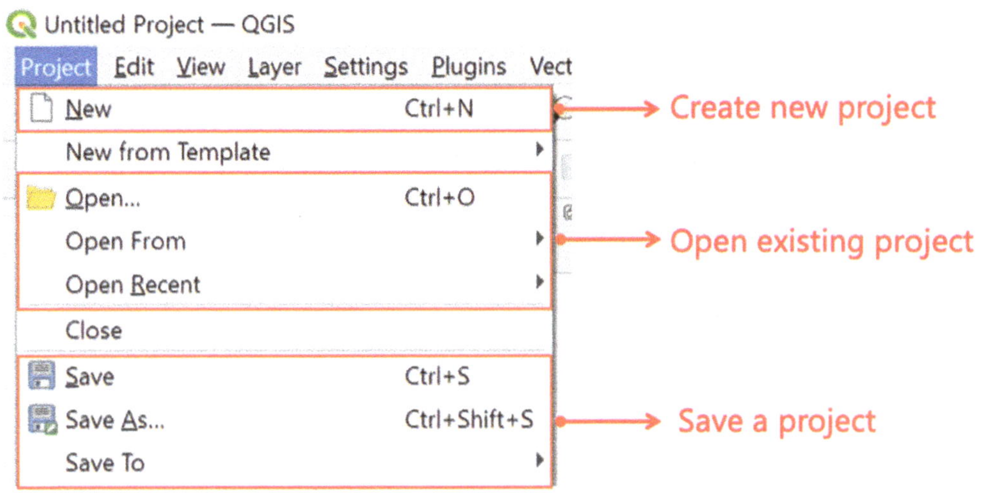

Source: Screenshot generated by Asian Development Bank using QGIS 3.34.14. 2025.

2.3.2 Importing Base Maps (Satellite Images)

Base maps serve as reference layers that provide geographic context for visualizing and interpreting geospatial information. They provide background information and enhance the readability of GIS projects. Base maps can include web layers, raster data, or multiple features.

The two most common methods to add base maps in QGIS are through (i) QuickMapServices and (ii) the XYZ Tiles plugins.

1. QuickMapServices

(i) In Module 1, Plugin Manager instructions are provided for installing plugins in QGIS, including QuickMapServices.

(ii) To add a base map to the layer, go to **Menu Toolbar | Web | QuickMapServices | Settings.** Select the **More services** tab in the dialog box, then click **Get contributed pack** (Figure 2.14). A dialog box will appear after the download is done; click **OK** and then **Save**.

Figure 2.14: Downloading the QuickMapServices Contributed Pack

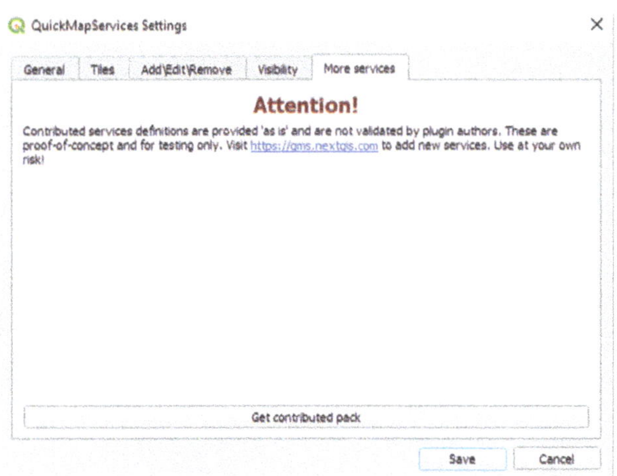

Source: Screenshot generated by Asian Development Bank using QGIS 3.34.14. 2025.

(iii) Go to **Menu Toolbar | Web | QuickMapServices** to access available web map layers. A wide range of base map options will be displayed (Figure 2.15).

Figure 2.15: Adding a Base Map Using the QuickMapServices Plugin

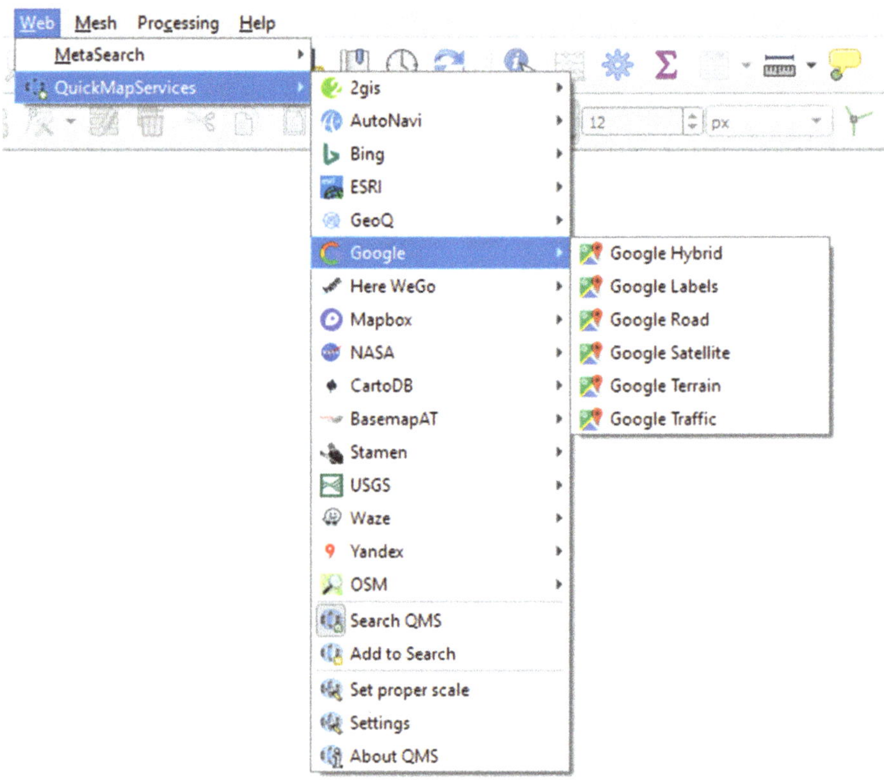

Source: Screenshot generated by Asian Development Bank using QGIS 3.34.14. 2025.

2. XYZ Tiles

a. Connecting to a Tile Service

One method for adding a base map using the XYZ Tiles plugin involves connecting to a tile service. Table 2.2 provides a selection of some of the commonly used Tile services.

Table 2.2: XYZ Tile Servers

Provider/Layer	URL
OpenStreetMap Monochrome	https://tile.openstreetmap.org/{z}/{x}/{y}.png
Google Maps	https://mt1.google.com/vt/lyrs=m&x={x}&y={y}&z={z}
Google Satellite	http://mt1.google.com/vt/lyrs=s&hl=en&x={x}&y={y}&z={z}
Google Hybrid (Google Map + Satellite)	https://mt1.google.com/vt/lyrs=y&x={x}&y={y}&z={z}

Source: QGIS. 2025. *QGIS Server 3.34 User Guide*. https://docs.qgis.org/3.34/pdf/en/QGIS-3.34-ServerUserGuide-en.pdf.

(i) In the Browser panel, right click **XYZ Tiles** and select **New Connection** (Figure 2.16). The XYZ Connection dialog box will appear (Figure 2.18).

Figure 2.16: Adding New Connections Using the XYZ Tiles Plugin

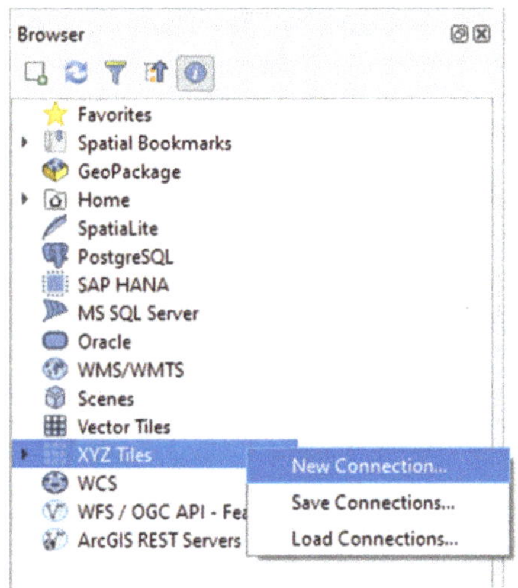

Source: Screenshot generated by Asian Development Bank using QGIS 3.34.14. 2025.

(ii) Provide a title for the XYZ map under **Name**, then type in the server's address under **URL** (Figure 2.17). Leave all the other settings at their defaults, and click **OK**.

Figure 2.17: Dialog Box for New XYZ Tile Connections

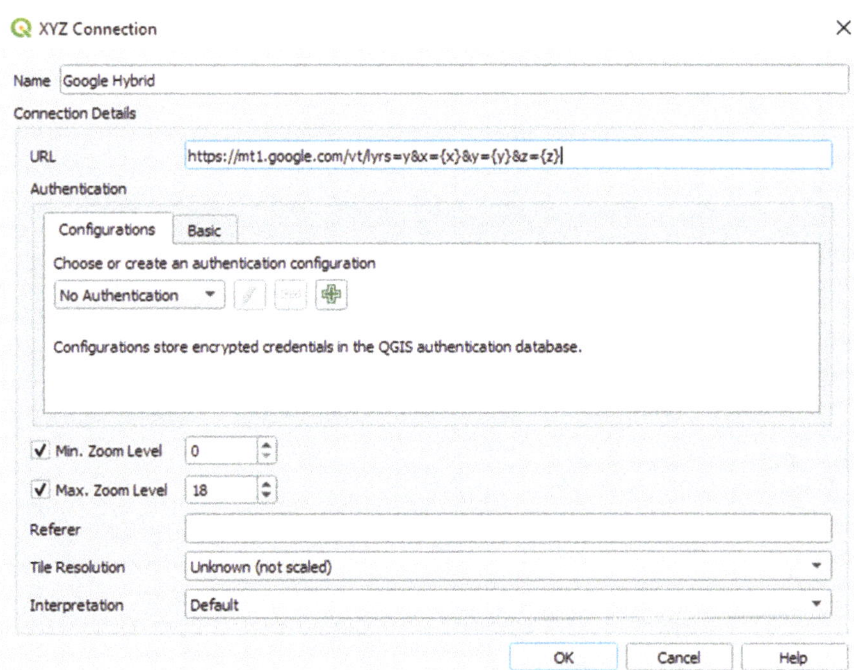

Source: Screenshot generated by Asian Development Bank using QGIS 3.34.14. 2025.

(iii) Verify that the new tile has been added to the list of connections. It should appear under the **XYZ Tiles** node in the **Browser** panel.

b. Python Script Setup

Numerous tile map services can also be set up simultaneously using the Python script posted at: https://raw.githubusercontent.com/klakar/QGIS_resources/master/collections/Geosupportsystem/python/qgis_basemaps.py.

(i) In the top menu, go to **Plugins** | **Python Console** (Figure 2.18).

Figure 2.18: Enabling the Python Console

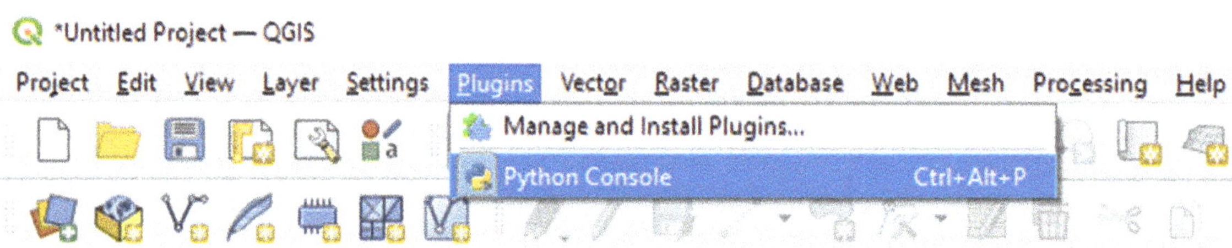

Source: Screenshot generated by Asian Development Bank using QGIS 3.34.14. 2025.

(ii) In the **Python Console**, paste the script in the panel located at the bottom of the workspace. Click ▶ to run the command (Figure 2.19).

Figure 2.19: Enabling and Running the Script in the Python Console

Source: Screenshot generated by Asian Development Bank using QGIS 3.34.14. 2025.

(iii) In the **Browser panel**, under **XYZ Tiles**, numerous selections of base maps are available (Figure 2.20).

Figure 2.20: XYZ Tiles

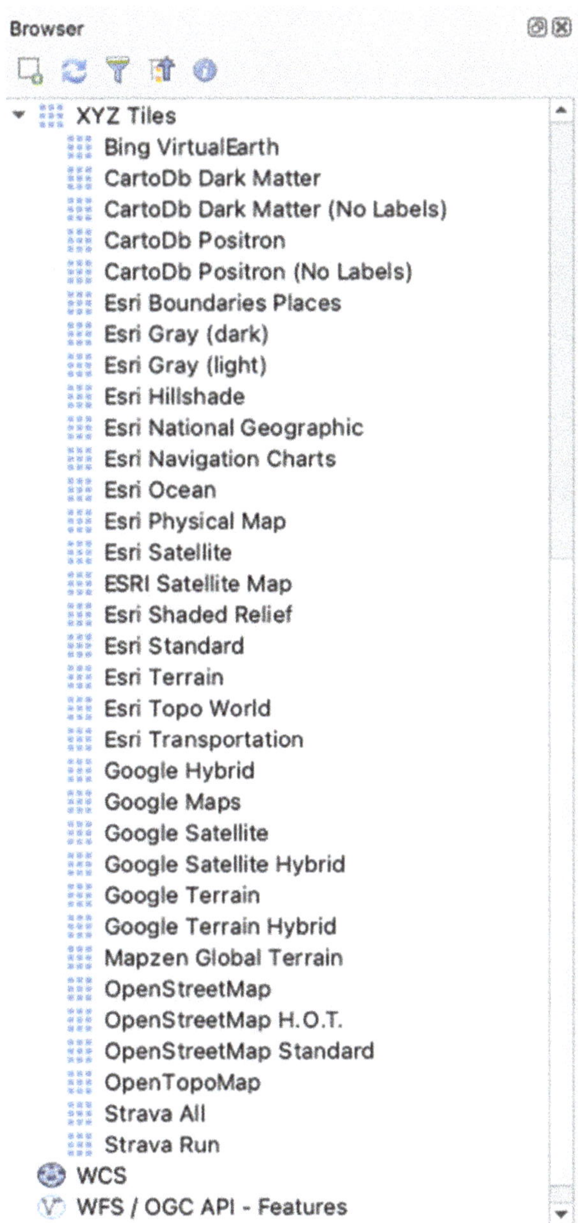

Source: Screenshot generated by Asian Development Bank using QGIS 3.34.14. 2025.

(iv) Add an Esri satellite map to the **map area** from XYZ Tiles. Right-click **Esri Satellite**, under the **XYZ Tiles** tab in the **Browser panel**, select **Add Layer to Project** (Figure 2.21).

Figure 2.21: Adding an Esri Satellite Map Using XYZ Tiles

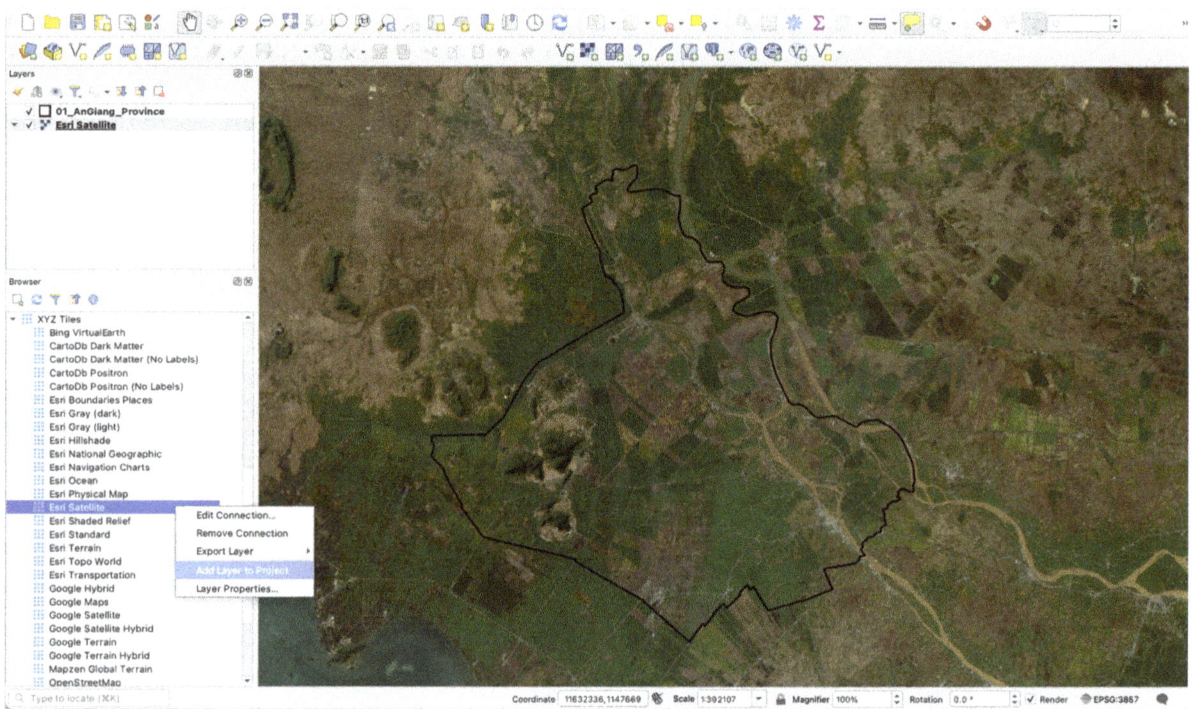

Sources: Screenshot generated by Asian Development Bank using QGIS 3.34.14. 2025.; Esri. XYZ Tiles Satellite Map. https://server.arcgisonline.com/ArcGIS/rest/services/World_Imagery/MapServer/tile/{z}/{y}/{x}.

2.3.3 Importing Map Layers

There are several methods of importing and adding map layers in a QGIS project, including the following:

(i) Go to the **Menu Toolbar** and click **Layer | Add Layer | Add Vector Layer** or **Add Raster Layer** (Figure 2.22). A dialog box titled "**Data Source Manager**" will appear.

(ii) In the Data Source Manager dialog box, click to browse through the files (Figure 2.23). The dialog box can also be quickly accessed through this icon in the Data Source Manager toolbar.

Figure 2.22: Importing a Map Layer from the Main Menu

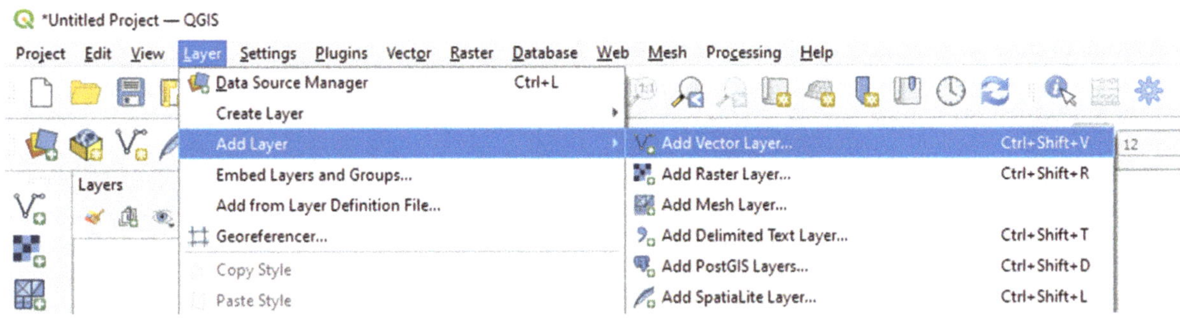

Source: Screenshot generated by Asian Development Bank using QGIS 3.34.14. 2025.

Figure 2.23: Importing a Vector File Using the Data Source Manager Dialog Box

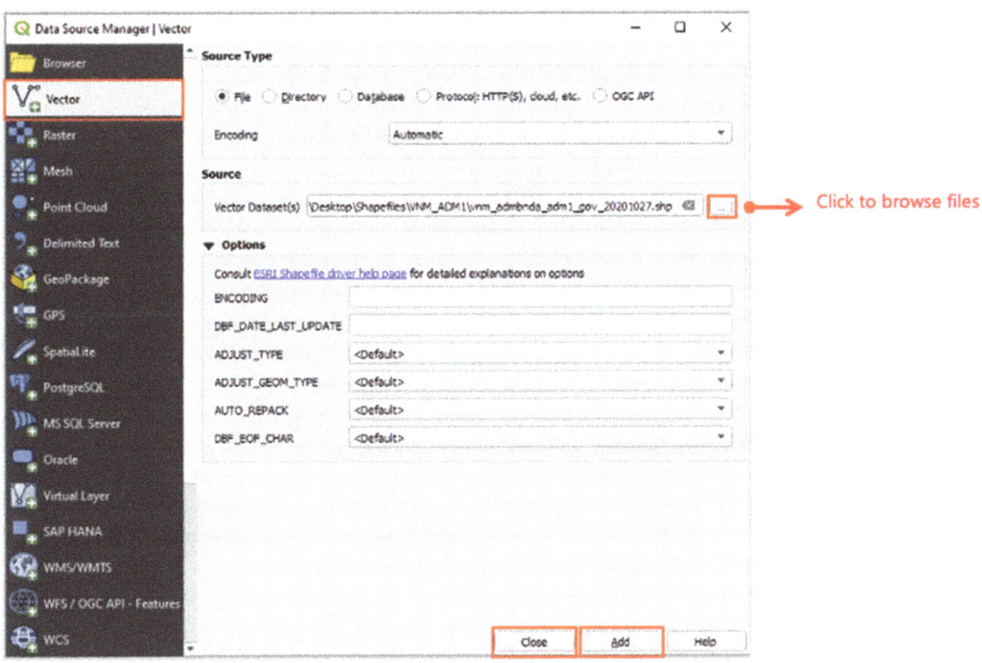

Source: Screenshot generated by Asian Development Bank using QGIS 3.34.14. 2025.

(iii) In the **Manage Layers Toolbar**, located in the leftmost part of the **Main Window**, click on the appropriate icon for adding a vector or raster layer (Figure 2.24).

Figure 2.24: Importing a Map Layer Using the Manage Layers Toolbar

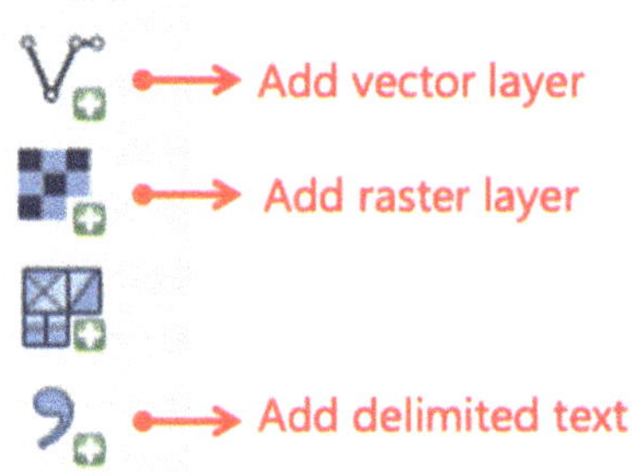

Source: Screenshot generated by Asian Development Bank using QGIS 3.34.14. 2025.

(iv) To add a layer to the project, go to the **Browser panel**, right-click the layer you want, and then click **Add Layer to Project** (Figure 2.25); or simply drag and drop the data into the **map area**. To add two or more layers at once, click and hold **Ctrl** and select the layers you want.

Figure 2.25: Importing a Map Layer to the Project from the Browser Panel

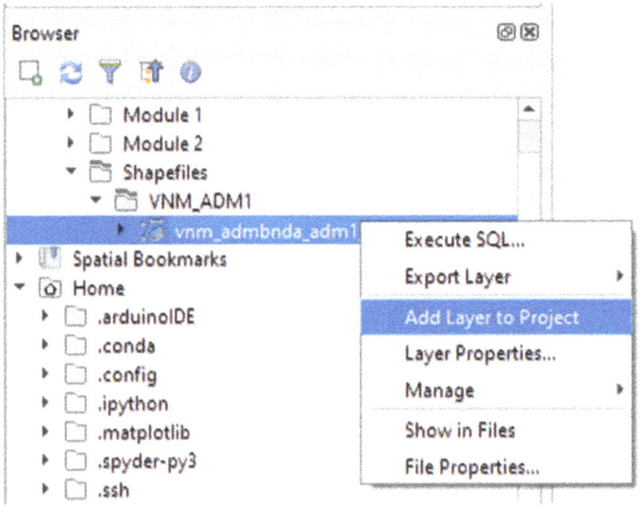

Source: Screenshot generated by Asian Development Bank using QGIS 3.34.14. 2025.

(v) Other options—such as **Delete File, Export Layer, Layer Properties**, and **File Properties—**are also available when you right-click on a layer. **Layer Properties** provides more information about the metadata and attributes, and includes tools for previewing the data. The **File Properties** dialog box displays information such as file type, creation date, and file location.

(vi) To facilitate quick access, a project folder can be added to the **Favorites** tab by right-clicking **Favorites | Add a Directory** (Figure 2.26).

Figure 2.26: Adding a Directory to Favorites via the Browser Panel

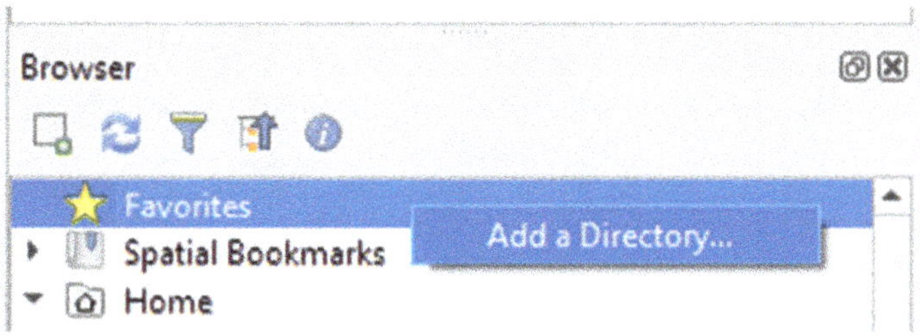

Source: Screenshot generated by Asian Development Bank using QGIS 3.34.14. 2025.

Below is a list of ways to import map vector layers relating to An Giang Province:

Import **Viet Nam—Subnational Administrative Boundaries** (ADM1) (polygon) by dragging and dropping it from the **Browser panel** into the **map area**.

Import **Points of Interest** (points) by clicking **Menu Toolbar | Layer | Add Layer | Add Vector Layer**.

Import **Roads** (lines) by clicking **Manage Layers Toolbar | Add Vector Layer**.

Import the **N10E104** (raster file) from the **Browser panel** by right-clicking it and selecting **Add Layer to Project**. Repeat similar steps for N10E105. Note that the land cover map is not automatically classified when added to the project. The method for classifying it will be discussed later in this module.

Add an Esri satellite map to the **map area** from **XYZ Tiles**. Right-click on **Esri Satellite**, under the **XYZ Tiles tab** in the **Browser panel**, and then select **Add Layer to Project**. Alternatively, you could select **Google Satellite** from the **XYZ Tiles tab** in the **Browser panel**, then drag and drop it into the **map area**.

2.3.4 Exporting Map Layers as New Layers

Exporting map layers as new layers in QGIS allows users to save a copy of the existing data with changes or in a different format. This is useful for organizing data, sharing specific features, or preparing layers for further analysis. The steps are as follows:

(i) Right-click on the layer you want to export in the **Layers panel**. Select **Export**, then **Save Features As** (Figure 2.27).

Figure 2.27: Exporting a Layer in QGIS

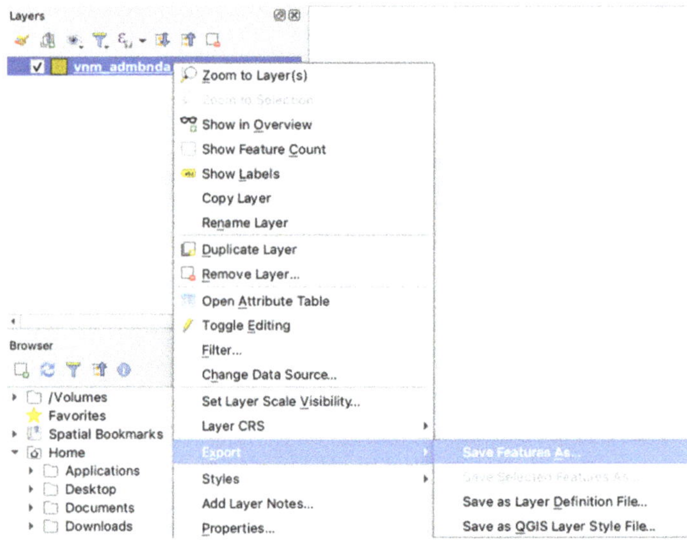

Source: Screenshot generated by Asian Development Bank using QGIS 3.34.14. 2025.

(ii) In the dialog box that appears, select a format under the **Format tab**—for instance, Esri Shapefile, GeoPackage, GeoJSON, or Keyhole Markup Language (KML) for vector data. Click **Browse** or this icon ... and set the **file name and location** where the new layer will be saved. Confirm or adjust the **CRS** if needed. Click **OK** to complete the export (Figure 2.28).

Figure 2.28: Configuring the Export Settings for a New Layer

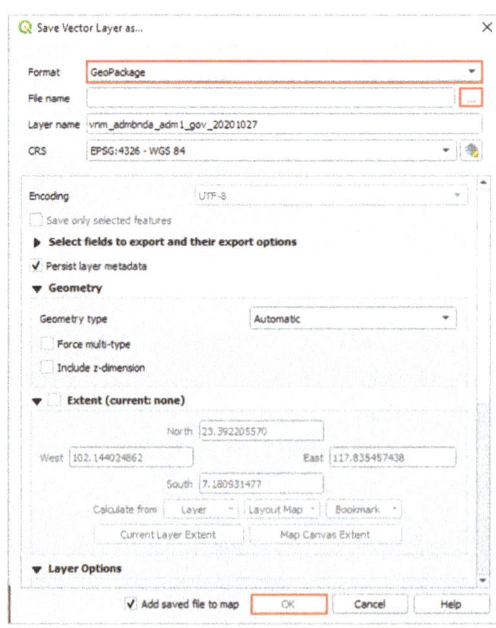

Source: Screenshot generated by Asian Development Bank using QGIS 3.34.14. 2025.

2.3.5 Interacting with Map Layers

1. Reordering Map Layers

Following the previous exercise, five active layers are now visible in the **map area** (Figure 2.29). These layers can be managed through the **Layers panel**, located to the left of the main window. The symbols in the legend indicate the differences among the point, line, polygon, and raster data layers.

Figure 2.29: The Layers Panel

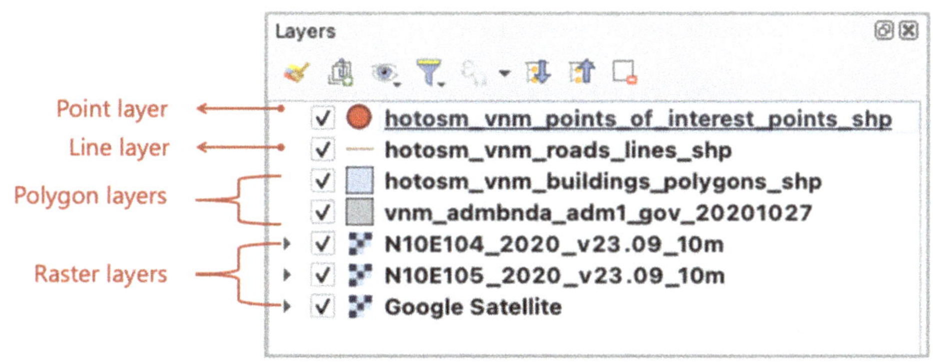

Source: Screenshot generated by Asian Development Bank using QGIS 3.34.14. 2025.

The order of the layers in the **Layers panel** can be rearranged by dragging the layers up and down using the mouse. The visibility of the map layers in the **map area** corresponds to the order set in the Layers panel. The visibility of the layers can be enabled and disabled by clicking the checkbox beside the layer symbol. A layer can be copied, renamed, duplicated, or removed from the project by right-clicking on the layer and selecting the function (Figure 2.30).

Figure 2.30: Copying, Renaming, Duplicating, and Removing a Map Layer

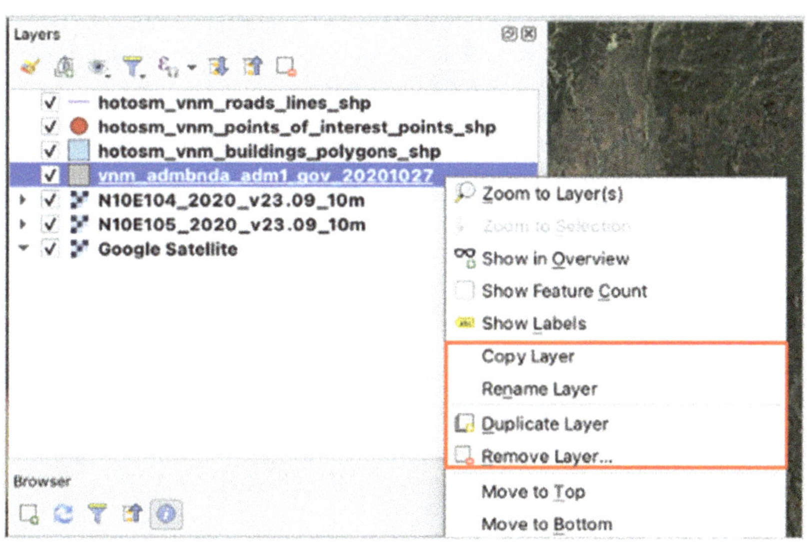

Source: Screenshot generated by Asian Development Bank using QGIS 3.34.14. 2025.

(i) Rename the Points-of-Interest layer under the category of **Landmarks,** the Lines layer under **Roads,** the Buildings (polygons) layer under **Buildings,** and the administrative boundary under **Boundaries.** To do this, right-click on the layer, and select **Rename.**

(ii) Arrange the layers according to the following line-up from top to bottom: Landmarks, Roads, Buildings, Boundaries, Land Cover (N10E104 and N10E105), and Google Satellite, by dragging the layers up and down in the **Layers panel.**

(iii) Switch off/disable the **Land Cover** (N10E104 and N10E105) layers in the **map canvas** by clicking the check mark next to the name of each layer. Note that the **Google Satellite map** is now visible.

2. Navigating the Map Area

Map navigation can be performed using the mouse (by dragging, scrolling, and clicking on the map area) or through the **Map Navigation toolbar** (Figure 2.31).

Figure 2.31: Map Navigation Toolbar

1 Pan Map *(moves the map view in any direction)*
2 Pan Map to Selection *(centers the map view on selected features)*
3 Zoom In *(increases the map magnification)*
4 Zoom Out *(decreases the map magnification)*
5 Zoom Full *(adjusts the map view to show the entire extent of all layers)*
6 Zoom to Selection *(zooms in to the extent of the selected features)*
7 Zoom to Layer *(adjusts map view to the full extent of the selected layer)*
8 Zoom to Native Resolution *(displays the map at the layer's original resolution)*
9 Zoom to Last *(returns to previous map view in navigation history)*
10 Zoom to Next *(moves forward to the next map view in navigation history)*
11 New Map View *(opens a new 2D map view)*
12 New 3D Map View *(creates a new 3D map view)*
13 New Spatial Bookmark *(saves the current map extent as bookmark)*
14 Show Spatial Bookmarks *(displays a list of all saved spatial bookmarks)*
15 Temporal Controller Panel *(manages temporal data and animations over time)*
16 Refresh *(updates the map view to reflect changes)*

Source: Screenshot generated by Asian Development Bank using QGIS 3.34.14. 2025.

An alternative method for zooming to the full extent of a layer is to right-click on the layer in the **Layers panel**, and then select **Zoom to Layer(s)** (Figure 2.32).

Figure 2.32: Zoom to Layer(s) in the Layers Panel

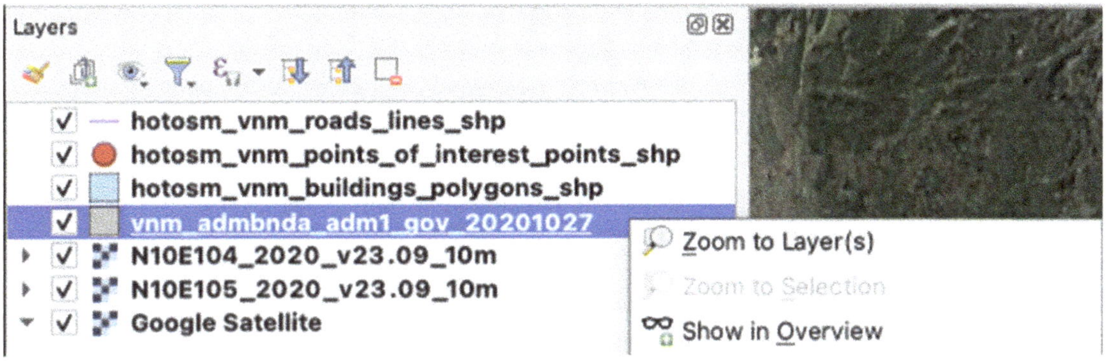

Source: Screenshot generated by Asian Development Bank using QGIS 3.34.14. 2025.

The zoom level can be set at a fixed scale by clicking the current map scale (formatted as x:y) in the **status bar**, located at the bottom of the main window (Figure 2.33).

Figure 2.33: Fixed Scales in the Status Bar

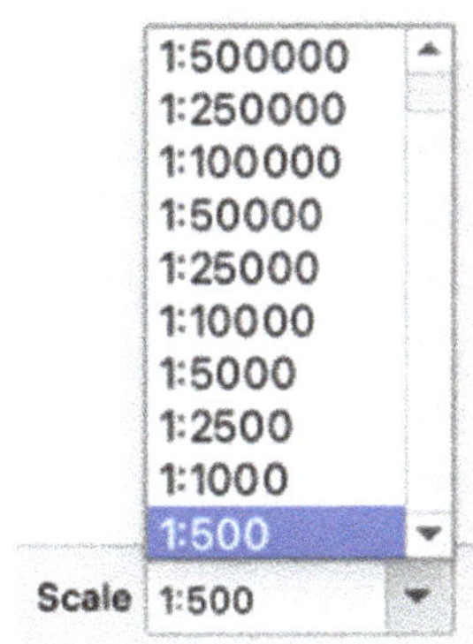

Source: Screenshot generated by Asian Development Bank using QGIS 3.34.14. 2025.

3. Identifying Data

Features within the map area can be identified by using the **identify tool**, located in the **Attributes toolbar**. This tool enables access to information on attributes and selected features.

To identify a point or polygon on the map, click the **identify tool**. The selected feature will be highlighted in the map area, and the **Identify Results dialog box** will be displayed to the right of the main window, containing information on the selected feature (Figure 2.34). This dialog box includes data retrieved from the **attribute table**, which will be discussed in detail in Module 4.

Figure 2.34: Finding Features Using the Identify Results Dialog Box

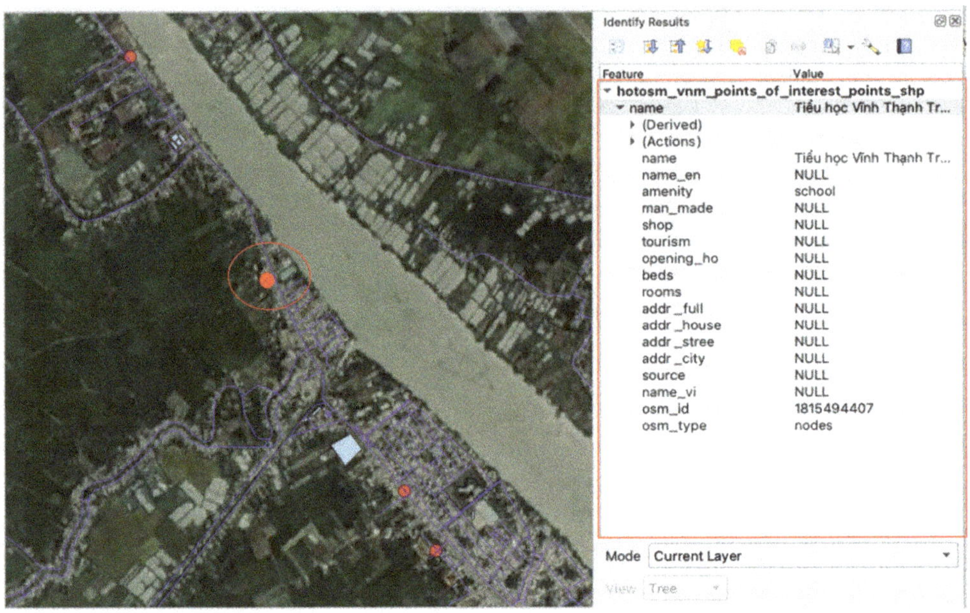

Source: Screenshot generated by Asian Development Bank using QGIS 3.34.14. 2025.

In the **Identify Results dialog box**, various methods are available for identifying features by area (polygon, freehand, radius) or through a single-click function. Explore the identifying methods shown in Figure 2.35.

Figure 2.35: Identifying Features by Area or by a Single-Click Function

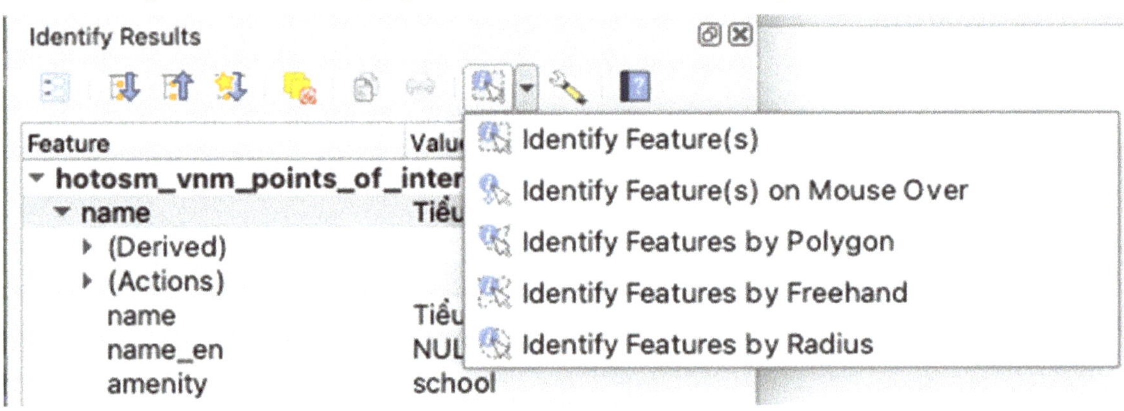

Source: Screenshot generated by Asian Development Bank using QGIS 3.34.14. 2025.

4. Checking Layer Properties

The **Layer Properties dialog box** provides general settings for managing the data layers. Each data layer's properties contain settings such as information, source, symbology, metadata, and labels. The Layer Properties settings for vector and raster layers are different in some ways and similar in others.

To open Properties for a layer, double-click on the layer in the **Layers panel** or right-click on **Properties** (Figure 2.36).

Figure 2.36: Opening Properties for a Layer

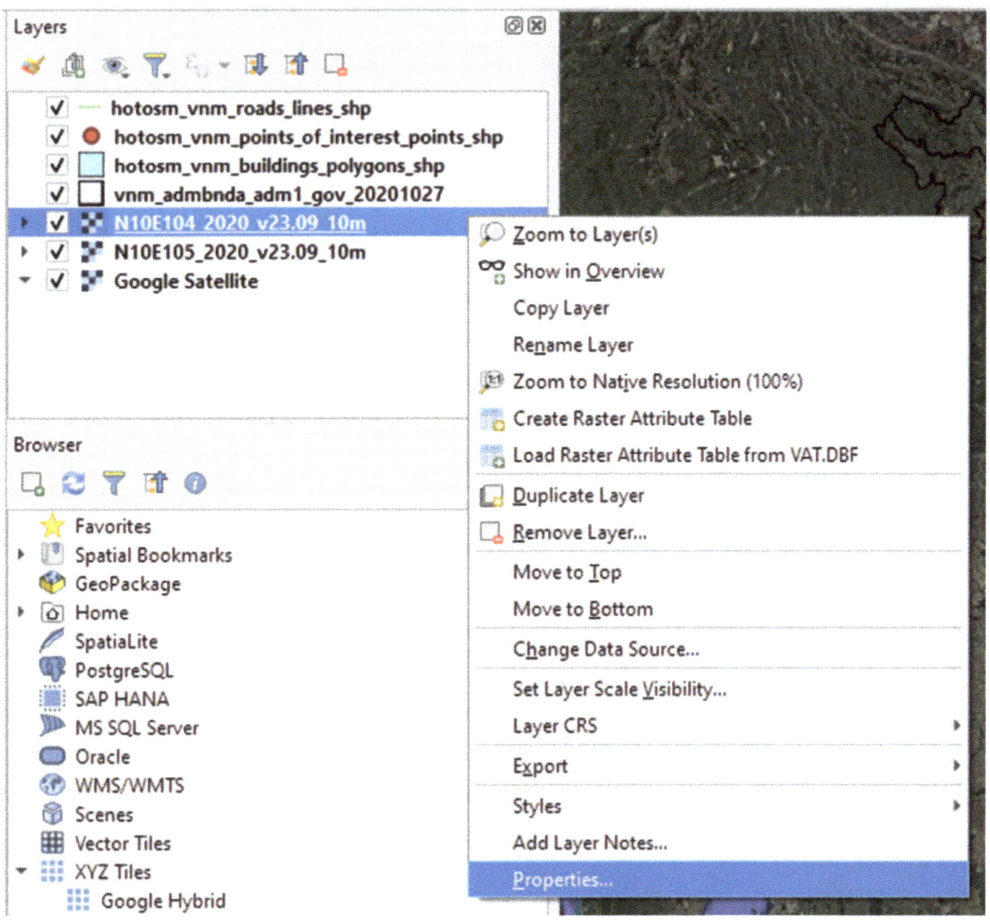

Source: Screenshot generated by Asian Development Bank using QGIS 3.34.14. 2025.

The common tab settings available for both vector and raster layers in QGIS include **Information, Source, Symbology, Rendering, Temporal, Metadata, Legend**, and **QGIS Server**. The **Layer Properties** interface for vector layers is shown in Figure 2.37, while the corresponding interface for raster layers is presented in Figure 2.38.

(i) The **Information** tab is read-only and provides information about the data type, the directory where the data layer was saved, CRS data, type of features, units, etc.
(ii) The **Source** tab provides options for configuring general layer settings, such as assigning a coordinate system to the layer (without reprojecting), creating a spatial index for an OGR-supported file format, and accessing the query builder.[1]
(iii) The **Symbology** tab is used to define or modify the visual style of the layer. The available options are different for vector and raster layers.
(iv) The **Metadata** section provides information and attributes concerning the layer.
(v) The **Rendering** section controls the range of scale at which the layer can be visible.
(vi) The **Legend** section provides advanced configuration options that determine how the layer appears in the Layers Panel.
(vii) The **QGIS Server** section enables the definition of the data layer's properties, such as the description, attributes, metadata-URL, and legend-URL.

Figure 2.37: Properties of a Vector Layer

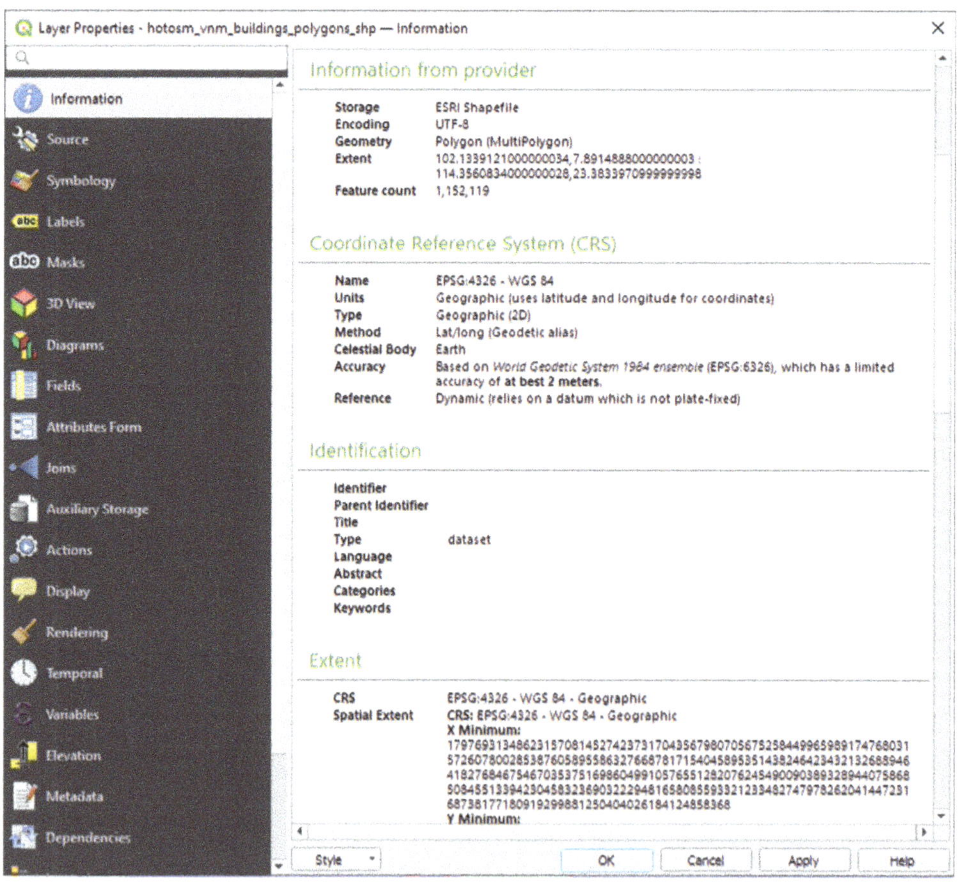

Source: Screenshot generated by Asian Development Bank using QGIS 3.34.14. 2025.

1 "OGR" used to stand for "OpenGIS Simple Features Reference Implementation," and then for "OGR Simple Features Library," but now only the abbreviation is used as the name.

Figure 2.38: Properties of a Raster Layer

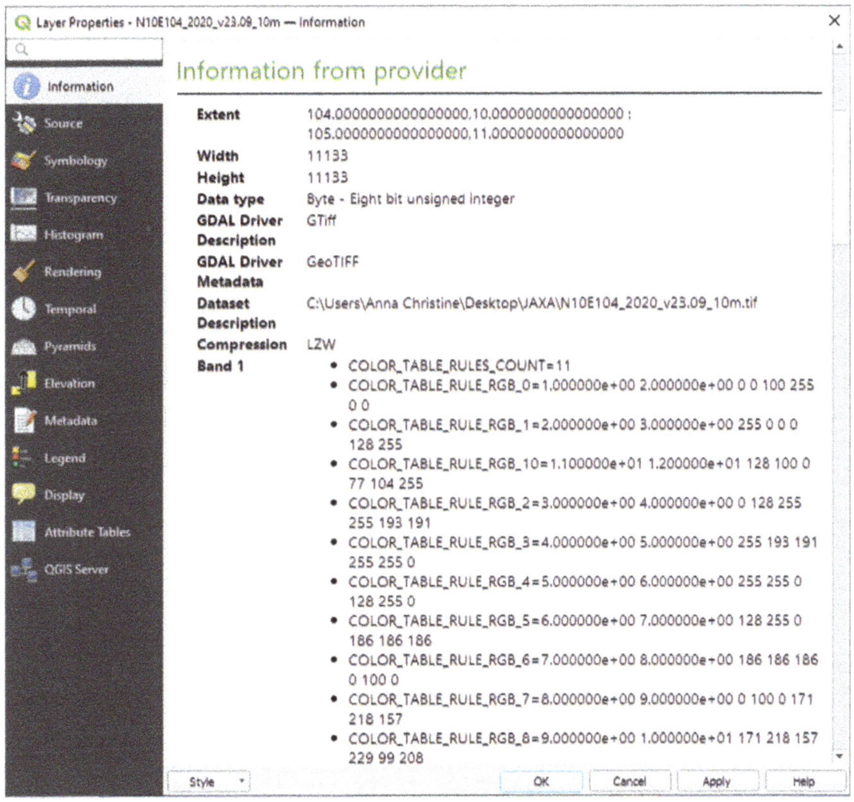

Source: Screenshot generated by Asian Development Bank using QGIS 3.34.14. 2025.

The key tabs in the **Vector Layer Properties** are **Labels** (used to configure labels in the layer), **Fields** (provides access to attribute-field information and tools for organizing, editing, and customizing fields), **Join** (enables linking the current vector layer to another vector layer, such as tabular data based on a common attribute, allowing additional data to be displayed and used), and **Diagram** (supports the addition of graphical overlays to a vector layer, such as pie charts or bar diagrams, to visually represent attribute data on the map).

The frequently used tabs of the **Raster Layer Properties** are **Transparency** (allows control over the visibility of the raster by setting the transparency levels or defining specific pixel values as transparent), **Pyramids** (used to create lower-resolution copies of a high-resolution raster layer), and **Histogram** (provides a visual representation of the distribution of pixel values in the raster layer, supporting analysis and adjustments to rendering settings).

2.4 Visualizing and Styling Data Layers

By default, QGIS automatically styles the data layers when they are imported. This default styling is useful for the initial visual inspection of the layers. But to make the data more intuitive, QGIS provides options for customizing the styling of each data layer. The styles can also be saved for future use.

There are two ways to open the **Layer Styling panel:**

(i) by right-clicking the map layer in the **Layers panel | Properties | Symbology**; and

(ii) by clicking the **Symbology** icon in the **Layers panel**. The Layer Styling panel will be displayed on the far left of the map display. This panel becomes available only when a vector or raster layer is loaded into the Layers panel. Although it reduces the visible area of the map display, it allows simultaneous visualization of both the map and the styling options, making it convenient for adjusting layer appearance.

The subsections below present exercises involving the styling of four types of data layers: point, line, polygon, and raster. They are imported from a previous activity (hyperlink exercise). As previously noted, each vector feature and raster layer offers distinct styling options. The different options allow for tailored visual representation.

Styling Vector Layers

2.4.1 Styling Points—Landmarks

(i) Continuing our previous exercise, select the **Points of Interest | Landmarks layer** in **the Layers panel**. Click to open the **Layer Styling panel** (Figure 2.39).

Figure 2.39: Styling Points in the Layers Panel

Source: Screenshot generated by Asian Development Bank using QGIS 3.34.14. 2025.

In the topmost tab of the Layer Styling panel, the Points of Interest layer is automatically selected. To shift to styling other layers, click the drop-down menu (Figure 2.40). If you hover the mouse over the layers, pop-up information about the CRS and directory of each layer will be displayed.

Figure 2.40: Selecting Features in the Layer Styling Panel

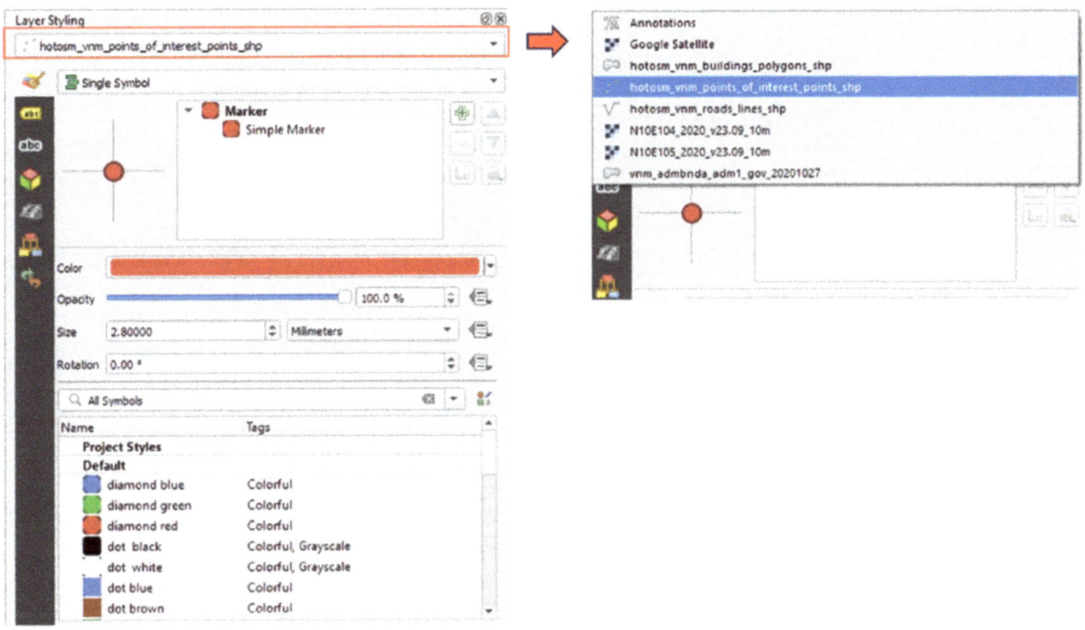

Source: Screenshot generated by Asian Development Bank using QGIS 3.34.14. 2025.

On the left side of the Layer Styling panel, there are seven tabs for rendering layers: Symbology, Labels, Mask, 3D View, Shading Renderer, Style Manager, and History. These tabs are commonly used for any features rendered in the Layer Styling panel. With the exception of the Symbology tab, the settings vary for each feature (points, lines, polygons) selected (Figure 2.41).

Figure 2.41: Tabs in the Layer Styling Panel (Vector Layers)

Source: Screenshot generated by Asian Development Bank using QGIS 3.34.14. 2025.

(ii) There are several ways to render point features (Figure 2.42). In this activity, the **Categorized** option will be used to identify the various establishments on the map.

Figure 2.42: Options for Rendering Point Features

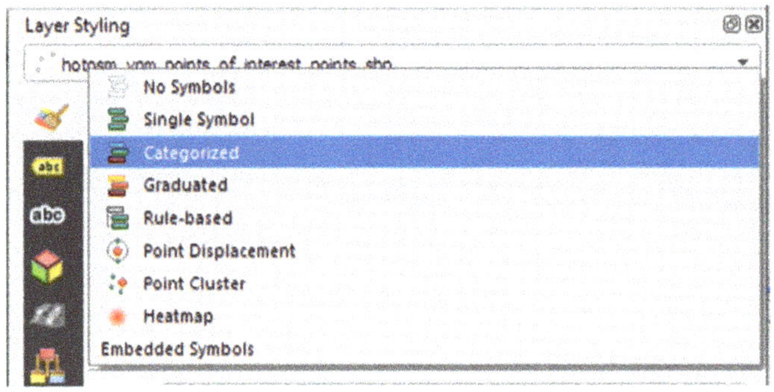

Source: Screenshot generated by Asian Development Bank using QGIS 3.34.14. 2025.

To understand the functions and view examples of other options for rendering points—such as No Symbols, Single Symbol, Graduated, Rule-based, or Point Displacement—refer to QGIS User Guide. 16. Working with Vector Data. 16.1.3.1. Features rendering.

(iii) In the Layer Styling panel, select **amenity** and click **Apply** to categorize the points. The amenity value defines the category of each point that is embedded in the attribute table. Note the classifications chosen under the Symbol, Value, and Legend tabs (Figure 2.43).

Figure 2.43: Types of the Renderers of Point Features

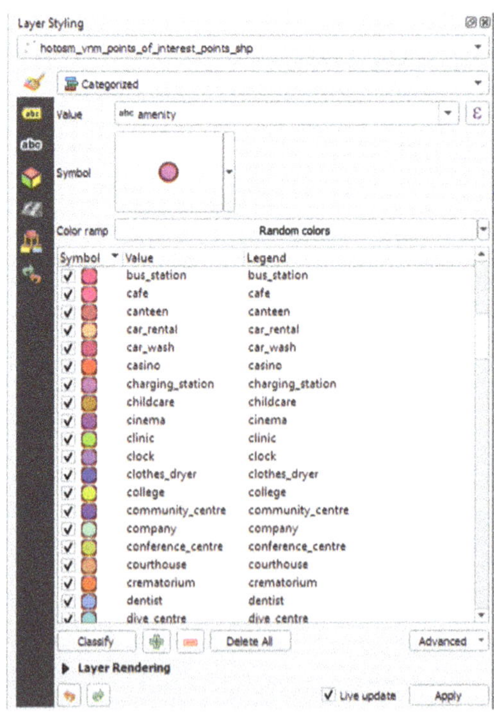

Source: Screenshot generated by Asian Development Bank using QGIS 3.34.14. 2025.

(iv) To further define the marker, shape, color, size, opacity, etc. of the point symbols, click on the **Symbol** tab, and explore the settings. Try selecting **Simple Marker** (Figure 2.45a) and explore the different Symbol Layer types, as well as the other settings (Figure 2.44b). Click **Apply** to save the selections. Click ◀ to go back to the main Layer Styling panel.

Figure 2.44: Symbol Settings in the Layer Styling Panel for Point Features

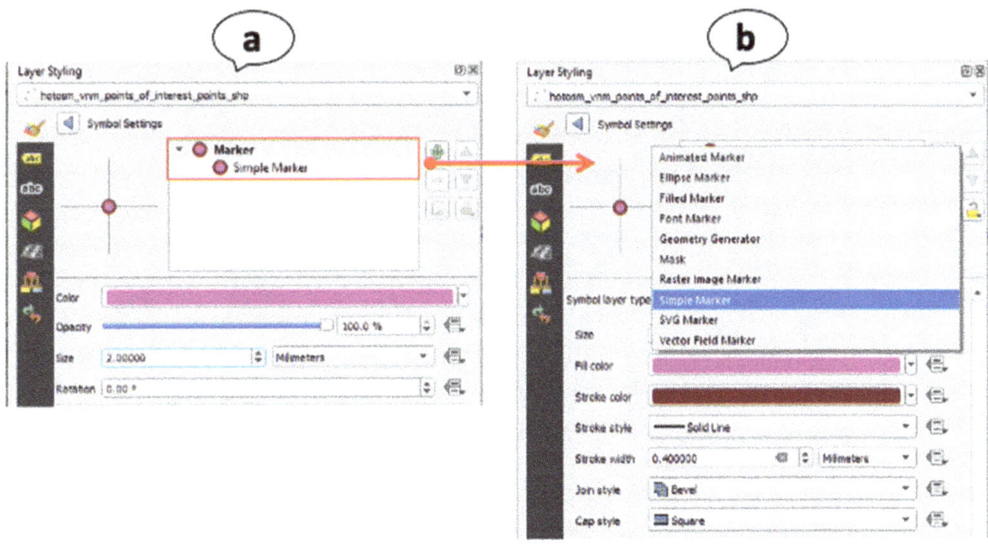

Source: Screenshot generated by Asian Development Bank using QGIS 3.34.14. 2025.

2.4.2 Styling Lines—Roads

(i) Switch on the **Roads layer** and **Google Satellite map**, and switch off the other layers. Select **Roads** in the Layers panel and click the Symbology icon. In the Layer Styling panel, select **Roads** and use the **Single Symbol**. Under the **Line** tab, select **Simple Line**. Change the **stroke width** to 0.200 millimeters, and the **stroke style** to a dashed line. Click **Apply** to finalize (Figure 2.45).

Figure 2.45: Styling Line—An Example of Roads

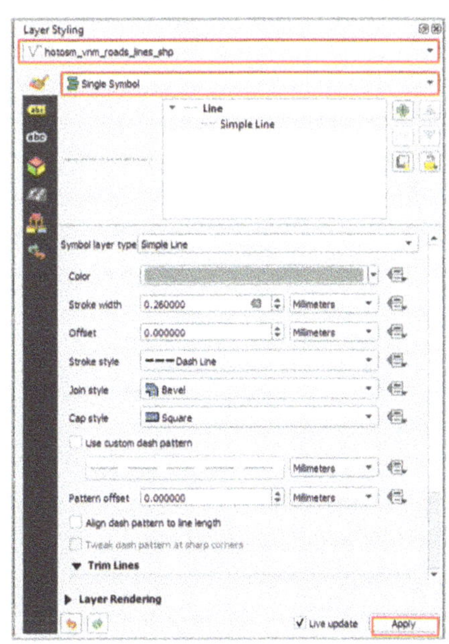

Source: Screenshot generated by Asian Development Bank using QGIS 3.34.14. 2025.

The layer should look like Figure 2.46.

Figure 2.46: Overlay of Road Data on Satellite Imagery of An Giang Province, Viet Nam

Source: Screenshot generated by Asian Development Bank using QGIS 3.34.14. 2025.

There are several methods for rendering line features (Figure 2.47). In this exercise, the **Single Symbol** option has been selected. This method applies a uniform, user-defined symbol to all features within the layer, thereby ensuring consistent visual representation across the dataset. To learn more about other renderer types, refer to QGIS User Guide. 12. Working with Vector Data. 12.1.3.1. Features rendering

Figure 2.47: Options for Line Rendering

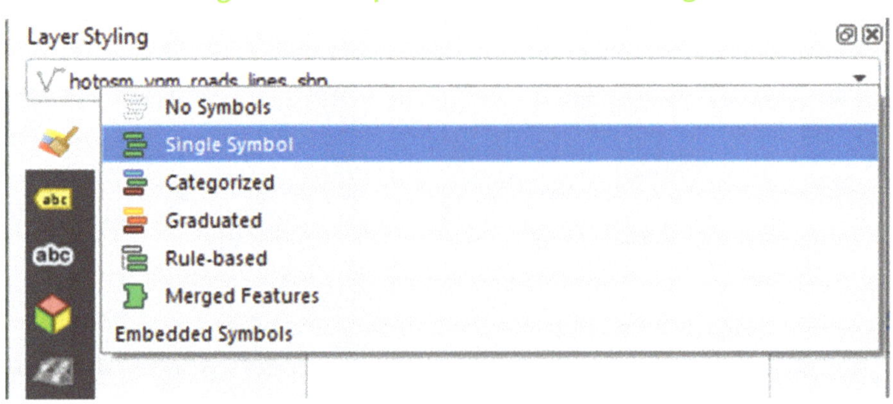

Source: Screenshot generated by Asian Development Bank using QGIS 3.34.14. 2025.

2.4.3 Styling Polygons—Administrative Boundaries

(i) Turn on the **boundary and road layers** in the **Layers panel**, and switch off the other layers. Select **Boundaries** in the Layers panel and click the **Symbology** icon.

(ii) In the **Layer Styling panel**, select **Boundaries** and use **Single Symbol**. Under the **Fill tab**, select **Simple Fill**. Then select **No Brush** under **Fill style** and black as **Stroke color**. Change the stroke width to 0.260 millimeters, and the stroke style to a solid line (Figure 2.48a).

Figure 2.48: Styling Polygons—An Example of Administrative Boundaries

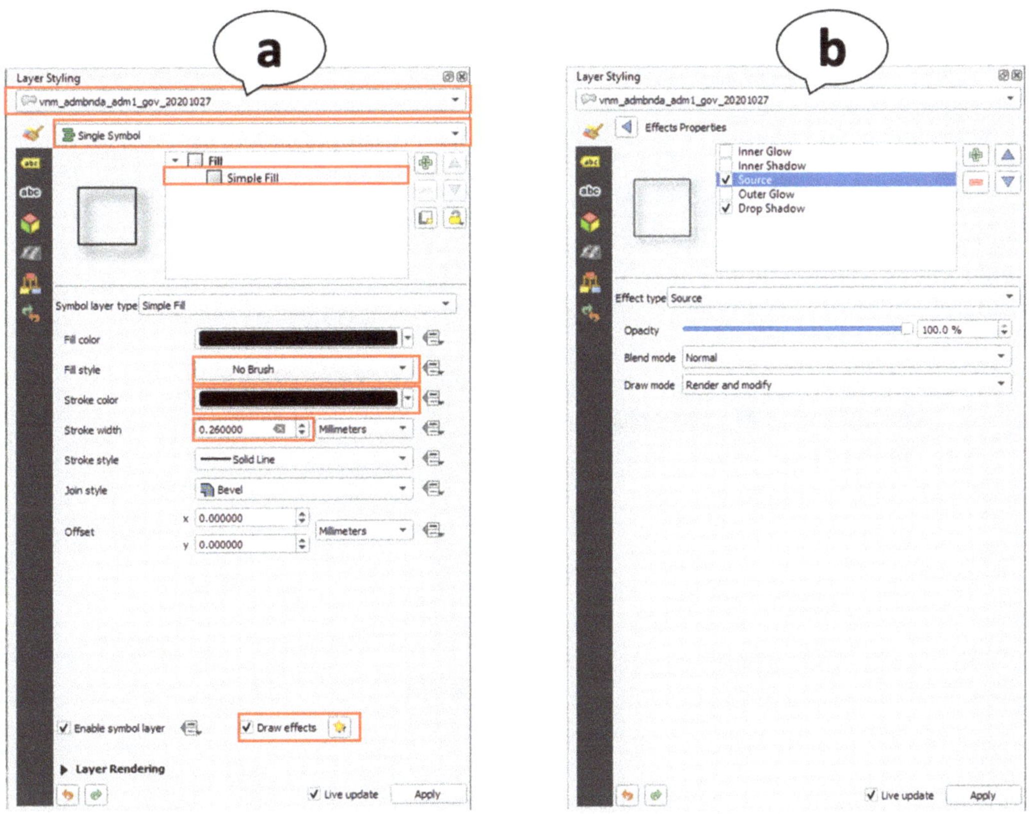

Source: Screenshot generated by Asian Development Bank using QGIS 3.34.14. 2025.

(iii) Select **Draw effects** and click the customizing effects icon (Figure 2.48a). Under **Effects Properties**, select **Source and Drop Shadow** (Figure 2.48b). These effects will provide a shadow for a more defined boundary line (Figure 2.50).

(iv) To add labels to the boundaries of a polygon, select **Single Labels** (Figure 2.49a). Select ADM1_EN for Value, which will represent the district boundaries in Viet Nam. Click the **ABC tab**, and select the font, style, size, color, etc. to make the label visible. Save the settings and click **Apply**.

Figure 2.49: Styling Polygons—Adding Labels

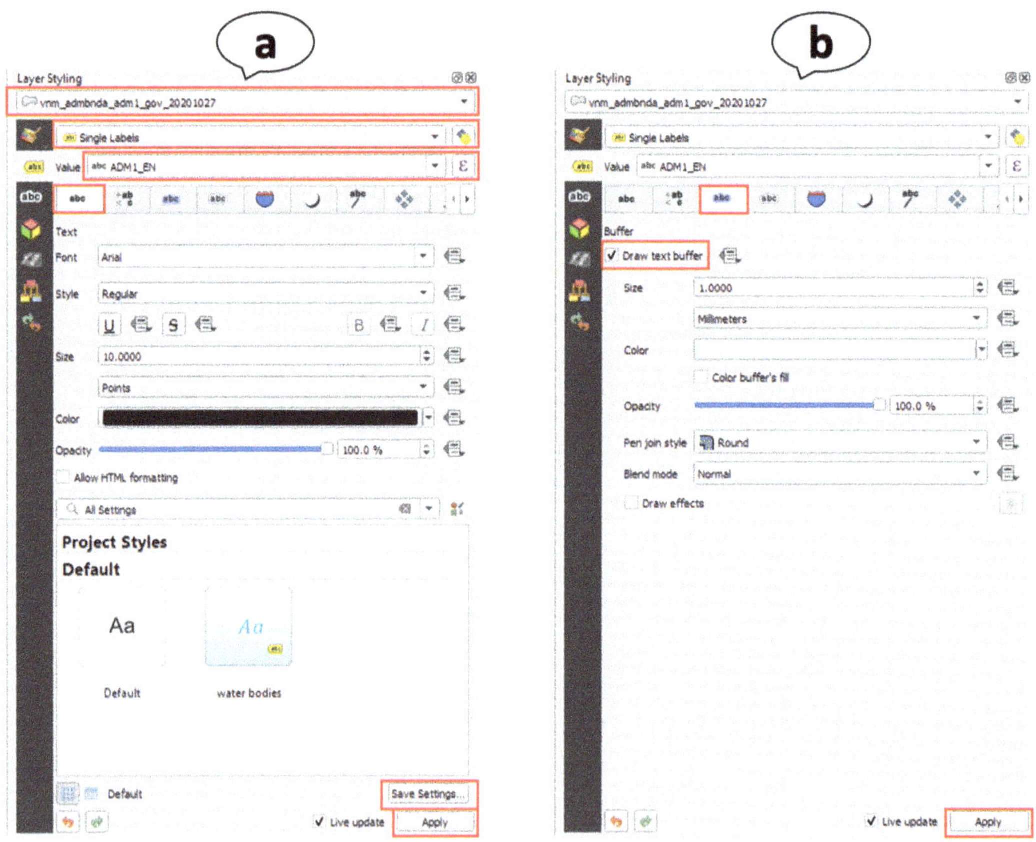

Source: Screenshot generated by Asian Development Bank using QGIS 3.34.14. 2025.

(v) To make the labels more visible, add a **Buffer** on them (Figure 2.49b). Click the abc tab. Check the option **Draw text buffer**. Change the size and color, as necessary. Then click **Apply**.

Finally, the boundary and road layers should look like Figure 2.50:

Figure 2.50: Administrative Boundary Layer—An Giang Province, Viet Nam

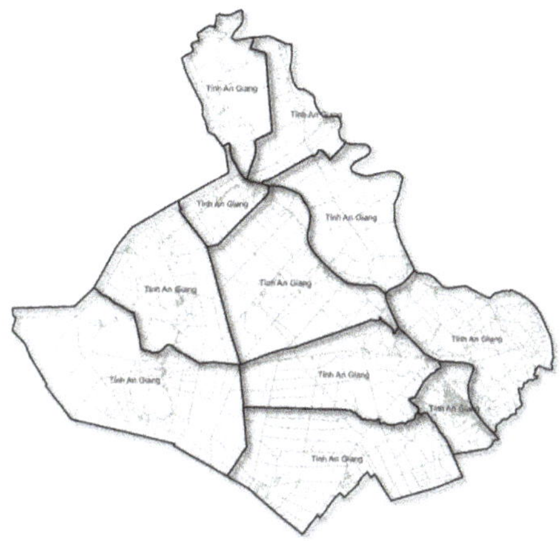

Note: The administrative boundaries are those of An Giang Province, Viet Nam, prior to its merger with Kien Giang Province in June 2025. As such, they do not reflect the newly configured administrative boundaries.

Source: Screenshot generated by Asian Development Bank using QGIS 3.34.14. 2025.

2.4.4 Styling Raster Layers—Land Cover Map

(i) Turn on the **Land Cover** layers. To check the legend of the raster layer, right-click on layer N10E104 and select **Layer Properties**. Go to the **Attribute Tables tab**, and select **New attribute table from current symbology**. A prompt will appear; select **Managed by the data provider** (Figure 2.51).

Figure 2.51: Opening the Raster Attribute Table

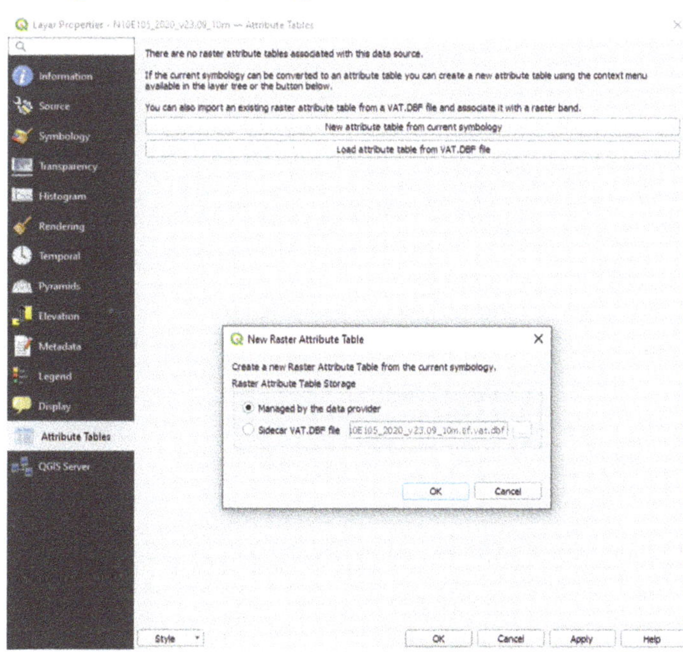

Source: Screenshot generated by Asian Development Bank using QGIS 3.34.14. 2025.

The land cover map of Viet Nam was downloaded from the Japan Aerospace Exploration Agency (JAXA) Earth Observation Research Center (EORC) to provide basic information for a variety of applications, such as biodiversity evaluation, countermeasures for disasters triggered by natural hazards, and urban planning. This 2020 map, with a 10-meter resolution, was created using Sentinel-2, Sentinel-1, and PALSAR-2 ScanSAR time series satellite data. It features land use and land cover (LULC) classifications for Viet Nam based on 12 categories. The entire country map comprises several raster tiles; however, for this exercise, only the two tiles covering An Giang Province were used.

The 12 LULC categories according to JAXA are listed in Figure 2.52:

Figure 2.52: The Categories for the Raster Attribute Table

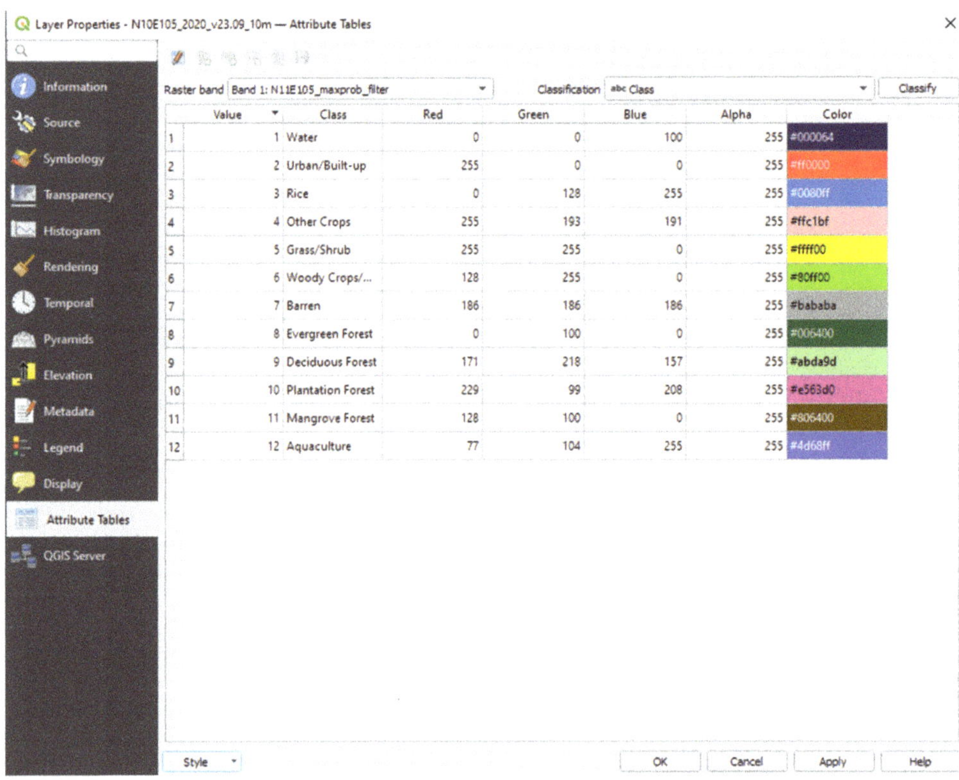

Source: Screenshot generated by Asian Development Bank using QGIS 3.34.14. 2025.

(ii) Go to the **Layers panel**, and open the **Layer Styling panel** by clicking on the icon: . Note that, in the Value section, the layer contains 255 values, or "bands." For the purpose of the map, only the bands corresponding to the classified land uses will be shown (Figure 2.53).

(iii) To classify the LULC map layer, select **Paletted/Unique values**, and then select **Classify**.

Figure 2.53: Default Grayscale Symbology for Single-Band Raster Values

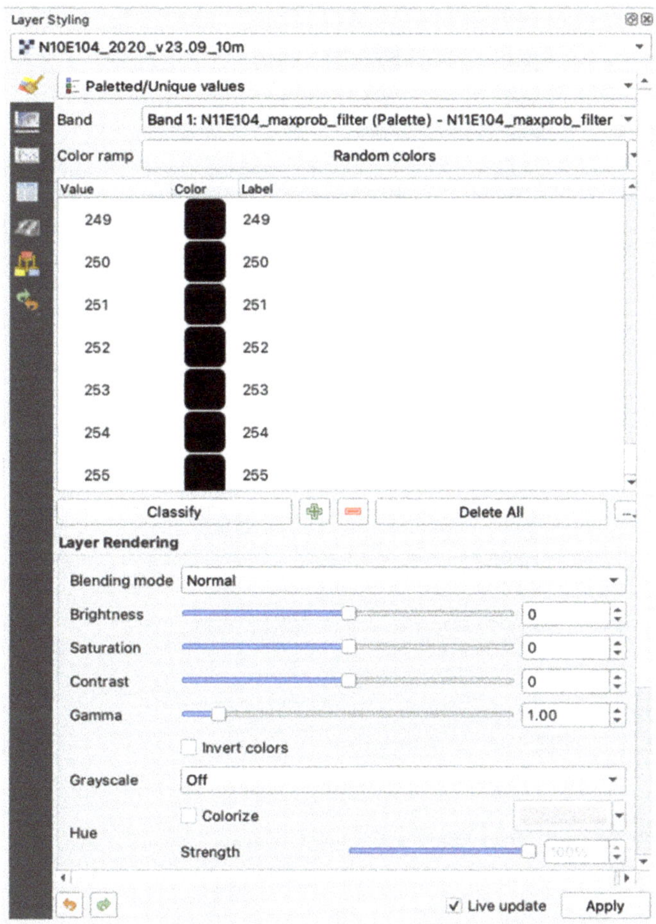

Source: Screenshot generated by Asian Development Bank using QGIS 3.34.14. 2025.

(iv) After classification, the map will display different colors corresponding to the land use classes. Right-click on the color of each class to change its color, opacity, or label. Change the label manually according to the classification provided in the **attribute table** (Figure 2.54).

Figure 2.54: Applying Unique Value-Color Palettes to Classify Land Cover Types

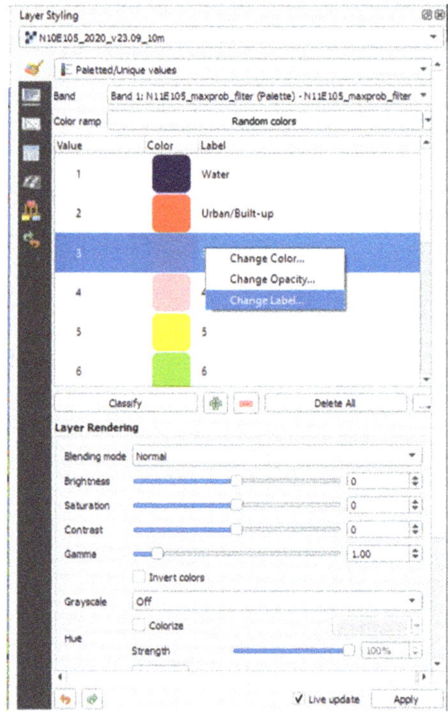

Source: Screenshot generated by Asian Development Bank using QGIS 3.34.14. 2025.

(v) The land cover map with the overlaid boundary layer should look like Figure 2.55. Note that the classifications are also depicted in the Layers panel.

Figure 2.55: Land Cover and Boundary-Layer Map of An Giang Province, Viet Nam

Note: The administrative boundaries are those of An Giang Province, Viet Nam, prior to its merger with Kien Giang Province in June 2025. As such, they do not reflect the newly configured administrative boundaries.

Source: Screenshot generated by Asian Development Bank using QGIS 3.34.14. 2025.

To further redefine the map, turn on the **base map (Google Satellite), boundary layer,** and **road layer**. Click on the **Layer rendering tab**, and select **Dodge** from among the **blending modes** (Figure 2.56). Explore other blending modes.

Figure 2.56: Blending the Satellite Data Layer with the Land Cover Map of An Giang Province, Viet Nam

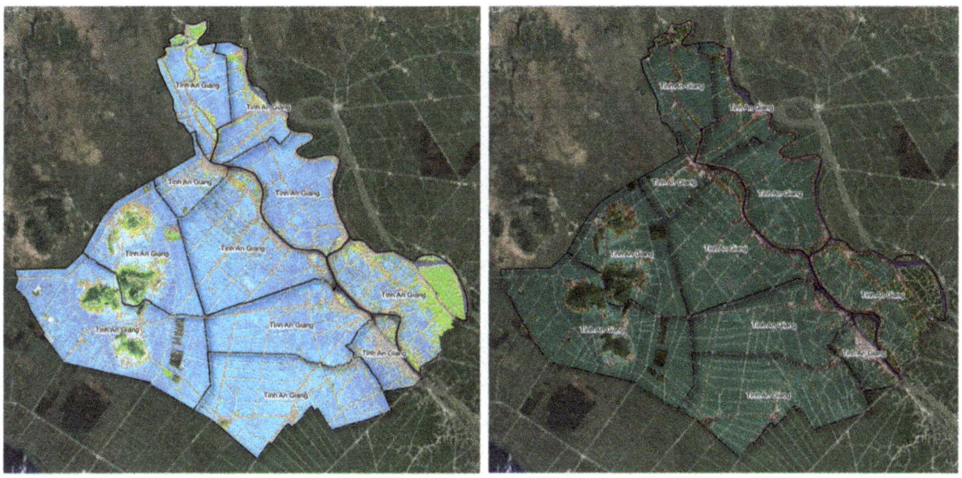

Notes: The administrative boundaries are those of An Giang Province, Viet Nam, prior to its merger with Kien Giang Province in June 2025. As such, they do not reflect the newly configured administrative boundaries. This image shows a land cover map of An Giang Province, Viet Nam, with different blending modes applied: dodge on the left and multiply on the right.

Source: Screenshot generated by Asian Development Bank using QGIS 3.34.14. 2025.

Figure 2.57: Blending Mode Settings in the Layer Styling Panel

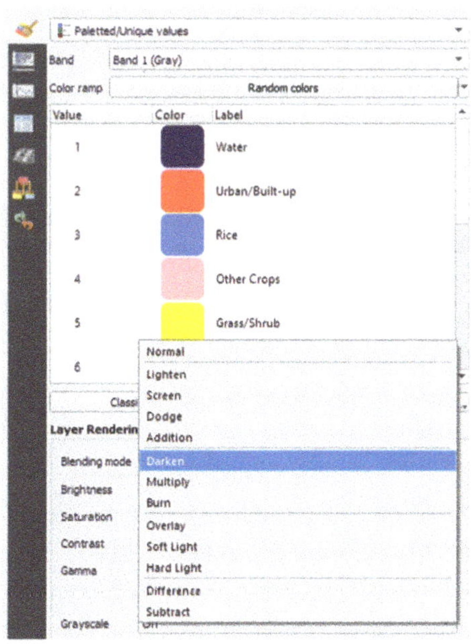

Source: Screenshot generated by Asian Development Bank using QGIS 3.34.14. 2025.

To learn more about other renderer types, refer to QGIS User Guide. 13. Working with Raster Data. 13.1.3.1. Band rendering.

2.5 Where to Find Data to Download

Listed below are links to free online archives and other reputable sources of global raster and vector data.

(i) European Space Agency WorldCover (2020 and 2021): https://esa-worldcover.org/en
(ii) ArcGIS Hub: http://opendata.arcgis.com/
(iii) Natural Earth: http://www.naturalearthdata.com/downloads/
(iv) US Geological Survey (USGS) EarthExplorer: https://earthexplorer.usgs.gov/
(v) GIS Geography OpenStreetMap: https://gisgeography.com/openstreetmap-download-osm-data/
(vi) National Aeronautics and Space Administration (NASA) Socioeconomic Data and Application Center: http://sedac.ciesin.columbia.edu/
(vii) OpenTopography: http://www.opentopography.org/
(viii) United Nations Environment Programme Environmental Data Explorer: http://geodata.grid.unep.ch/
(ix) NASA Earth Observations: http://neo.sci.gsfc.nasa.gov/
(x) Sentinel Satellite Data: https://scihub.copernicus.eu/dhus/
(xi) IPUMS Terra: https://terra.ipums.org/
(xii) Food and Agriculture Organization of the United Nations GeoNetwork: https://www.fao.org/land-water/databases-and-software/geonetwork/en/
(xiii) Geofabrik.de (OSM data and shapefiles): https://www.geofabrik.de/data/download.html
(xiv) DIVA-GIS: https://www.diva-gis.org/gdata
(xv) SoilGrids: https://soilgrids.org/
(xvi) USGS Global Food-and-Water Security-Support Analysis Data: https://www.usgs.gov/centers/western-geographic-science-center/science/global-food-and-water-security-support-analysis
(xvii) GADM Maps and Data: https://gadm.org/

The list above is not exhaustive. It is largely drawn from GIS Geography, at https://gisgeography.com/best-free-gis-data-sources-raster-vector/, in addition to various archives that were added.

Module 3 covers basic raster analysis, with section 3.2 focusing on the process of downloading raster data from open-access platforms.

The following exercise demonstrates how to download raster data from ESA WorldCover. This raster file, specifically an LULC map, is commonly used in agricultural applications such as area frame mapping and sample design. LULC maps identify land cover types and support more accurate and efficient agricultural surveys and planning.

Access to the data requires registration for a free account at https://viewer.esa-worldcover.org/worldcover/. This account is necessary to enable the full download features and access to additional services provided by the platform.

(i) Go to the WorldCover viewer, at https://viewer.esa-worldcover.org/worldcover/. Log into the viewer using the Terrascope credentials.
(ii) In the map settings, located on the right-hand side of the interface, regional boundaries can be enabled to help locate the area of interest (AOI) and make it visible (Figure 2.58).

Figure 2.58: Enabling Administrative Boundaries in the WorldCover Map Settings

Source: European Space Agency. Layers: WorldCover V2 2021. https://viewer.esa-worldcover.org/worldcover/.

(iii) Navigate to the area of interest using the map; or search for a location, such as *An Giang, Viet Nam*, using the search bar (Figure 2.59).

Figure 2.59: Navigating to the Area of Interest—An Giang Province, Viet Nam

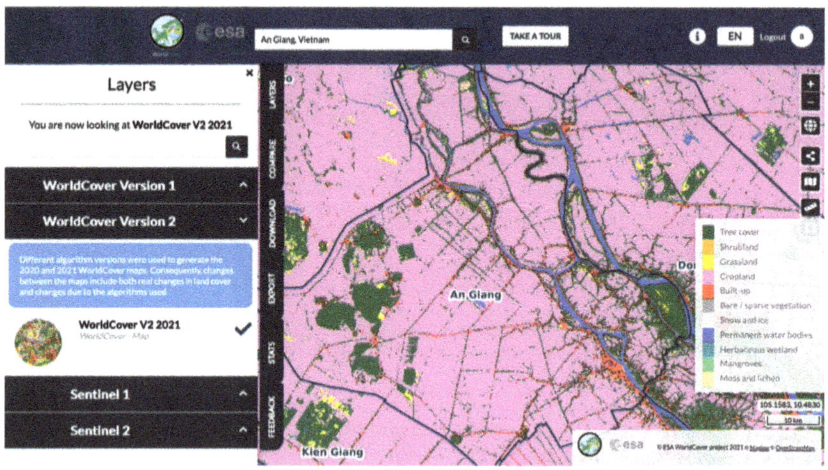

Source: European Space Agency. Layers: WorldCover V2 2021. https://viewer.esa-worldcover.org/worldcover/.

(iv) Open the **Download** tab in the viewer. Several options are available, including *WorldCover Version 1* (2020) and *WorldCover Version 2* (2021). For this exercise, select Version 2, which provides a classified land cover map with a 10-meter resolution. There are also multiple options for defining the extent of the download. For this exercise, choose **Draw area of interest**, and draw a bounding box around the borders of An Giang Province (Figure 2.60).

Figure 2.60: Defining the Area of Interest for Downloading

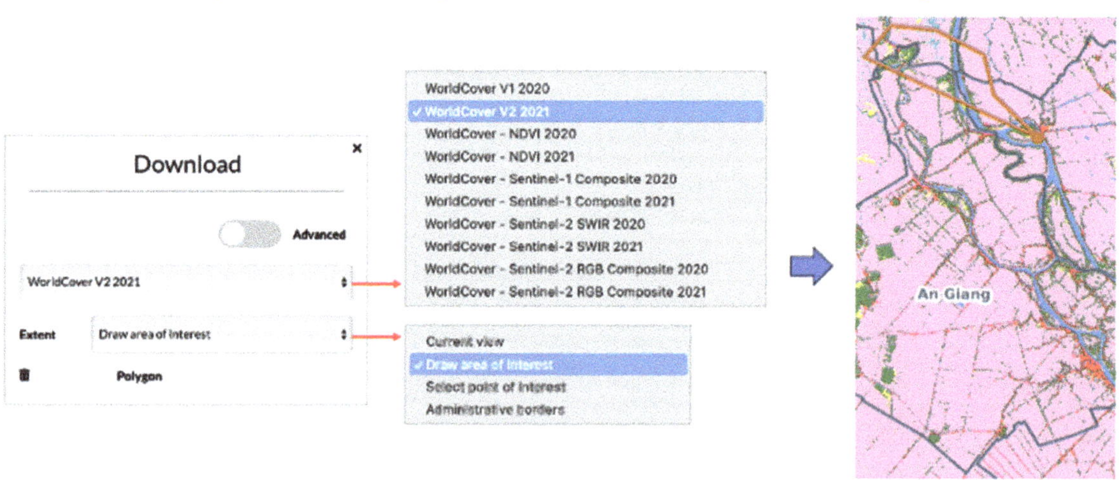

Source: European Space Agency. Layers: WorldCover V2 2021. https://viewer.esa-worldcover.org/worldcover/.

(v) As mentioned above, the province is covered by two raster tiles. At the bottom of the page, click **Next**, and select both tiles for downloading (Figure 2.61).

Figure 2.61: Selecting the Raster Tiles for Downloading

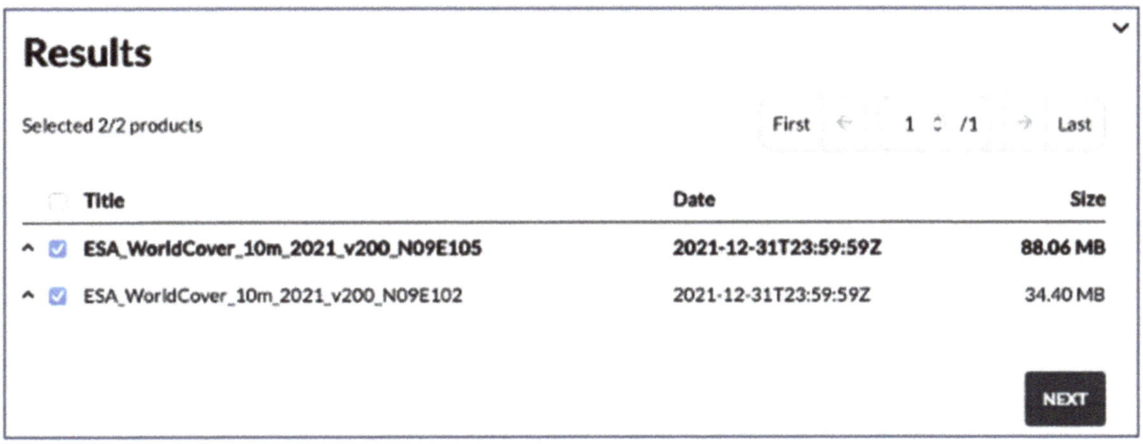

Source: European Space Agency. Layers: WorldCover V2 2021. https://viewer.esa-worldcover.org/worldcover/.

(vi) The platform will generate a ZIP file containing the selected tiles and will initiate the download (Figure 2.62).

Figure 2.62: Downloading the Raster Tiles

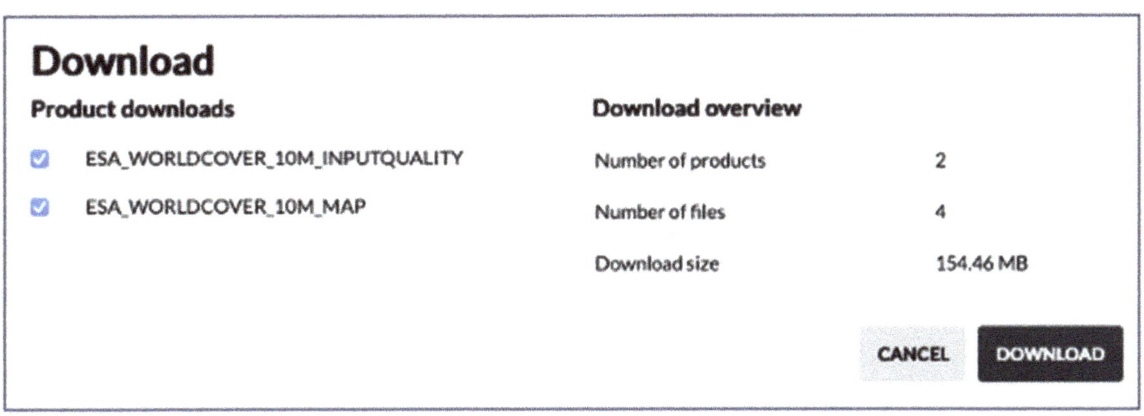

Source: European Space Agency. Layers: WorldCover V2 2021. https://viewer.esa-worldcover.org/worldcover/.

Module 3: Working with Raster Data in QGIS

The growing availability of high-resolution, frequently updated, and openly accessible geospatial data has created new opportunities for agricultural monitoring and policy development. Remote sensing technologies now offer timely insights into land use patterns, crop distribution, and environmental conditions that are key inputs for designing effective sampling strategies and targeting support to vulnerable agricultural communities.

To fully leverage these datasets, it is important to understand how raster data is managed and analyzed within a geographic information system (GIS) environment. This module introduces the core concepts and tools for working with raster data in QGIS, including importing, clipping, and converting raster layers to align with vector formats. These foundational skills will be essential in later modules, where land cover maps are used to support spatial analysis and sampling design.

The topics covered in this module include:

- 3.1 Introduction to Raster Data
  - 3.1.1 What is Raster Data?
  - 3.1.2 Types of Raster Data
- 3.2 Preparing Raster Data in QGIS
  - 3.2.1 Downloading Raster Data
  - 3.2.2 Importing Raster Data
  - 3.2.3 Clipping Raster Layers to Match the Size of the Study Area
  - 3.2.4 Saving Raster Data
- 3.3 Raster to Vector Conversion
  - 3.3.1 Converting Land Use and Land Cover Raster to Polygon Features
  - 3.3.2 Spatial Intersection: Transferring Attributes by Using Overlays
  - 3.3.3 Dissolving Polygonized Features by Land Use Category
  - 3.3.4 Reclassifying Digital Number Values as Land Use Categories Using the Field Calculator
  - 3.3.5 Using the Field Calculator to Reclassify Land Use and Land Cover Categories

By the end of this module, you will be able to prepare and manipulate raster datasets in QGIS. These skills will be essential for conducting spatial analysis and supporting sampling designs in the succeeding modules, particularly those that make use of land cover maps for agricultural survey applications.

3.1 Introduction to Raster Data

3.1.1 What is Raster Data?

Raster data is a type of geospatial data represented as a grid of cells (also called "pixels"), with each cell containing a value that corresponds to a specific geographic location. This format is particularly suited for representing continuous surfaces and spatial phenomena such as elevation, temperature, vegetation indices, or land cover (Lim 2016).

Unlike vector data that represents discrete features using points, lines, and polygons, raster data captures internal variations across a surface by assigning values to each cell. This structure allows for a detailed spatial analysis of gradients and patterns within geographic features (Campbell and Shin 2011).

Raster datasets are organized as two-dimensional arrays, with rows and columns forming a matrix. The resolution of a raster is defined by the size of its cells, with smaller cells yielding higher spatial detail. Raster data is commonly sourced from satellite imagery, aerial photography, digital elevation models (DEMs), and scanned maps. Raster datasets also include metadata such as coordinate systems, pixel dimensions, and geographic extent, which are essential for accurate georeferencing and spatial analysis (Lim 2016; Esri, n.d.-a).

3.1.2 Types of Raster Data

Raster data can be categorized by how it represents information, by its source, or its intended use.

1. Based on Data Representation

- **Continuous raster data** represent phenomena that vary smoothly across space, such as elevation, temperature, or rainfall. Each pixel contains a numeric value reflecting a measurement of a precise location.
- **Discrete raster data** encode categorical information, such as land cover types (e.g., forest, urban, cropland). Each pixel holds a code representing a specific class or category.

2. Based on Source or Use

- **Satellite imagery** is captured by platforms such as Landsat or Sentinel; the datasets are used for land cover classification, crop monitoring, and environmental analysis.
- **DEMs** are specialized rasters representing terrain elevation; they are useful for slope analysis, watershed delineation, and terrain modeling.
- **Digital orthophotos** are aerial photographs that have been corrected for distortion and georeferenced. They are commonly used for visual interpretation and base mapping.
- **Drone imagery** is high-resolution imagery from unmanned aerial vehicles (UAVs), increasingly used in precision agriculture and localized monitoring.
- **Scanned maps and binary raster files** are digitized versions of historical or paper maps, often used for archival purposes or legacy data integration.

3. Based on Content Type

- **Thematic raster data** represent classified or measured phenomena such as pollution levels, or vegetation indices.
- **Spectral raster data** contain multiband data from remote sensing platforms, used for analyzing reflectance across different wavelengths.
- **Imagery raster data** include aerial or satellite photographs used primarily for visual interpretation and mapping.

4. Common Raster Formats

- **GeoTIFF (.tiff or .tif)** is a widely used format that supports georeferencing and metadata embedding.
- **JPEG (.jpg or .jpeg) and PNG (.png)** are both suitable for visualization but typically lack spatial reference information.
- **NetCDF (.nc)** is designed for multidimensional scientific data, commonly used in climate and environmental modeling.

Table 3.1 summarizes raster data sources, detailing update frequency, spatial and temporal resolution, accessibility, and examples of their relevance or application in agricultural contexts.

Table 3.1: Raster Data Sources

Source	Update Frequency	Spatial Resolution	Temporal Resolution	Accessibility	Agricultural Relevance
Esri, National Geospatial Open Data, https://livingatlas.arcgis.com/esri_ngda_datasets/	Regular	Variable	Variable	Open access (some restrictions)	Land cover, infrastructure, environmental layers
Sentinel Hub, Copernicus Data Space Ecosystem, https://www.sentinel-hub.com/explore/copernicus-data-space-ecosystem/	5–10 days	10–60 m	5–10 days	Free with registration	Soil moisture, crop health, land use
EarthStat, http://www.earthstat.org/	Static	~10 km	Static	Open access	Crop yields, harvested areas, land use
FAO GeoNetwork, https://www.fao.org/land-water/databases-and-software/geonetwork/en/	Irregular	Variable	Irregular	Open access	Land cover, soil, climate layers
FAO GAEZ https://gaez.fao.org/	Periodic	~10 km	Static or scenario-based	Open access	Land suitability, crop potential

continued on next page

Table 3.1 *continued on next page*

Source	Update Frequency	Spatial Resolution	Temporal Resolution	Accessibility	Agricultural Relevance
JAXA, https://earth.jaxa.jp/en/data/	Daily to monthly	10 m–1 km	Daily to monthly	Open access	Forest monitoring, precipitation, vegetation indices
NASA, LAADS DAAC, https://ladsweb.modaps.eosdis.nasa.gov/learn/view-data/	Daily	250 m–1 km	Daily	Free with NASA Earthdata login	Vegetation indices, temperature, drought monitoring
NASA EOSDIS, https://www.earthdata.nasa.gov/about/esdis/eosdis	Daily to monthly	250 m–1 km	Daily to monthly	Free with NASA Earthdata login	Vegetation, soil moisture, land surface temperature
OpenStreetMap, Raster Tiles, https://wiki.openstreetmap.org/wiki/Raster_tiles	Frequent (community-driven)	Tile-based (256×256 px)	Variable	Open access	Contextual mapping, base layers
OpenTopography, https://opentopography.org/	Project-based	Under a meter to 30 m	Irregular	Open access	Terrain modeling, watershed analysis
USGS, EarthExplorer, https://earthexplorer.usgs.gov/	Daily to seasonal	15 m–1 km	16 days (Landsat), daily (MODIS)	Free with registration	Land cover, crop monitoring, hydrology

~ = approximately; EOSDIS = Earth Observing System Data and Information System; FAO = Food and Agriculture Organization of the United Nations; GAEZ = Global Agro-Ecological Zones; JAXA = Japan Aerospace Exploration Agency; km = kilometer; LAADS DAAC = Level-1 and Atmosphere Archive & Distribution System Distributed Active Archive Center; m = meter; MODIS = moderate resolution imaging spectroradiometer; NASA = National Space and Aeronautics Administration; px = pixel, USGS = United States Geological Survey.

Note: The table content is drawn from various raster open data sources accessed in August 2025.

Source: Asian Development Bank.

3.2 Preparing Raster Data in QGIS

This section outlines the process of downloading raster data from open-access platforms, using the Copernicus Data Space Ecosystem as an example. To illustrate the workflow, Sentinel-2 imagery was selected due to its suitability for agricultural and environmental applications. Sentinel-2 provides multispectral data with spatial resolutions ranging from 10 to 20 meters, covering visible, near-infrared (NIR), and shortwave infrared (SWIR) bands. These datasets are commonly used for vegetation monitoring, land cover classification, and the calculation of indices such as the Normalized Difference Vegetation Index (NDVI) and Normalized Difference Water Index (NDWI).

Visit https://dataspace.copernicus.eu and register for a free account. This account is required to access full download features and other services.

3.2.1 Downloading Raster Data

(i) After logging in, navigate to the **Copernicus Browser**, a web-based interface for searching, previewing, and downloading satellite imagery.
(ii) Define the area of interest (AOI), and use the map interface to locate the AOI. This can be done by typing the location name (e.g., "An Giang Province"); manually drawing a polygon; or uploading a shapefile, Keyhole Markup Language (KML) file, or GeoJSON file that defines the AOI (Figure 3.1).

Figure 3.1: Defining the Area of Interest

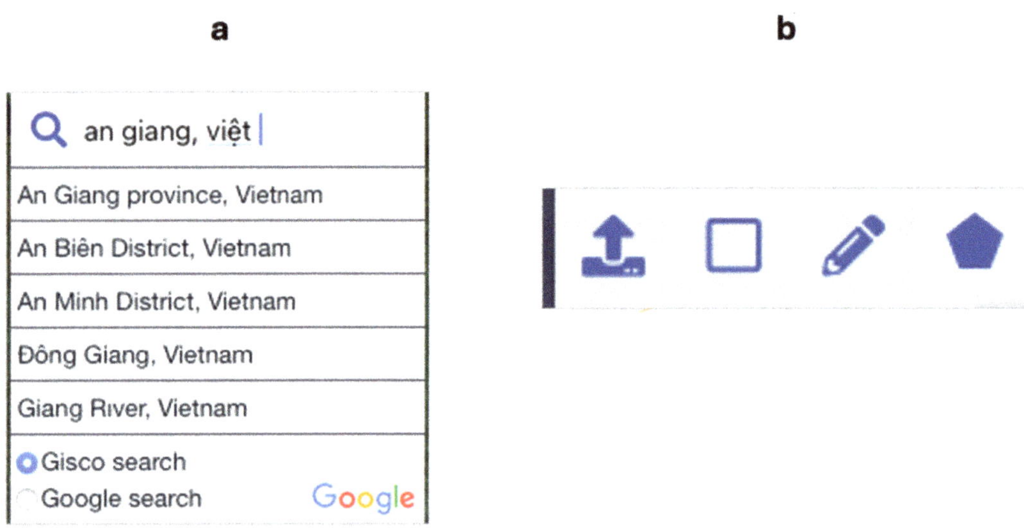

Note: The area of interest (AOI) can be defined by using a search location function (as shown in image "a"), by manually drawing a polygon (image "b"), or by uploading a file that defines the AOI.
Source: European Space Agency. Copernicus Browser. https://browser.dataspace.copernicus.eu (accessed 11 August 2025).

(iii) Apply search filters. Select the desired dataset, such as **Sentinel-2 L2A**, which provides atmospherically corrected imagery. Set a cloud-cover threshold (e.g., less than 20%) to filter for clearer scenes. Adjust the date range to target the specific time period of interest, such as **July 2025** (Figure 3.2).

Figure 3.2: Applying Search Filters

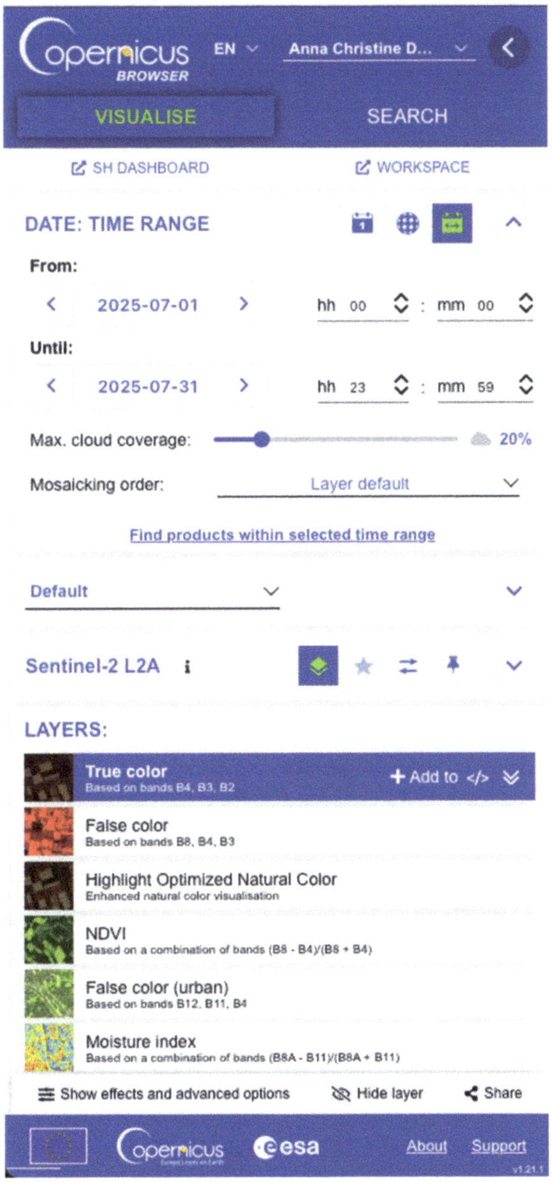

Source: European Space Agency. Copernicus Browser. https://browser.dataspace.copernicus.eu (accessed 11 August 2025).

(iv) The browser will display a list of matching satellite scenes, as shown in Figure 3.3. Preview each product to evaluate the image quality, cloud coverage, and metadata.

Figure 3.3: Reviewing Available Scenes

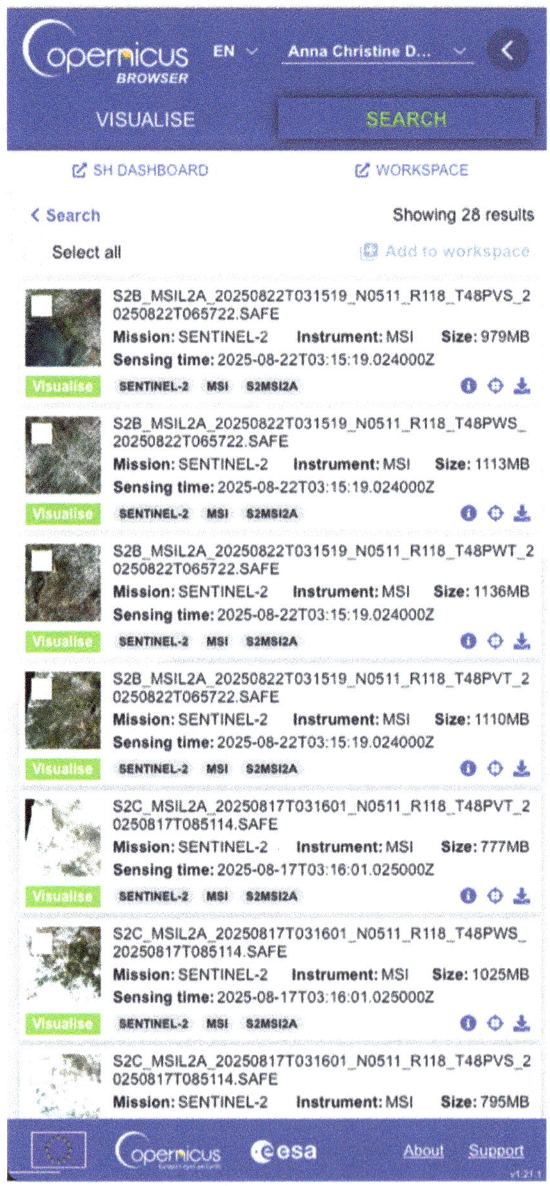

Source: European Space Agency. Copernicus Browser. https://browser.dataspace.copernicus.eu (accessed 11 August 2025).

(v) Ensure that the entire AOI is covered by the selected scenes. Larger areas may span multiple raster tiles, so it is important to download all the relevant tiles to achieve complete spatial coverage. As shown in Figure 3.4, the study area is covered by three raster tiles.

Figure 3.4: Checking Tile Coverage

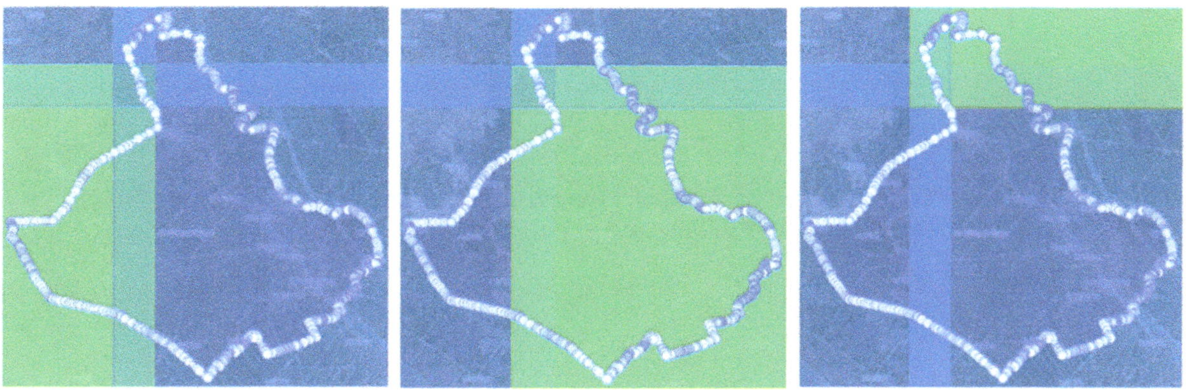

Source: European Space Agency. Copernicus Browser. https://browser.dataspace.copernicus.eu (accessed 11 August 2025).

(vi) Click the download icon next to each selected product (Figure 3.5). Choose the GeoTIFF format, which is compatible with most GIS software, and then save the image in the target folder.

Figure 3.5: Downloading the Data

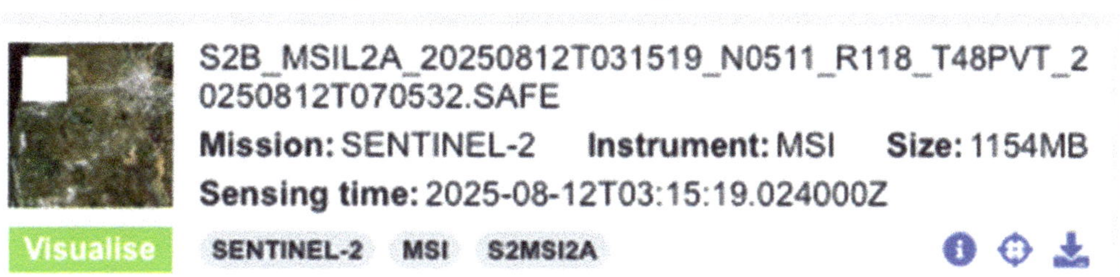

Source: European Space Agency. Copernicus Browser. https://browser.dataspace.copernicus.eu (accessed 11 August 2025).

(vii) The downloaded files can now be opened in QGIS and processed using the available tools for visualizing spectral bands, computing vegetation indices (e.g., NDVI), and for generating maps or statistics for further analysis.

In addition to Sentinel-2 imagery, this section demonstrates how to access thematic raster data by retrieving the **Viet Nam 2020 land use and land cover (LULC) map** from the Japan Aerospace Exploration Agency (JAXA). This dataset, referenced in earlier chapters, serves as a practical example of how to integrate land cover information into spatial analysis workflows.

JAXA requires user registration before downloading. Click the link below or the one provided for user registration, and follow the instructions for creating an account. Registration is free, and typically requires an email confirmation. https://www.eorc.jaxa.jp/ALOS/en/lulc/registration.htm.

(i) Go to the official JAXA dataset page (Figure 3.6), at: https://www.eorc.jaxa.jp/ALOS/en/dataset/lulc/lulc_vnm_v2309_e.htm.

(ii) Review the dataset information. The page provides details on the 2020 high-resolution LULC map for Viet Nam, including classification categories, satellite sources (Sentinel-1, Sentinel-2, ALOS-2), and the use of convolutional neural networks for mapping. The dataset includes 12 land cover classes, such as water, rice paddy, urban, forest types, and aquaculture.

Figure 3.6: Viet Nam 2020 Land Use and Land Cover Dataset Information

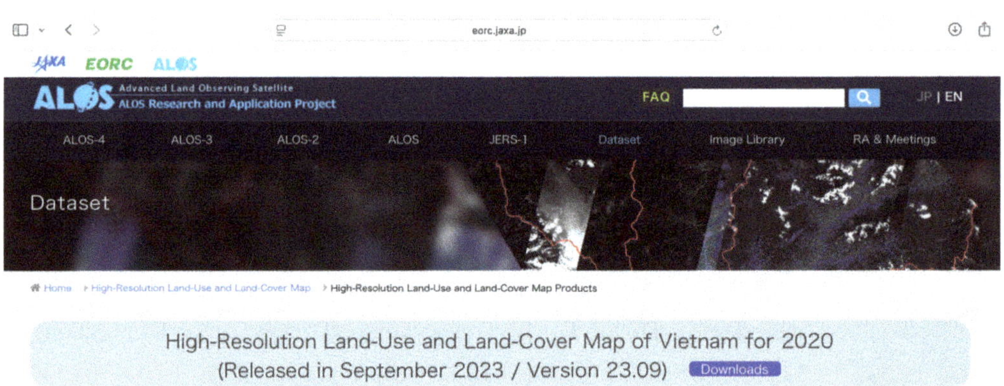

Source: Japan Aerospace Exploration Agency (JAXA), Earth Observation Research Center. JAXA High-Resolution Land Use and Land Cover Map. https://www.eorc.jaxa.jp/ALOS/en/dataset/lulc/lulc_vnm_v2309_e.htm (accessed 7 October 2025).

(iii) Scroll down to the **Downloads section**, and look for the file labeled: "2020VNM_v23.09 – GeoTIFF format of 10m equivalent resolution (474MB)."

(iv) Download this GeoTIFF file. After logging in, return to the dataset page and click the download link for the 10 m GeoTIFF file. The file name will resemble: "N08E104_2020_v23.09_10m.tif." This naming convention indicates the tile's geographic coverage (e.g., 8–11° N, 104–105° E).

3.2.2 Importing Raster Data

(i) To view the coverage of An Giang Province, add the vector layer **AnGiang_Province** to the map canvas. Change the symbology to a simple line style to ensure the visibility of the other layers beneath it.

(ii) The extent of An Giang Province is covered by two raster tiles: **N10E104_2020_v23.09_10m and N10E105_2020_v23.09_10m**. These can be added by selecting both files from the folder and dragging them from the **Browser panel** into the **map canvas**. Alternatively, the layers can be added by clicking the **Layers panel** in the main menu, selecting **Add Layer | Add Raster Layer**. The process is similar to that for adding a vector layer. The raster layer will appear in the **Layers panel** and on the map canvas (Figure 3.7).

Figure 3.7: Adding a Raster Layer to QGIS

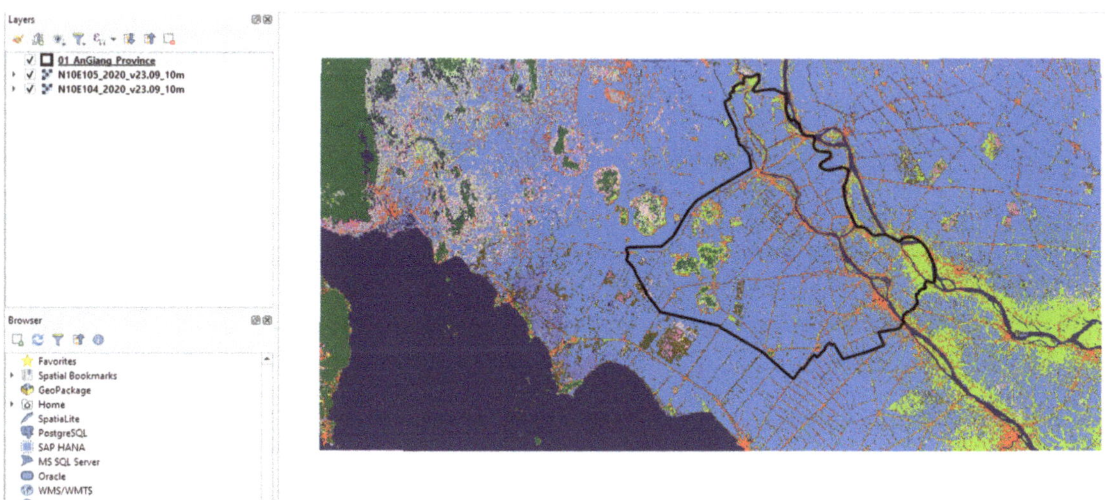

Source: Screenshot generated by the Asian Development Bank using QGIS 3.34.14. 2025.

3.2.3 Clipping Raster Layers to Match the Size of the Study Area

(i) To prepare the raster data for clipping, begin by merging the two raster tiles that cover An Giang Province: **N10E104_2020_v23.09_10m and N10E105_2020_v23.09_10m**. Go to **Raster | Miscellaneous | Merge**, then select both raster files as input layers. Specify the output file location or leave it as temporary and click **Run** (Figure 3.8).

Figure 3.8: Merging Raster Layers

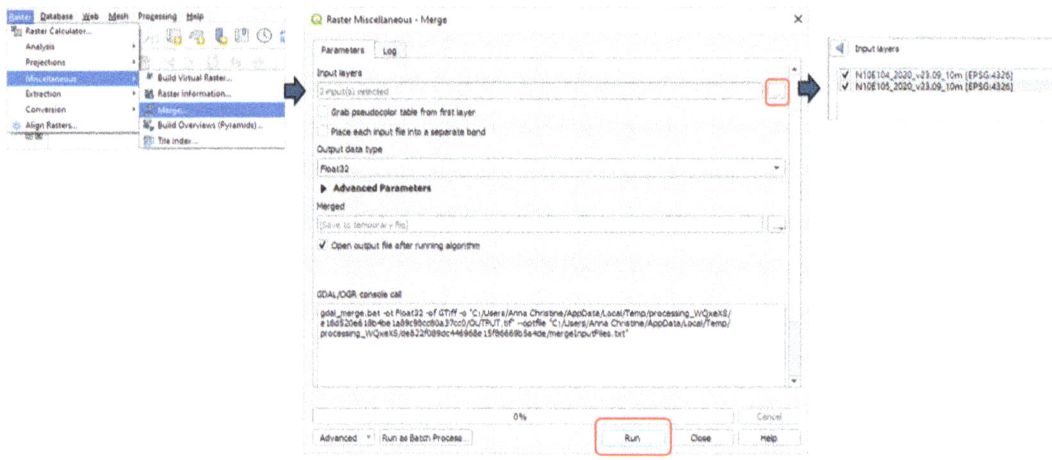

Source: Screenshot generated by the Asian Development Bank using QGIS 3.34.14. 2025.

The merged raster layer will be added to the map canvas.

(ii) To clip a raster layer using a study area boundary, go to **Raster | Extraction | Clip Raster by Mask Layer**. In the dialog box that appears, select the merged raster layer as the input. Choose the vector mask layer representing the study area (e.g., the An Giang Province shapefile). Specify the output file location or leave it as temporary. Click **Run** to execute the process (Figure 3.9).

Figure 3.9: Clipping Raster by Mask Layer

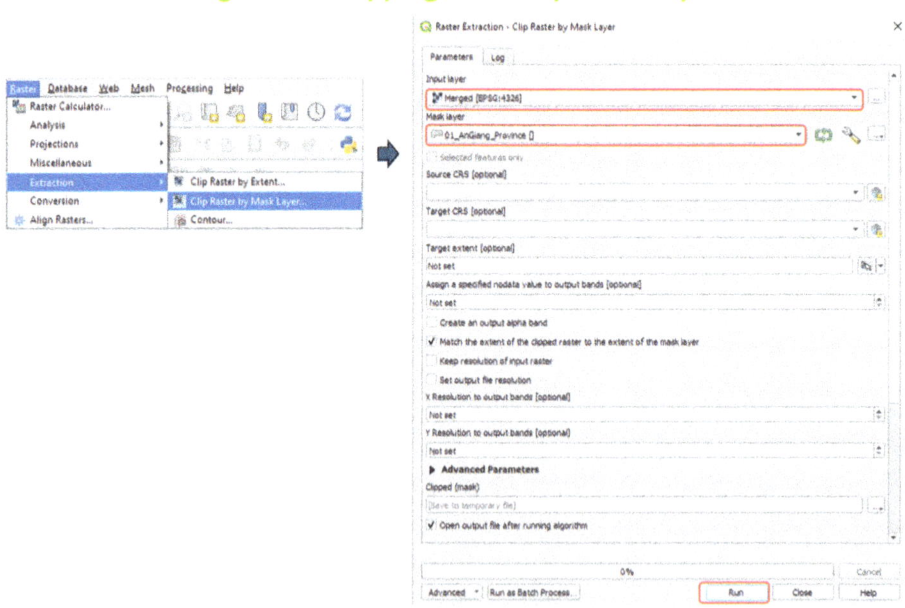

Source: Screenshot generated by the Asian Development Bank using QGIS 3.34.14. 2025.

The clipped raster will be automatically added to the map canvas (Figure 3.10).

The option **Clip Raster by Extent** trims the raster using a rectangular area defined by coordinate boundaries, while **Clip Raster by Mask Layer** uses the shape of a vector layer to crop the raster more precisely to a specific region. The mask method is preferred when working with irregular areas.

Figure 3.10: Clipped Raster Layer

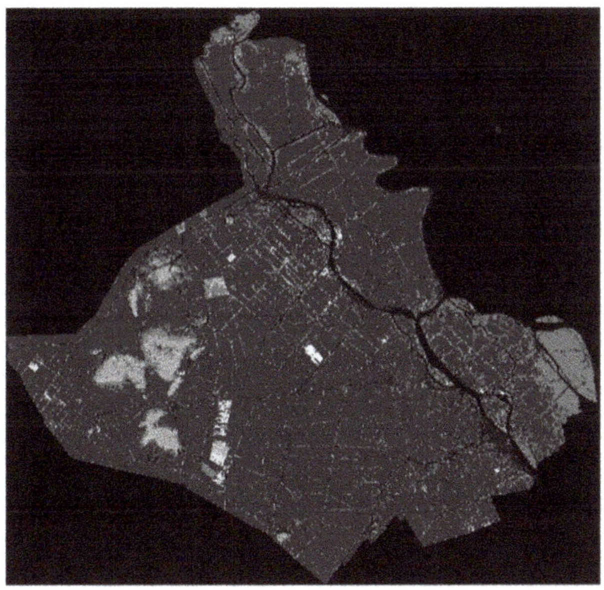

Source: Screenshot generated by the Asian Development Bank using QGIS 3.34.14. 2025.

(iii) The resulting raster will appear in grayscale. To apply a pseudocolor scheme, right-click on the clipped (mask) raster layer and select **Properties**. Go to the **Symbology** tab and change the **Render type** from **Singleband gray** to **Paletted/Unique values**. Click **Classify**. Double-click on the color for the value 0, and set its opacity to 0% (Figure 3.11).

Figure 3.11: Applying Pseudocolor Symbology

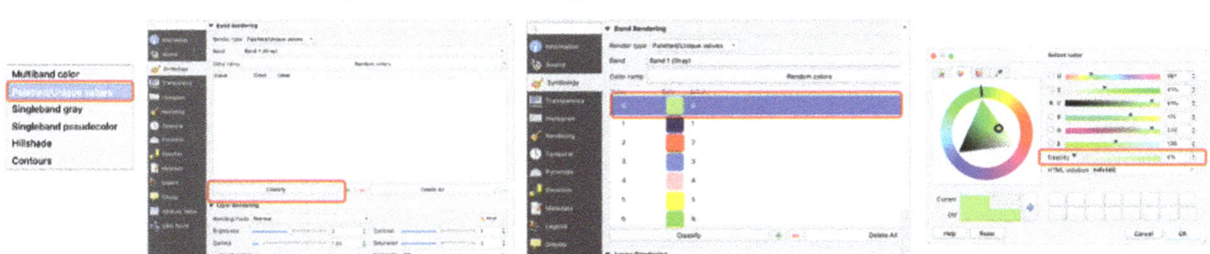

Set opacity to 0%

Click Classify

Double-click on the color of the 0 Value

Source: Screenshot generated by the Asian Development Bank using QGIS 3.34.14. 2025.

Values from 1 to 12 will now be visible, representing the **12 LULC classes** in the study area, as illustrated in Figure 3.12. Note that the color assigned to each land use type may vary among users, depending on individual settings or how the pseudocolor scheme is applied. These color assignments can also be manually adjusted to accommodate specific visualization requirements.

Figure 3.12: Land Use and Land Cover in An Giang Province, Viet Nam

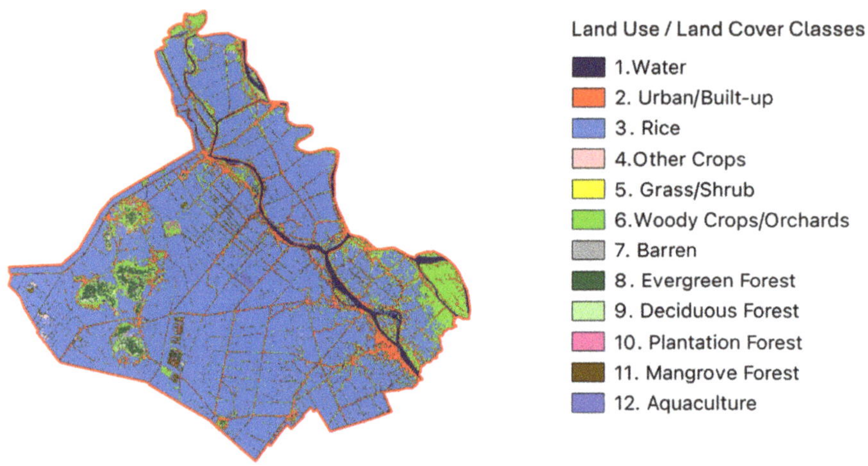

Source: Screenshot generated by the Asian Development Bank using QGIS 3.34.14. 2025.

3.2.4 Saving Raster Data

(i) To save the clipped raster layer, right-click on the raster layer in the **Layers panel** and select **Export | Save As...**. (Figure 3.13).

Figure 3.13: Saving the Raster Layer

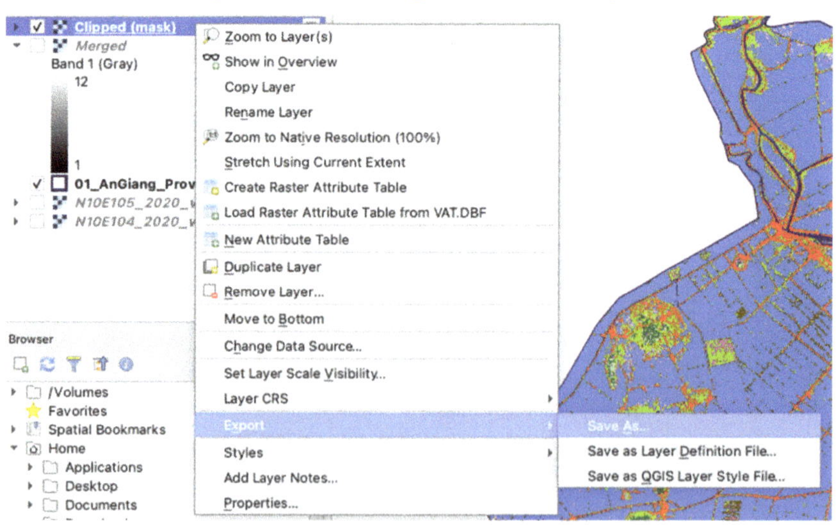

Source: Screenshot generated by the Asian Development Bank using QGIS 3.34.14. 2025.

(ii) In the **"Save Raster Layer As"** dialog box, which then appears, configure the settings as follows: Select **GeoTIFF** as the format. Click … to set the folder location where you want to save your file, and enter **AnGiang_Landuse** as the file name. If the layer already has a defined **coordinate reference system (CRS)** and no reprojection is needed, this setting can be left unchanged. Choose the appropriate **Extent** to export (the default is the full layer), and set the desired **Resolution**, or keep the default setting. If needed, specify a **No data value**, and adjust any **additional options**, such as compression. Once all parameters are set, click **OK** to save the raster as a new .tif file (Figure 3.14).

Figure 3.14: Configuring the Export Parameters

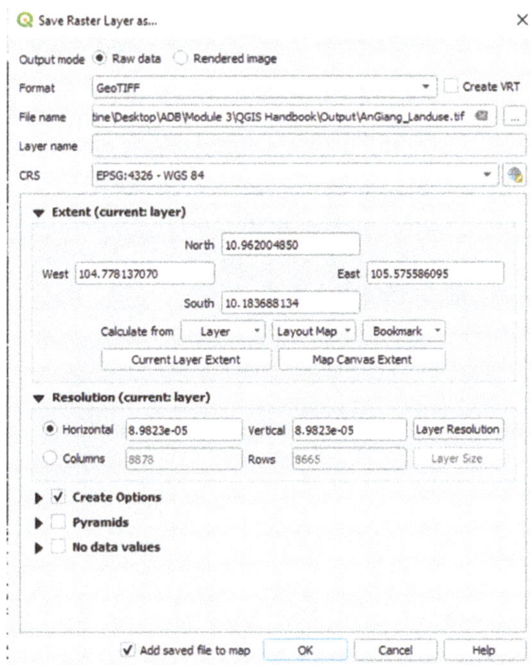

Source: Screenshot generated by the Asian Development Bank using QGIS 3.34.14. 2025.

Note that when exporting a raster layer, even if it has been styled using **Paletted/Unique values** or **Singleband pseudocolor**, the resulting file will typically revert to the default **Singleband gray** rendering when reloaded. This is because symbology settings are not preserved in the exported raster file unless explicitly saved through additional styling files or project settings.

Reminder:

Keep the file **AnGiang_Landuse.tiff** saved and readily accessible to ensure continuity in the succeeding modules. This output will serve as a key input for further analyses and exercises.

3.3 Raster to Vector Conversion

Converting a raster to polygon features transforms pixel-based data into vector shapes, allowing for the delineation of clear and discrete boundaries suitable for spatial analysis. This is useful in agricultural sampling to accurately target specific areas like crop fields.

3.3.1 Converting Land Use and Land Cover Raster to Polygon Features

For this exercise, use the clipped (mask) layer or the AnGiang_Landuse raster layer containing the 12 LULC categories from section 3.2.4.

(i) To convert the raster into the vector format, go to **Raster | Conversion | Polygonize (Raster to Vector)**. Alternatively, use the **Processing Toolbox** and search for **Polygonize (Raster to Vector)**.

(ii) In the tool dialog box, set the **Input layer** by selecting the raster **AnGiang_Landuse** layer. Specify a **Field name** for the attribute that will store the raster values: in this example, that will be field DN. "DN" stands for "Digital Number," a term commonly used in the context of raster data—particularly satellite imagery—in which each pixel is assigned a DN value representing the intensity or reflectance in a specific spectral band. Choose a file name for the output or save it as a temporary layer (Figure 3.15).

Figure 3.15: Converting a Raster into a Vector Layer

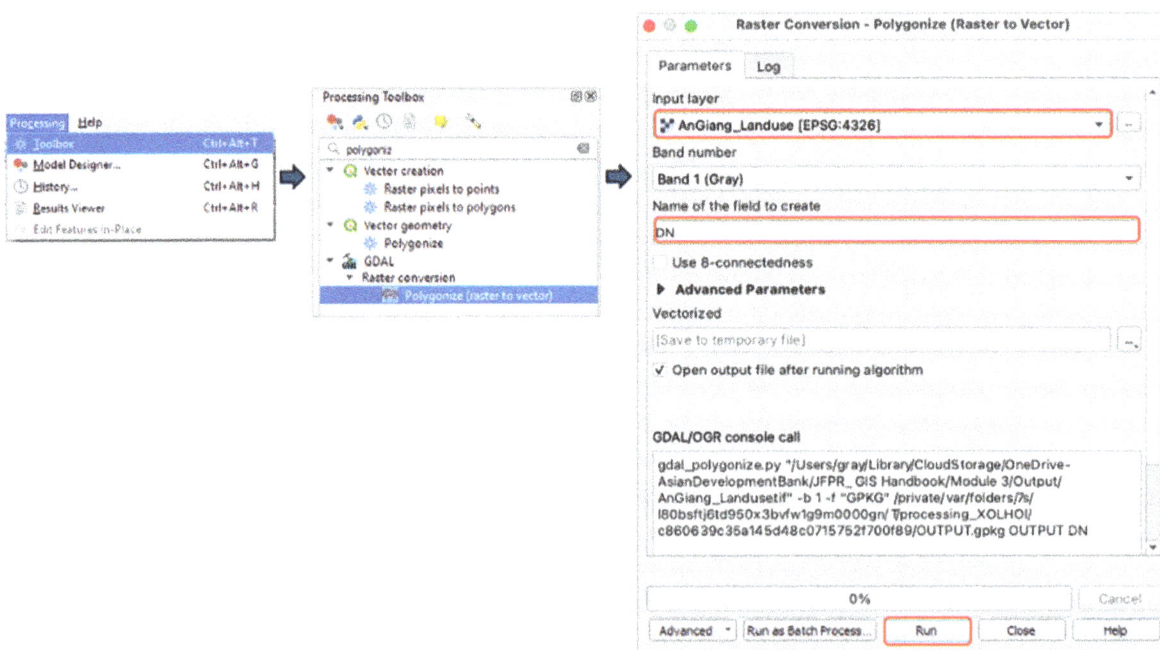

Source: Screenshot generated by the Asian Development Bank using QGIS 3.34.14. 2025.

(iii) Click **Run** to generate the polygon layer.
(iv) Open the **attribute table** to confirm that each polygon has a corresponding land cover value (Figure 3.16).

Figure 3.16: Vectorized Raster Layer

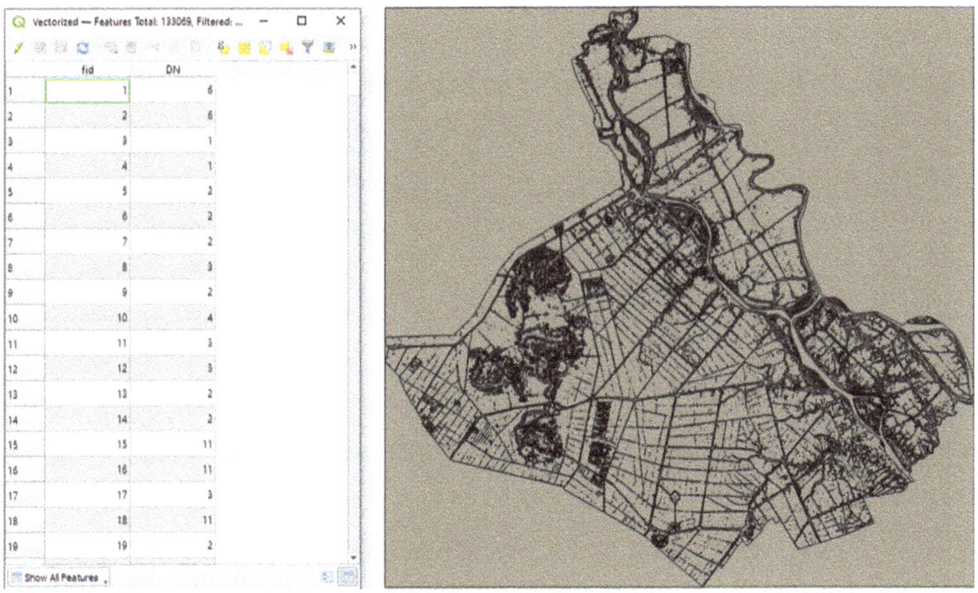

Source: Screenshot generated by the Asian Development Bank using QGIS 3.34.14. 2025.

3.3.2 Spatial Intersection: Transferring Attributes by Using Overlays

The process of spatial intersection enables features from one layer to inherit attribute information from another, based on their shared geographic space. In the context of agricultural survey sampling, this step is essential for linking land use classifications derived from raster data with administrative boundaries such as districts or provinces. By overlaying these datasets, each land use feature can be associated with the administrative units it intersects with, enabling more precise stratification, localized analysis, and informed decision-making. The steps for spatial intersection are as follows:

(i) Ensure that the polygonized or **vectorized** raster layer (containing the **fid** and **DN** attributes) from the previous step is already loaded into the QGIS workspace before proceeding.
(ii) Add the vector layer **AnGiang_District** by dragging and dropping it from the **Browser panel** into the **map area**. Alternatively, you can add the layer by clicking the **Layer** tab in the **menu bar**, then selecting **Add Layer** and **Add Vector Layer**.
(iii) Go to **Vector | Geoprocessing Tools | Intersection** (Figure 3.17).

Figure 3.17: Vector Intersection Tool

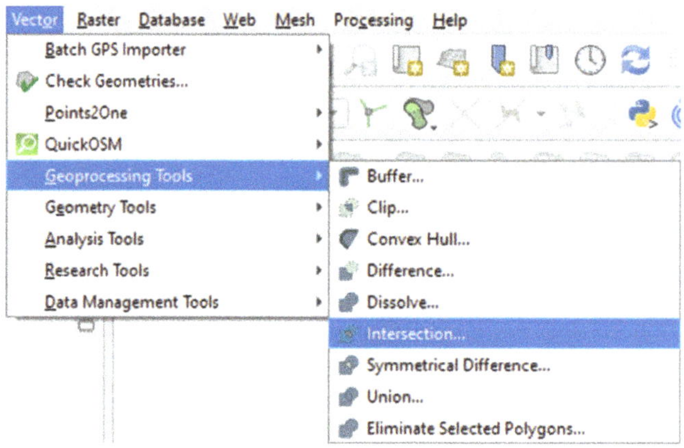

Source: Screenshot generated by the Asian Development Bank using QGIS 3.34.14. 2025.

(iv) In the tool dialog box that appears, set the **Input layer** to the **polygonized** or **Vectorized** land use layer, and the **Overlay layer** to the administrative boundary layer, such as the **AnGiang_District** layer in this example. This configuration ensures that the land use polygons will intersect with the administrative boundaries, allowing the resulting features to inherit attribute information from both layers. Specify the output file location and file name or leave the field blank to create a temporary output. Ensure that the desired fields from the overlay layer (e.g., DN, ADM1) are selected for inclusion in the output, or leave the selection at its default setting to include all available fields. Once all the parameters are set, click Run to execute the intersection process (Figure 3.18).

Figure 3.18: Configuring the Intersection Parameters

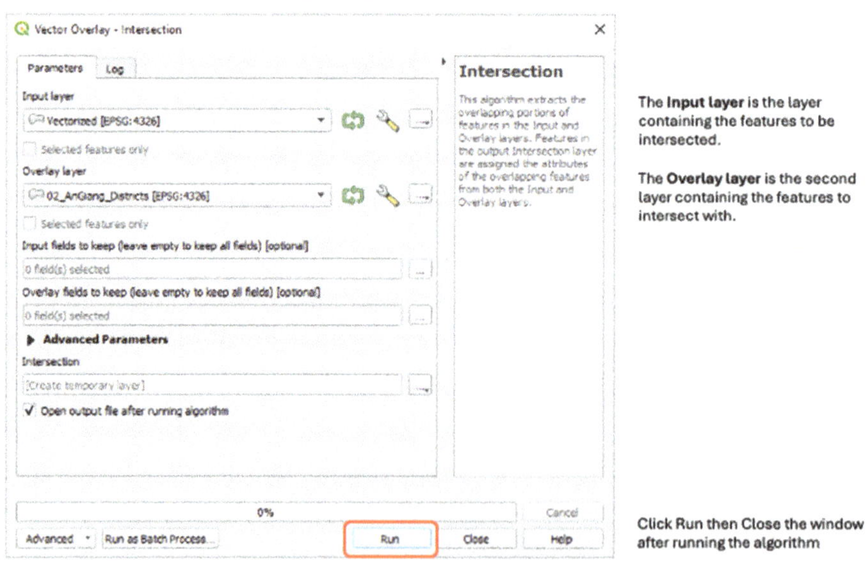

Source: Screenshot generated by the Asian Development Bank using QGIS 3.34.14. 2025.

(v) Right-click on the **Intersection layer** and select **Open the Attribute Table**. The new layer will contain approximately 135,000 features, each representing a spatial intersection between the land use polygons and the administrative boundaries (Figure 3.19).

Figure 3.19: Attribute Table After Spatial Intersection

Intersection — Features Total: 135091, Filtered: 135091, Selected: 0

	fid	DN	Shape_Leng	Shape_Area	ADM2_EN	ADM2_VI	ADM2_PCODE	ADM1_EN	ADM1_PCODE
1	1	2	0.69484591989	0.01866079879	An Phu	Huyện An Phú	VN80505	An Giang	VN805
2	2	2	0.69484591989	0.01866079879	An Phu	Huyện An Phú	VN80505	An Giang	VN805
3	3	2	0.69484591989	0.01866079879	An Phu	Huyện An Phú	VN80505	An Giang	VN805
4	4	3	0.69484591989	0.01866079879	An Phu	Huyện An Phú	VN80505	An Giang	VN805
5	5	2	0.69484591989	0.01866079879	An Phu	Huyện An Phú	VN80505	An Giang	VN805
6	6	4	0.69484591989	0.01866079879	An Phu	Huyện An Phú	VN80505	An Giang	VN805
7	7	3	0.69484591989	0.01866079879	An Phu	Huyện An Phú	VN80505	An Giang	VN805
8	8	3	0.69484591989	0.01866079879	An Phu	Huyện An Phú	VN80505	An Giang	VN805
9	9	2	0.69484591989	0.01866079879	An Phu	Huyện An Phú	VN80505	An Giang	VN805
10	10	2	0.69484591989	0.01866079879	An Phu	Huyện An Phú	VN80505	An Giang	VN805
11	11	11	0.69484591989	0.01866079879	An Phu	Huyện An Phú	VN80505	An Giang	VN805
12	12	11	0.69484591989	0.01866079879	An Phu	Huyện An Phú	VN80505	An Giang	VN805
13	13	3	0.69484591989	0.01866079879	An Phu	Huyện An Phú	VN80505	An Giang	VN805
14	14	11	0.69484591989	0.01866079879	An Phu	Huyện An Phú	VN80505	An Giang	VN805
15	15	2	0.69484591989	0.01866079879	An Phu	Huyện An Phú	VN80505	An Giang	VN805
16	16	6	0.69484591989	0.01866079879	An Phu	Huyện An Phú	VN80505	An Giang	VN805
17	17	11	0.69484591989	0.01866079879	An Phu	Huyện An Phú	VN80505	An Giang	VN805
18	18	12	0.69484591989	0.01866079879	An Phu	Huyện An Phú	VN80505	An Giang	VN805

Show All Features

Source: Screenshot generated by the Asian Development Bank using QGIS 3.34.14. 2025.

3.3.3 Dissolving Polygonized Features by Land Use Category

The **Dissolve** function merges adjacent polygons that share the same land use category, thereby simplifying the dataset. The **Processing Toolbox** also includes a **Dissolve** tool under the **Geospatial Data Abstraction Library (GDAL)** section. Both are used to merge adjacent polygons that share the same land use category. The **Vector Geometry Dissolve** tool provides more control, allowing dissolution by multiple fields and the retention of detailed attribute data. **GDAL Dissolve**, based in the GDAL library, is optimized for speed and handles large datasets efficiently, though it offers fewer options for managing attributes. For agricultural applications, **Vector Geometry Dissolve** is useful when working with detailed land cover data, while GDAL Dissolve is better suited for large-scale maps where speed is more important than attribute complexity.

(i) To dissolve features, go to **Vector | Geoprocessing Tools | Dissolve** (Figure 3.20). Alternatively, use the **Processing Toolbox** and search for **Dissolve**.

Figure 3.20: Vector Dissolve Tool

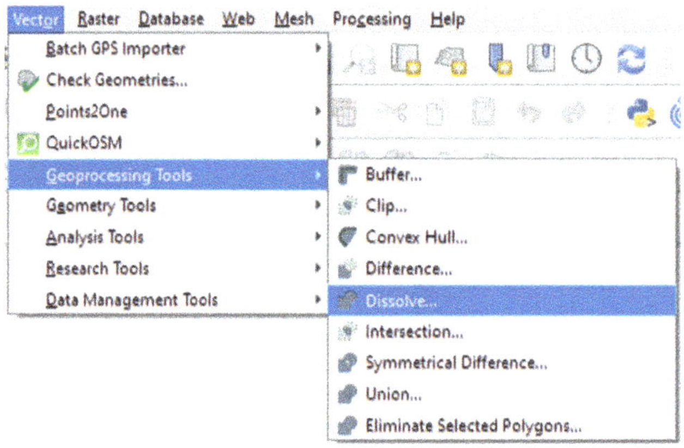

Source: Screenshot generated by the Asian Development Bank using QGIS 3.34.14. 2025.

(ii) In the tool dialog box that appears, select the **Intersection** layer from the previous step as the input layer. In the **Dissolve field(s)** section, select both the field representing land use categories (e.g., DN) and the district ID (e.g., ADM2_PCODE). Including both fields ensures that adjacent polygons are grouped not only by land use type, but also by their administrative locations. This prevents features with the same land use classification from being merged across district boundaries, thus preserving the spatial and administrative context necessary for location-specific analysis and reporting. Choose a file name and location for the output or save it as a temporary layer. Once all the parameters are set, click **Run** to execute the dissolution process (Figure 3.21).

Figure 3.21: Configuring the Dissolve Parameters

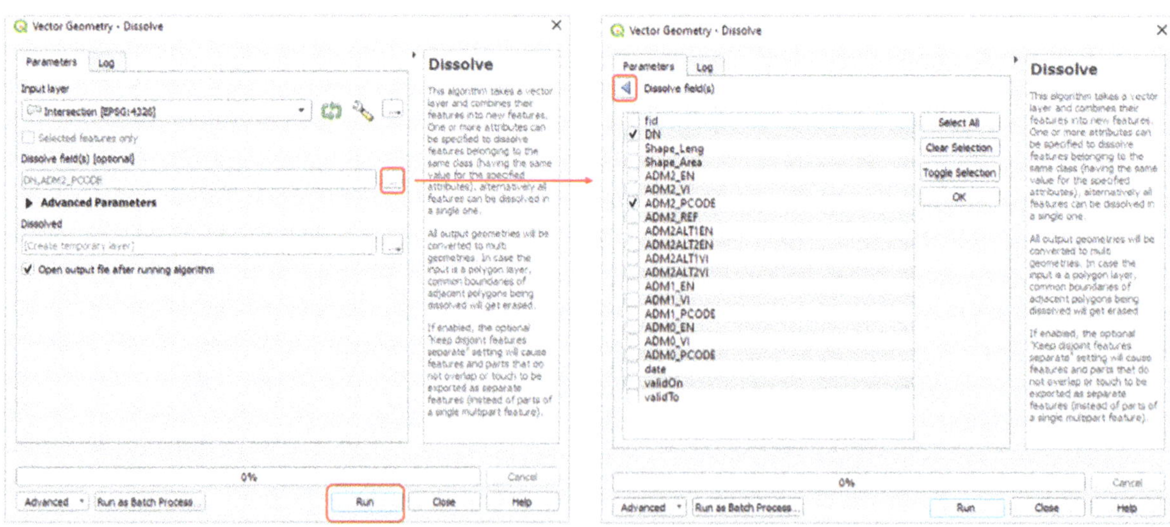

Source: Screenshot generated by the Asian Development Bank using QGIS 3.34.14. 2025.

(iii) Open the **attribute table** and inspect the map to confirm that the polygons belonging to the same category have been successfully merged (Figure 3.22). Note that the number of features has been reduced from approximately 135,000 to 134, in a process similar to the merging of the LULC categories.

Figure 3.22: Attributes of Merged Polygon Features

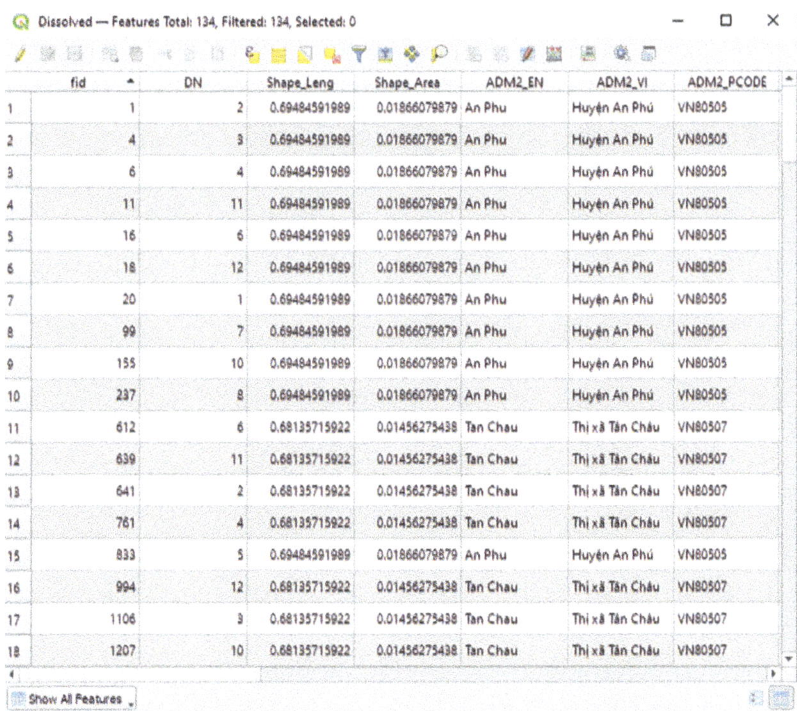

Dissolved — Features Total: 134, Filtered: 134, Selected: 0

	fid	DN	Shape_Leng	Shape_Area	ADM2_EN	ADM2_VI	ADM2_PCODE
1	1	2	0.69484591989	0.01866079879	An Phu	Huyện An Phú	VN80505
2	4	3	0.69484591989	0.01866079879	An Phu	Huyện An Phú	VN80505
3	6	4	0.69484591989	0.01866079879	An Phu	Huyện An Phú	VN80505
4	11	11	0.69484591989	0.01866079879	An Phu	Huyện An Phú	VN80505
5	16	6	0.69484591989	0.01866079879	An Phu	Huyện An Phú	VN80505
6	18	12	0.69484591989	0.01866079879	An Phu	Huyện An Phú	VN80505
7	20	1	0.69484591989	0.01866079879	An Phu	Huyện An Phú	VN80505
8	99	7	0.69484591989	0.01866079879	An Phu	Huyện An Phú	VN80505
9	155	10	0.69484591989	0.01866079879	An Phu	Huyện An Phú	VN80505
10	237	8	0.69484591989	0.01866079879	An Phu	Huyện An Phú	VN80505
11	612	6	0.68135715922	0.01456275438	Tan Chau	Thị xã Tân Châu	VN80507
12	639	11	0.68135715922	0.01456275438	Tan Chau	Thị xã Tân Châu	VN80507
13	641	2	0.68135715922	0.01456275438	Tan Chau	Thị xã Tân Châu	VN80507
14	761	4	0.68135715922	0.01456275438	Tan Chau	Thị xã Tân Châu	VN80507
15	833	5	0.69484591989	0.01866079879	An Phu	Huyện An Phú	VN80505
16	994	12	0.68135715922	0.01456275438	Tan Chau	Thị xã Tân Châu	VN80507
17	1106	3	0.68135715922	0.01456275438	Tan Chau	Thị xã Tân Châu	VN80507
18	1207	10	0.68135715922	0.01456275438	Tan Chau	Thị xã Tân Châu	VN80507

Show All Features

Source: Screenshot generated by the Asian Development Bank using QGIS 3.34.14. 2025.

3.3.4 Reclassifying Digital Number Values as Land Use Categories Using the Field Calculator

Reclassifying DN values as descriptive land use categories is an essential step in making spatial data more useful. For agricultural surveys and sampling frame development, this process helps distinguish different surface types, such as cropland, water bodies, or urban areas, allowing for more accurate stratification and better-targeted data collection.

(i) Right click on the **Dissolved** layer in the previous step, and select **Open Attribute Table** (Figure 3.23).

Figure 3.23: Opening the Attribute Table

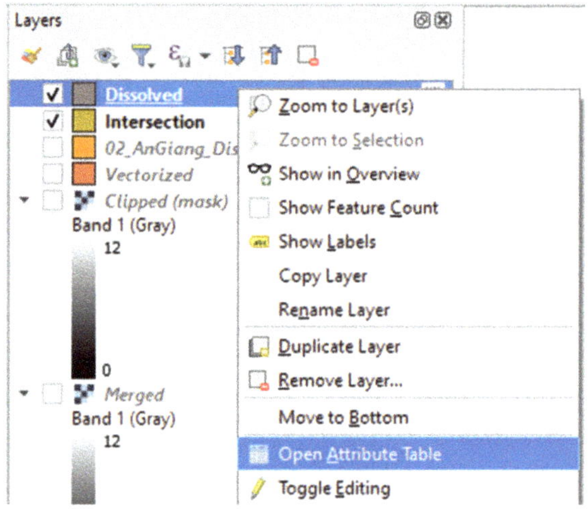

Source: Screenshot generated by the Asian Development Bank using QGIS 3.34.14. 2025.

(ii) In the **attribute table**, open the **Field Calculator** (Figure 3.24). In the **Field Calculator** dialog box, check the option **Create a new field**. Set the **Output field name** (e.g., LANDUSE) and choose the **Output field type** as **Text (string)**. Set the **Output field length** to **30** to accommodate longer category names. Next, enter the **CASE** expression to map the DN values onto descriptive land use categories, based on the classification shown in Figure 3.11. Click **OK** (Figure 3.24).

Figure 3.24: Function Configuration in the Field Calculator Using the CASE Expression

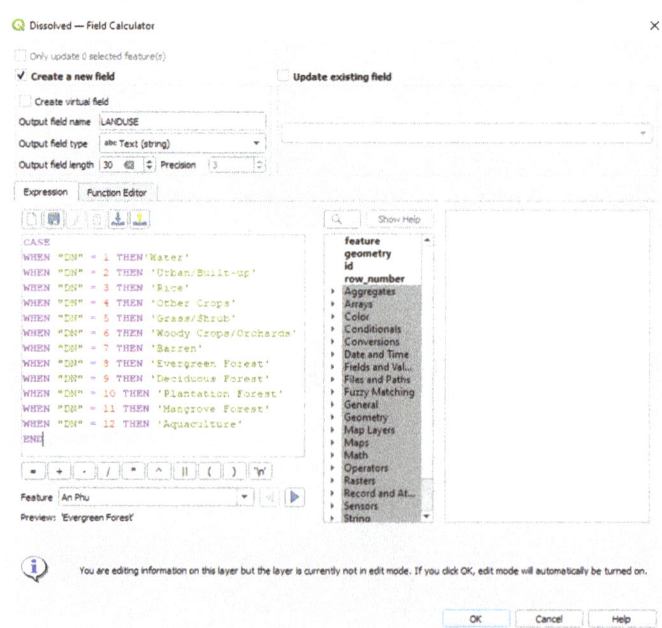

Source: Screenshot generated by the Asian Development Bank using QGIS 3.34.14. 2025.

The resulting attribute table will contain a new field named "**LANDUSE**," which will show the corresponding land use category for each **DN** value (Figure 3.25).

Figure 3.25: Updated Attribute Table with LANDUSE Categories Derived from DN Codes

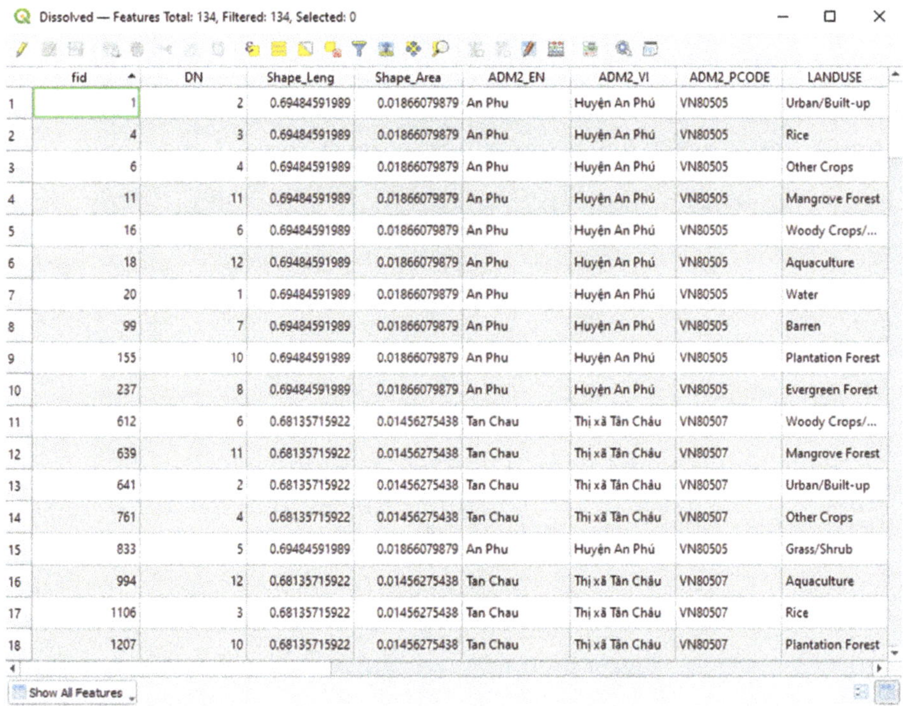

Dissolved — Features Total: 134, Filtered: 134, Selected: 0

	fid	DN	Shape_Leng	Shape_Area	ADM2_EN	ADM2_VI	ADM2_PCODE	LANDUSE
1	1	2	0.69484591989	0.01866079879	An Phu	Huyện An Phú	VN80505	Urban/Built-up
2	4	3	0.69484591989	0.01866079879	An Phu	Huyện An Phú	VN80505	Rice
3	6	4	0.69484591989	0.01866079879	An Phu	Huyện An Phú	VN80505	Other Crops
4	11	11	0.69484591989	0.01866079879	An Phu	Huyện An Phú	VN80505	Mangrove Forest
5	16	6	0.69484591989	0.01866079879	An Phu	Huyện An Phú	VN80505	Woody Crops/...
6	18	12	0.69484591989	0.01866079879	An Phu	Huyện An Phú	VN80505	Aquaculture
7	20	1	0.69484591989	0.01866079879	An Phu	Huyện An Phú	VN80505	Water
8	99	7	0.69484591989	0.01866079879	An Phu	Huyện An Phú	VN80505	Barren
9	155	10	0.69484591989	0.01866079879	An Phu	Huyện An Phú	VN80505	Plantation Forest
10	237	8	0.69484591989	0.01866079879	An Phu	Huyện An Phú	VN80505	Evergreen Forest
11	612	6	0.68135715922	0.01456275438	Tan Chau	Thị xã Tân Châu	VN80507	Woody Crops/...
12	639	11	0.68135715922	0.01456275438	Tan Chau	Thị xã Tân Châu	VN80507	Mangrove Forest
13	641	2	0.68135715922	0.01456275438	Tan Chau	Thị xã Tân Châu	VN80507	Urban/Built-up
14	761	4	0.68135715922	0.01456275438	Tan Chau	Thị xã Tân Châu	VN80507	Other Crops
15	833	5	0.69484591989	0.01866079879	An Phu	Huyện An Phú	VN80505	Grass/Shrub
16	994	12	0.68135715922	0.01456275438	Tan Chau	Thị xã Tân Châu	VN80507	Aquaculture
17	1106	3	0.68135715922	0.01456275438	Tan Chau	Thị xã Tân Châu	VN80507	Rice
18	1207	10	0.68135715922	0.01456275438	Tan Chau	Thị xã Tân Châu	VN80507	Plantation Forest

Show All Features

Source: Screenshot generated by the Asian Development Bank using QGIS 3.34.14. 2025.

(iii) To identify and remove features with missing land use classifications, begin by sorting through the **LANDUSE field** in the attribute table in reverse order by **clicking the column header**. This will help reveal the features with NULL values. Select all such features by highlighting them on the right side of the attribute table.

These NULL entries typically result from features that did not indicate common attributes during the intersection process. To maintain data integrity, these features should be removed. Before deleting any features, however, be sure to enable the **Toggle Editing** mode for the layer. Once editing is activated, you can proceed to delete the selected features by clicking the **Delete Selected Features** tab in the attribute table toolbar (Figure 3.26).

Figure 3.26: Deleting Features with NULL Values in the Attribute Table

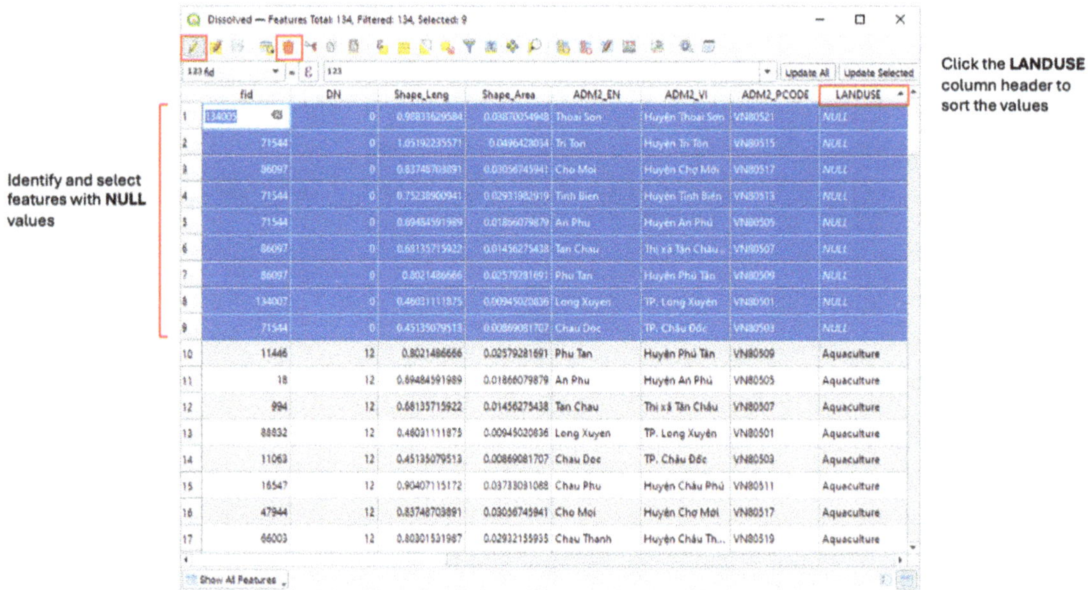

Source: Screenshot generated by the Asian Development Bank using QGIS 3.34.14. 2025.

After this step is completed, the resulting layer should contain **125 features**, representing valid land use classifications. Finally, disable the **Toggle Editing** mode and **Save** the changes to preserve the updated layers.

(iv) To export or save, right-click on **Dissolved layer** in the **Layers panel**. Select **Export**, then choose **Save Features As**. In the dialog box that appears, set the **Format** to **GeoPackage** or **.gpkg**. Click **Browse** to assign the file name **AnGiang_Landuse**, and specify the location where the new file will be stored. Once all the settings have been configured, click **OK** to complete the export or save operation.

For further guidance, refer to Module 2, section 2.3.4 ("Exporting Map Layers as New Layers") for detailed instructions and a step-by-step description of the procedures.

Reminder:

Keep the **AnGiang_Landuse.gpkg** file saved and readily accessible. Keeping this file ensures continuity in data processing and interpretation throughout the subsequent modules, particularly modules 4 and 5.

3.3.5 Using the Field Calculator to Reclassify Land Use and Land Cover Categories

This section further simplifies the LULC classifications by reclassifying the original 12 categories into four broader classes. It builds on the output from the previous section, in which a descriptive LANDUSE field was added, and polygons with similar attributes were dissolved. Using the **Field Calculator**, this step groups the detailed LULC classes into generalized categories to support streamlined analysis. For this exercise, use the **Vectorized** (AnGiang_Landuse.gpkg) layer generated in **Section 3.3.4** as the input.

(i) To begin, right-click the **Dissolved** layer and select **Open Attribute Table**. Then, launch the **Field Calculator**. In the dialog box, check the option **Create a new field**. Set the **Output field name** (e.g., LULC_reclass) and choose **Integer (32 bit)** as the **Output field type**.

(ii) Next, enter the **Reclassification Expression** using a CASE expression to map the original 12 classes onto the four new categories (Figure 3.27).

Figure 3.27: Reclassifying Categories Using the Field Calculator

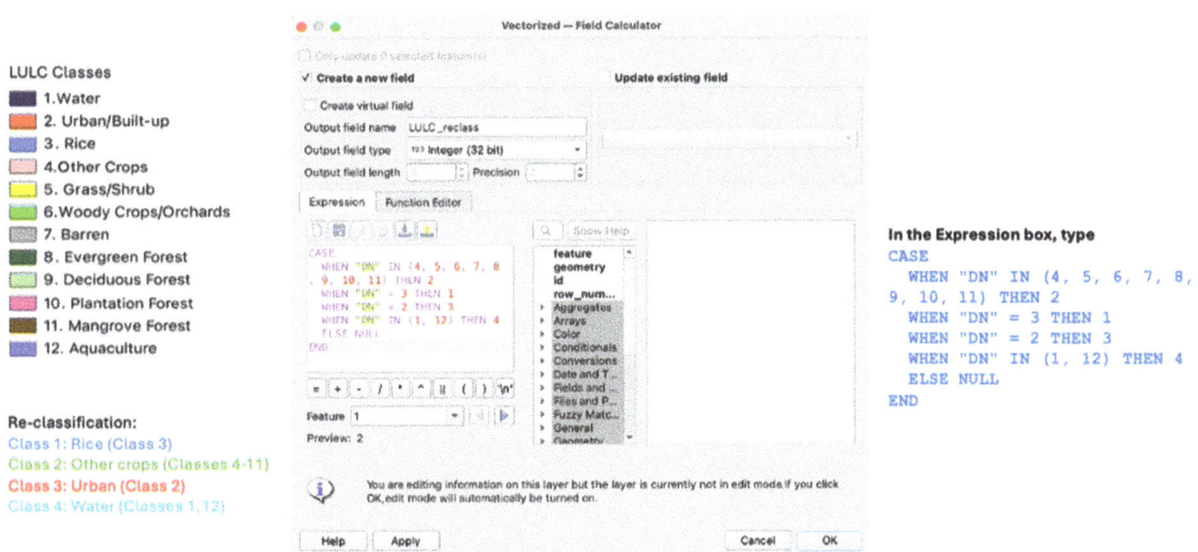

Source: Screenshot generated by the Asian Development Bank using QGIS 3.34.14. 2025.

(iii) Click **OK** to apply the calculation.

A new field will be added to the attribute table containing the reclassified values.

Module 4: Attribute Tables

This module focuses on attribute tables in QGIS, showing how to manage them effectively. Attribute tables are a fundamental component of vector layers in QGIS, providing a structured way to store and manipulate data associated with geographic features. Familiarity with the functionalities of attribute tables is crucial for leveraging QGIS in agricultural statistics, as it enables precise data analysis and visualization.

This module covers:

4.1 Overview of the Attribute Table
4.2 Selecting, Editing, and Filtering Features
4.3 Adding a New Field, Renaming, and Organizing Fields
4.4 Calculating a Variable (with Area Calculation as an Example)
4.5 Merging Tabular Data with Administrative Boundaries
4.6 Creating an Overlay Histogram on a Map

By the end of this module, you will be able to use attribute tables effectively in QGIS—an essential skill for analyzing agricultural statistics and for applying an advanced geographic information system (GIS) and remote sensing methods in later modules. These concepts will be illustrated throughout this module.

To understand the concepts of the attribute table for a vector layer in QGIS, this module utilizes the land use and land cover map (LULC) of An Giang Province, Viet Nam. This map is derived from the Japan Aerospace Exploration Agency (JAXA) Earth Observation Research Center (EORC) LULC map of Viet Nam for 2020, version 23.09 (Government of Japan, n.d.). Before proceeding with the exercises in this module, it is important to complete the raster data preprocessing steps outlined in **Module 3**. These steps include downloading, importing, clipping, and saving the raster data, followed by its conversion into vector format and initial classification. If Module 3 has been completed, the resulting vectorized and processed file **AnGiang_Landuse.gpkg** should be available, and will serve as the input for the activities in this module.

4.1 Overview of the Attribute Table

(i) Add the vector land use layer **AnGiang_Landuse.gpkg** in QGIS by dragging and dropping it from the **Browser panel** into the **map area**. A land use map appears with multiple polygons and uniform colors. By default, QGIS does not automatically classify data layers when they are added to a project (Figure 4.1).

Figure 4.1: Land Use and Land Cover Map of An Giang Province, Viet Nam (Vector Layer)

Sources: Screenshot generated by Asian Development Bank using QGIS 3.34.14. 2025; Government of Japan, Japan Aerospace Exploration Agency, Earth Observation Research Center. High-Resolution Land Use and Land Cover Map of Vietnam for 2020 (Released in September 2023 / Version 23.09).
https://www.eorc.jaxa.jp/ALOS/en/dataset/lulc/lulc_vnm_v2309_e.htm (accessed 14 February 2025).

(ii) To get an overview of the content of the layer, right-click the layer in the **Layers panel** and select **Open Attribute Table** (Figure 4.2).

Figure 4.2: Opening the Attribute Table

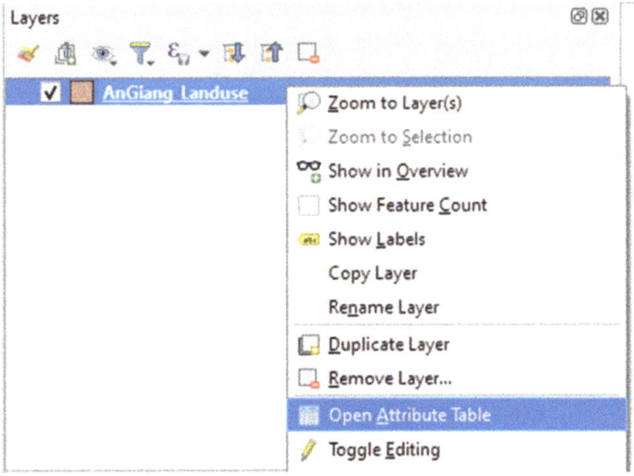

Source: Screenshot generated by Asian Development Bank using QGIS 3.34.14. 2025.

Alternatively, click the **Open Attribute Table** icon, , in the **Attributes toolbar** (Figure 4.3).

Figure 4.3: The Attribute Table Icon in the Attributes Toolbar

Source: Screenshot generated by the Asian Development Bank using QGIS 3.34.14. 2025.

(iii) The **Open Attribute Table** drop-down menu offers four options for viewing and interacting with the attribute table of a vector layer (Figure 4.4).

Figure 4.4: The Attribute Table Drop-Down Menu

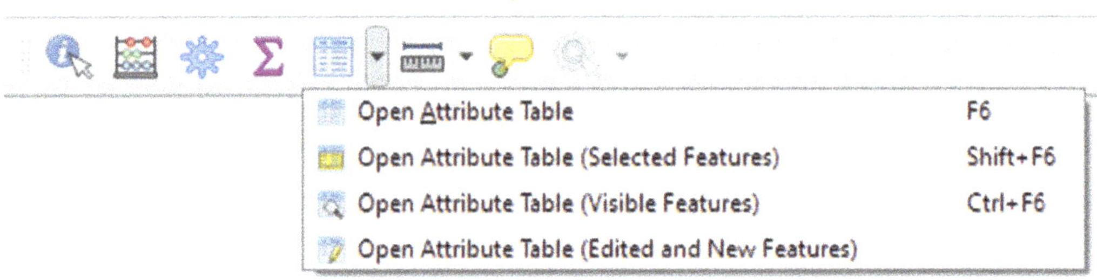

Source: Screenshot generated by the Asian Development Bank using QGIS 3.34.14. 2025.

They are as follows:

- **Open Attribute Table.** This option opens the attribute table for the entire layer, displaying all the features and their attributes.
- **Open Attribute Table (Selected Features).** This option opens the attribute table, but only shows the features that are currently selected on the map. This is useful for focusing on specific features without the distraction of the entire dataset.
- **Open Attribute Table (Visible Features).** This option opens the attribute table and displays only the features that are currently visible in the map view. This can be helpful when working with large datasets, as it limits the table to the features you are currently viewing.
- **Open Attribute Table (Edited and New Features).** This option opens the attribute table and shows only the features that meet a specified filter criterion—features that have been edited or newly added during the current session.

The **attribute table** is like a spreadsheet that provides an overview of the content of a vector layer (Figure 4.5). Each row in the attribute table represents one feature in the layer, denoted by a Feature ID. The features are real-world things such as roads, buildings, and rivers that are modeled using points, lines, polylines, and polygons, as determined by the type of layer selected.

In the example, the features are polygons. Each polygon feature has related attributes such as district ID (ADM2_PCODE), district name (ADM2_EN), province ID (ADM1_PCODE), province name (ADM1_EN), and LANDUSE, which are arranged in the columns of the table. Attribute data can be in the form of a whole number, decimal number, text, date, date and time, string list, or a Boolean that describe a feature. The column name is "Field Attributes," and the column is limited to alphanumeric characters. We will cover the functions in the attribute table in the following sections and exercises.

Figure 4.5: The Attribute Table

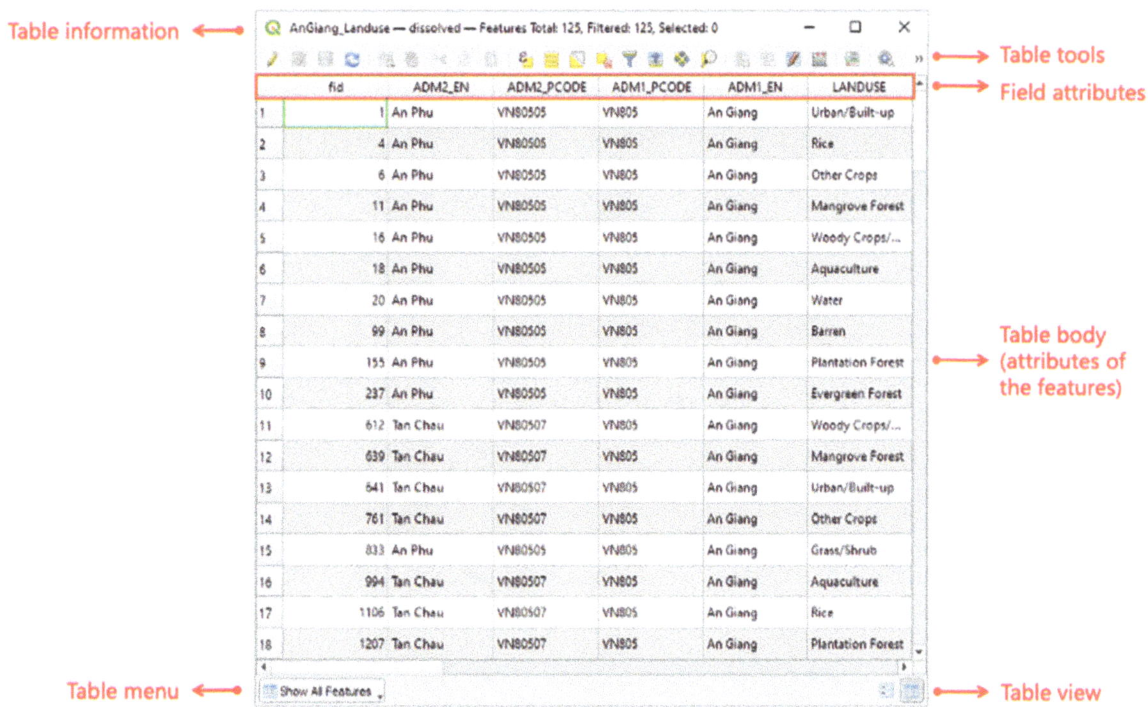

Source: Screenshot generated by the Asian Development Bank using QGIS 3.34.14. 2025.

4.2 Selecting, Editing, and Filtering Features

The attribute table window contains toolbars that can be used to manage the features and their attributes. These toolbars can be accessed through the corresponding buttons (Figure 4.6). You can see the pop-up name of each button in QGIS by hovering your mouse pointer over it. Some of the icons are inactivated, however, and can only be activated when other icons or features are selected.

Figure 4.6: Buttons at the Top of the Attribute Table

Source: Screenshot generated by the Asian Development Bank using QGIS 3.34.14. 2025.

Feature selection in QGIS is useful because it allows you to delete a layer, add specific data to a layer, or create a new layer altogether. There are several ways to select features in QGIS:

(i) In the attribute table, the icons are used to select particular features, deselect features, select all features, and other related functions (Figure 4.7):

Figure 4.7: Selection Icons in the Attribute Table

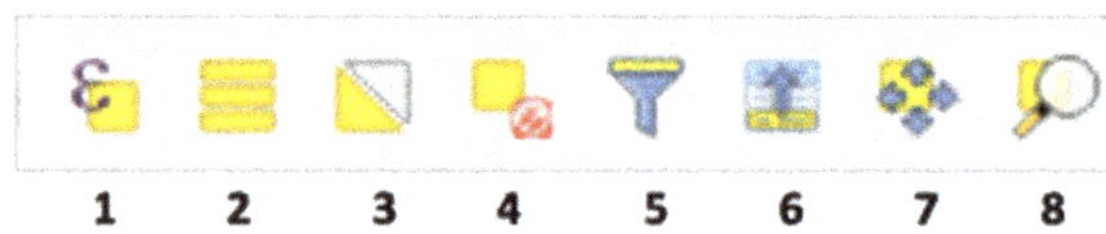

1	**Select features using an expression**
2	**Select all**
3	**Invert selection**
4	**Deselect all features from the layer**
5	**Select or filter features using forms**
6	**Move selection to top**
7	**Pan map to the selected rows**
8	**Zoom map to the selected rows**

Source: Screenshot generated by the Asian Development Bank using QGIS 3.34.14. 2025.

(ii) Use the **Selection toolbar** and its drop-down menu, as shown in Figure 4.8. To activate, go to **Menu Toolbar** | **View** | **Toolbars** | **Selection Toolbar**).

Figure 4.8: The Selection Toolbar

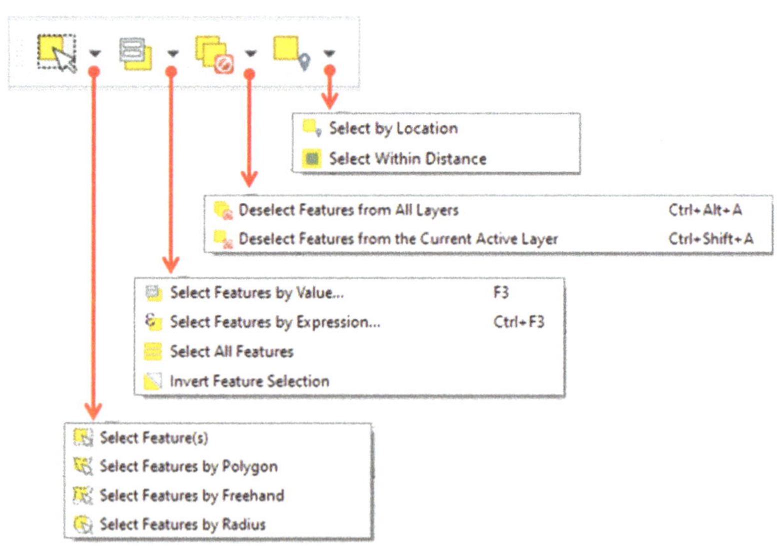

Source: Screenshot generated by the Asian Development Bank using QGIS 3.34.14. 2025.

In the following exercise, we will cover the different selections and filtering methods using the functions listed in the attribute table. Some of these functions, such as **Select Features by Expression** or **Select All Features**, can also be found in the **Selection toolbar** (Figure 4.8).

(i) Select a feature by clicking on the right end of a row in the attribute table. Upon selection, the corresponding polygon will be highlighted in yellow in the **map view**. In the example shown in Figure 4.9, the attribute table indicates that the land cover type of the highlighted polygon is classified as "Woody Crops/Orchards." Now, identify the land cover type of your selected polygon.

Figure 4.9: Selecting Features in the Attribute Table

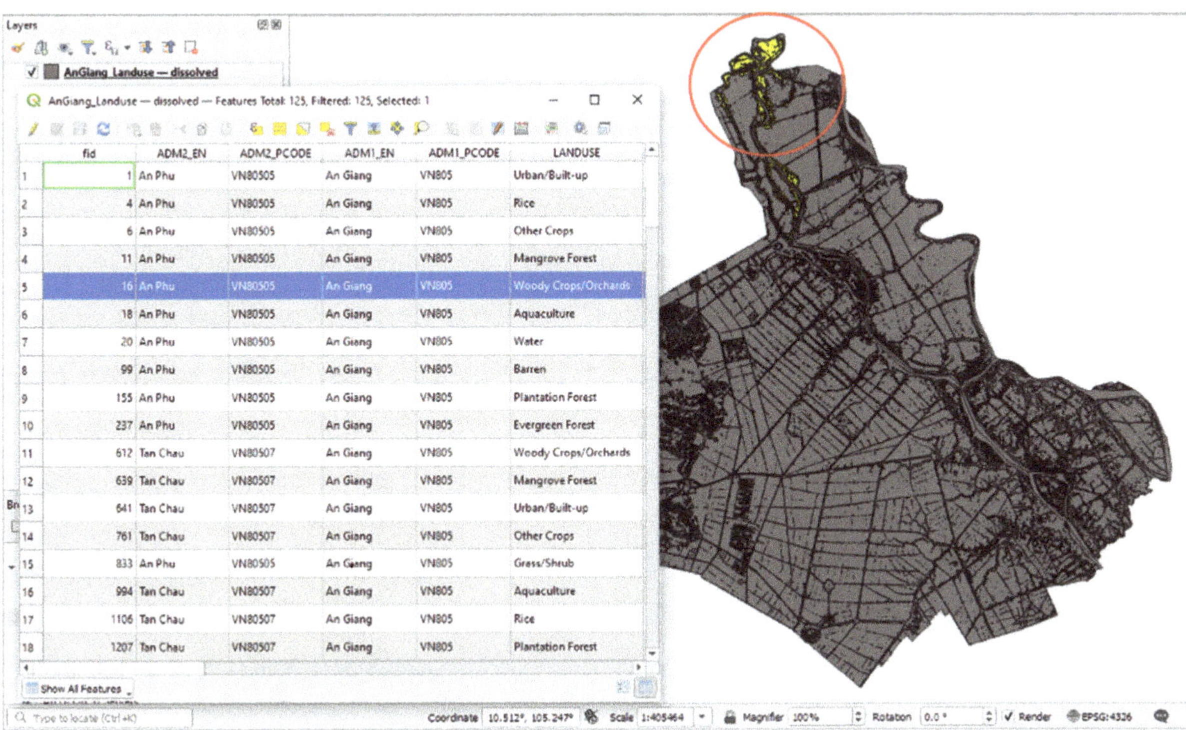

Source: Screenshot generated by the Asian Development Bank using QGIS 3.34.14. 2025.

(ii) Click or the **Select Features by Expression** tool to select several features at once using an expression. Type or copy-and-paste the expression "LANDUSE = 'Rice'" (Figure 4.10). This expression will select all the Rice-related features and polygons. While typing the expression, note that field names and function parameters automatically appear to enable faster writing of the expression. When you done, click **Select Features** to process the expression.

Figure 4.10: Selecting Features by Expression

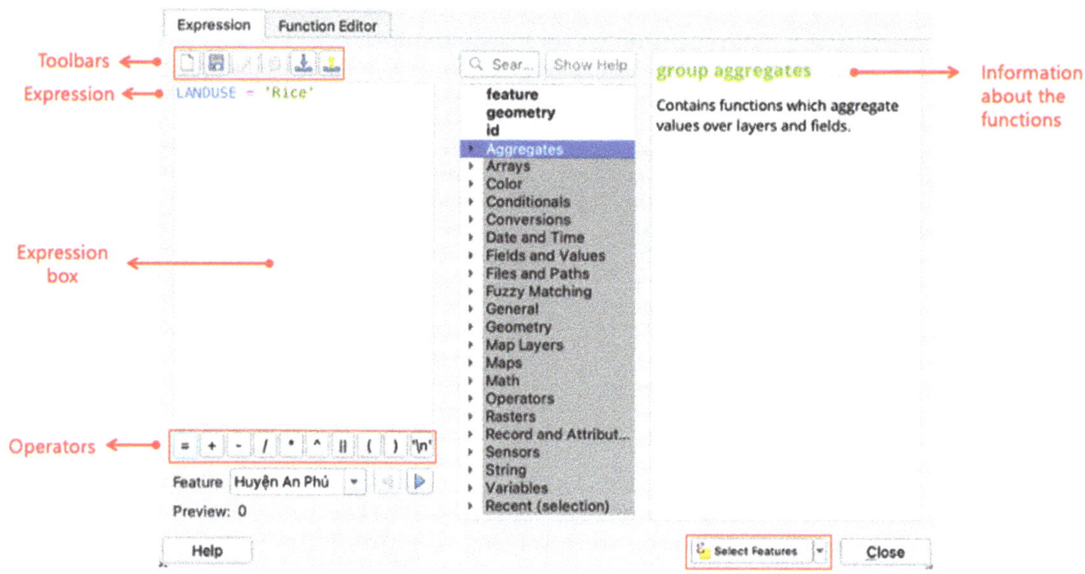

Source: Screenshot generated by the Asian Development Bank using QGIS 3.34.14. 2025.

(iii) Note that all polygons assigned to Rice categories are now highlighted in yellow on the map and highlighted blue in the attribute table (Figure 4.11).

Figure 4.11: Filtered and Selected Features

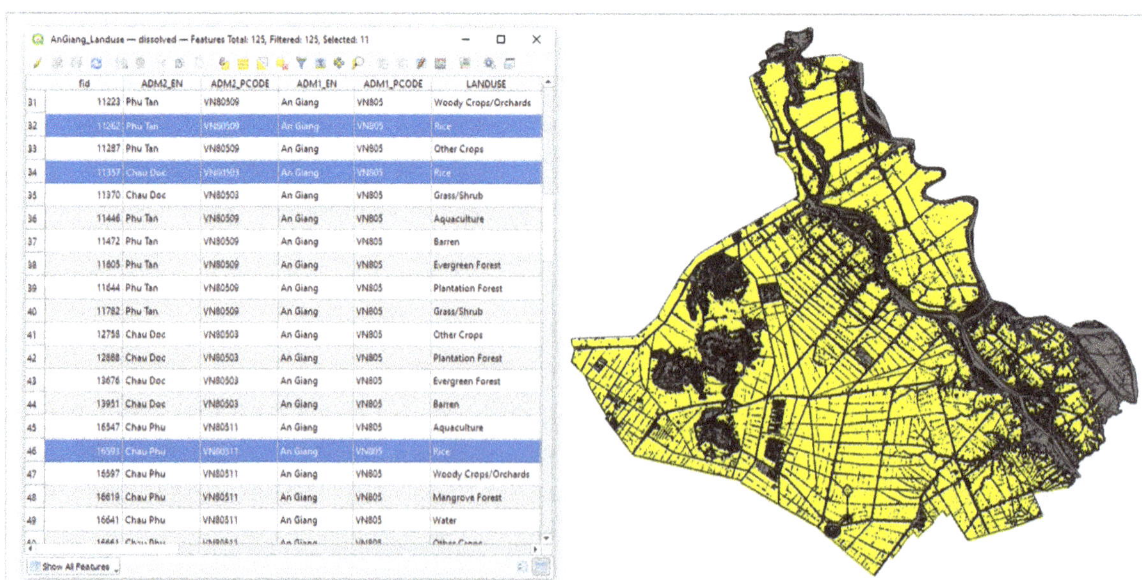

Source: Screenshot generated by the Asian Development Bank using QGIS 3.34.14. 2025.

At the top of the **attribute table**, the number of those features that were filtered or selected is displayed. The number of features selected is also displayed in the **status bar**, at the bottom of your main window.

(iv) Another way to filter and select features in the attribute table is to use the filter function. In the attribute table, click , the icon representing **Filter/Select features using form**. The filter form icon will then be highlighted at the bottom-left of the panel, indicating that the panel is in the filter format.

Let us select features that are identified with **Rice** land use in the lowlands. Type "Rice" in the LANDUSE field, and type "VN80511" in the **district ID** or **ADM2_PCODE** field. The **district ID** here is chosen at random for demonstration purposes, and you may select any district of interest for your own analysis. Note that the Contains button in the LANDUSE field and the district ID field are automatically activated, while other buttons remain excluded. Click **Select Features** (Figure 4.12).

Figure 4.12: Selecting Features Using the Filter Form

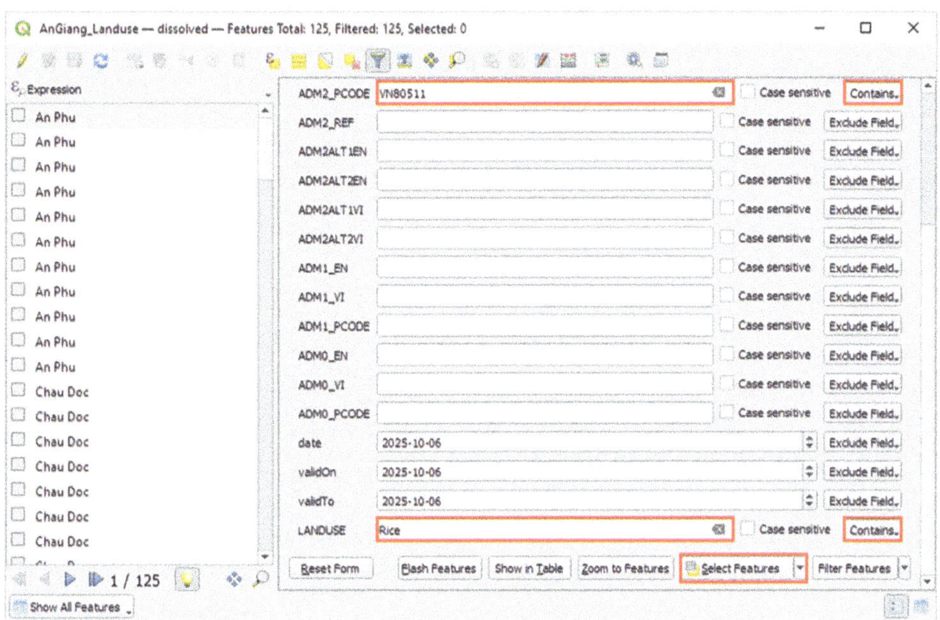

Source: Screenshot generated by the Asian Development Bank using QGIS 3.34.14. 2025.

(v) To view the selected features in the attribute table, click to switch in the table format. To show only the selected features in the attribute table, click **Show Selected Features**, at the bottom right of the attribute table (Figure 4.13).

Figure 4.13: Showing Selected Features Only in the Attribute Table

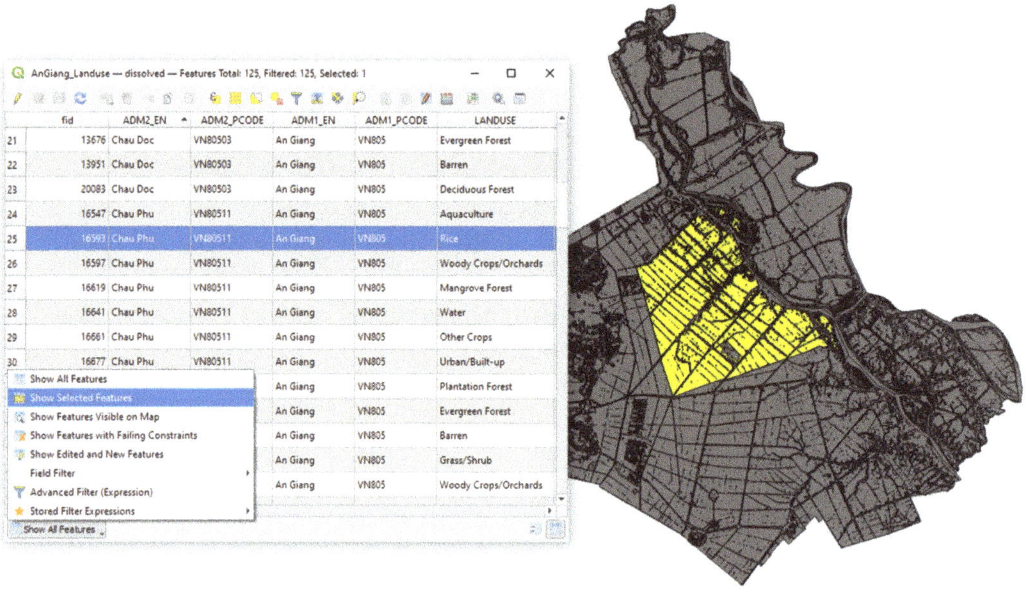

Source: Screenshot generated by the Asian Development Bank using QGIS 3.34.14. 2025.

(vi) To export the selected features into another file, right-click on the layer, select **Export** and **Save Selected Features As** (Figure 4.14).

Figure 4.14: Exporting and Saving Selected Features

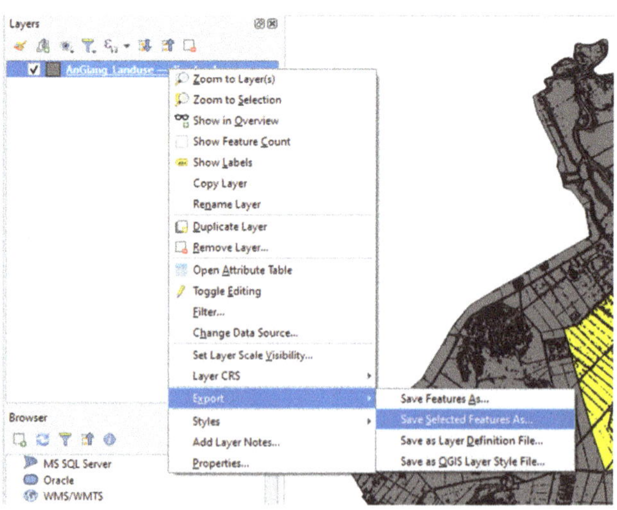

Source: Screenshot generated by the Asian Development Bank using QGIS 3.34.14. 2025.

(vii) In the **Save Vector Layer panel**, select the file format (e.g., Esri shapefile, GeoPackage . Click ... to find your project file folder. Ensure that the "Save only selected features" option is checked. Under "Select fields to export," click **Select All** to ensure that all the attributes of the features are included in the new layer; or, alternatively, select only the relevant attributes that you intend to retain for subsequent processing. Click **OK** (Figure 4.15).

Figure 4.15: Saving the Selected Vector Layer as a New Layer

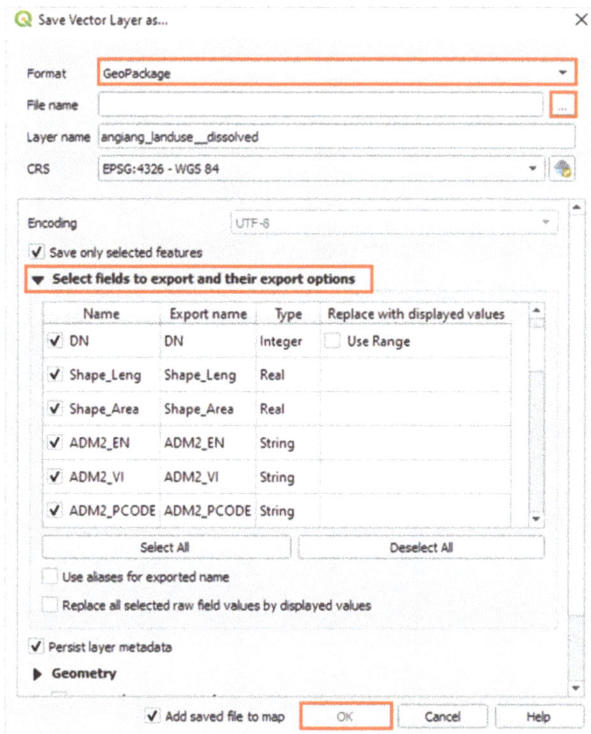

Source: Screenshot generated by the Asian Development Bank using QGIS 3.34.14. 2025.

4.3 Adding a New Field, Renaming, and Organizing Fields

(i) In the attribute table, enable the editing functionalities by clicking on the **Toggle editing mode** icon, . Note that other buttons will then be enabled, such as: "Add feature," "Paste features from clipboard," "New field," and "Delete field." To turn off the editing functionalities, click again.

(ii) To add a **new field attribute**, click on the **New field** icon, . A small pop-up window will appear (Figure 4.16).

Type the name of the field columns and select the type of attributes you wish to add. An attribute can be a whole number, decimal number, text, date, date and time, string list, or Boolean. "Length" refers to the total number of characters of the feature attribute, including the number of digits after the decimal places.

Figure 4.16: Creating a New Field Attribute

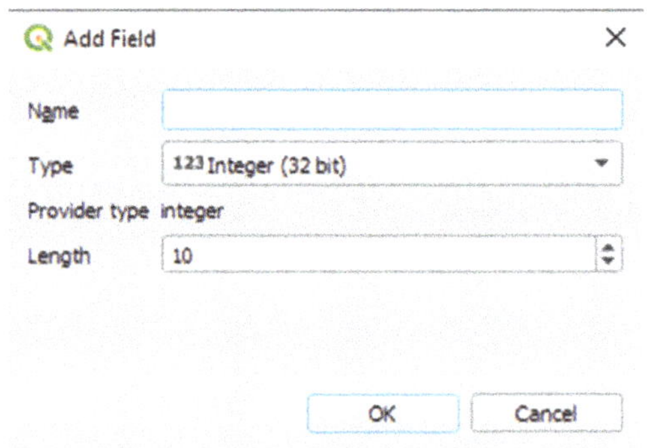

Source: Screenshot generated by the Asian Development Bank using QGIS 3.34.14. 2025.

(iii) To organize columns, click , the icon for the **Organize Table columns** tool. Organize the fields in the following order—fid, district name (ADM2_EN), district code (ADM2_PCODE), province name (ADM1_EN), province ID (ADM1_PCODE), and LANDUSE—by clicking and dragging the fields up and down (Figure 4.17). Then click **OK** to finish.

Figure 4.17: Organizing the Columns of the Attribute Table

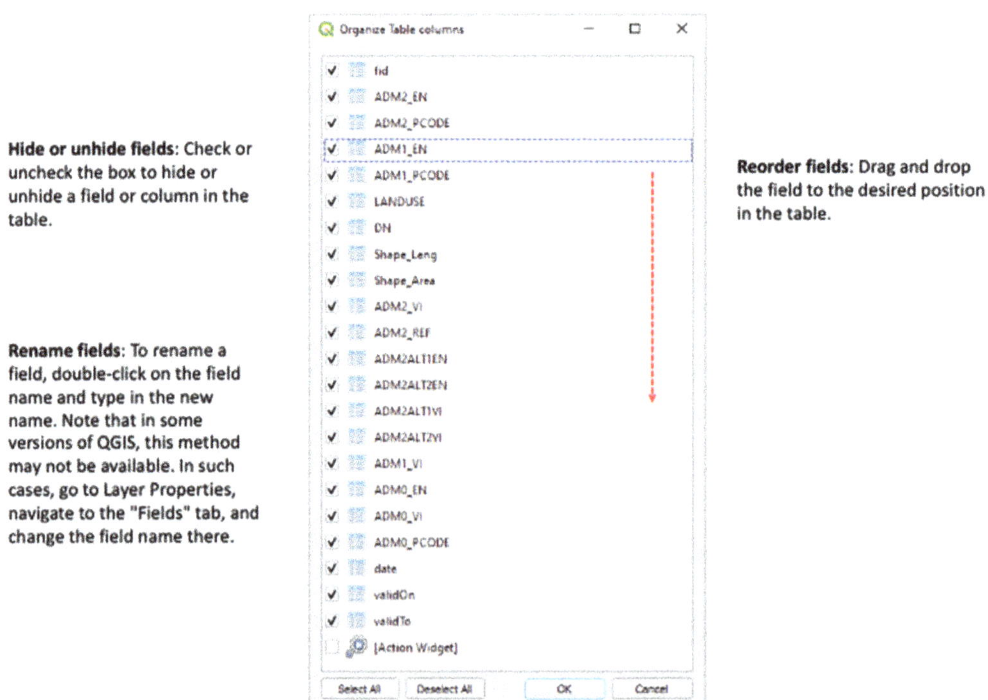

Source: Screenshot generated by the Asian Development Bank using QGIS 3.34.14. 2025.

(iv) To edit the names of the field attributes, right-click on the layer in the **Layers Panel** | **Properties** | **Fields** (Figure 4.18). To rename fields, click , the icon for the Toggle editing mode, double-click on the field name, and type the new name. Change "field id" ("fid") to "ID."

Click ▲ to reverse the position of the features in the rows. This function also applies in the attribute table. Click **Apply** and **OK** to save your edits.

Figure 4.18: Editing the Column Names of the Attribute Table

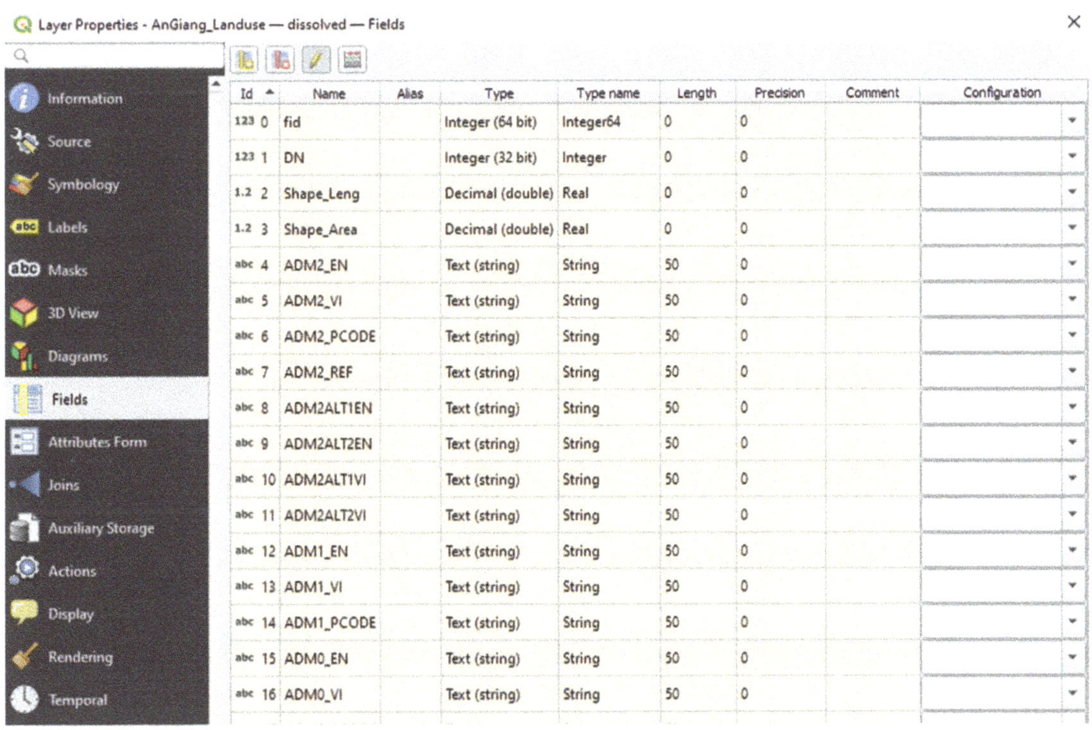

Source: Screenshot generated by the Asian Development Bank using QGIS 3.34.14. 2025.

4.4 Calculating a Variable (with Area Calculation as an Example)

There are two ways to calculate areas in the attribute table using the **open field calculator**. They use the expressions **$area** and **area($geometry).**

- **$area:** calculates the area of the current feature, considering the project's ellipsoid and area unit settings. If an ellipsoid is set, the area will be ellipsoidal; otherwise, it will be planimetric.
- **area($geometry):** computes the area of a given geometry polygon object using planimetric calculations based on the geometry's spatial reference system (SRS). The units of the returned area match the SRS units, in contrast to the $area, which adapts to the project's ellipsoid and area unit settings.

A more detailed discussion on area calculation is provided in Module 7, section 7.4, which focuses on this topic.

In the next exercise, we will calculate the areas of each polygon feature in the land-use map. Note that the current coordinate reference system (CRS) of the map is set to a geographic CRS, EPSG:4326 - WGS 84 - Geographic. For linear calculations such as area calculations, it is recommended that a Projected CRS (x, y) expressed in linear units be used, rather than a Geographic CRS (latitude, longitude), which is expressed in degrees. Hence, there will be a need to reproject the layer according to the recommended projection of the location before we calculate any areas.

(i) To reproject the vector layer, go to **main menu**, click **Vector** | **Data Management Tools** | **Reproject Layer** (Figure 4.19).

Figure 4.19: To Start Reprojecting a Vector Layer

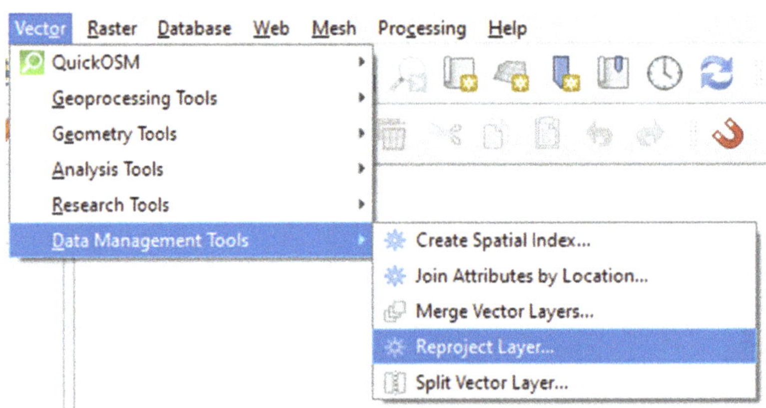

Source: Screenshot generated by the Asian Development Bank using QGIS 3.34.14. 2025.

(ii) Set the target CRS for Viet Nam to **EPSG:32648 / UTM** 48N. If the target CRS does not appear in the drop-down menu, click the **CRS widget,** , to open the **CRS selector**. In the CRS selector, use the **Filter tab** to search for the CRS. You can research the appropriate CRS for the locality; for instance, in our example concerning An Giang Province, it is EPSG:32648. Note that it will be highlighted on the map. Select this CRS. You may save the reprojected layer as a temporary file or as a file in your file folder. Click on **Open output file after running the algorithm**, and then click **Run** to reproject the layer (Figure 4.20).

Figure 4.20: To Finish Reprojecting a Vector Layer

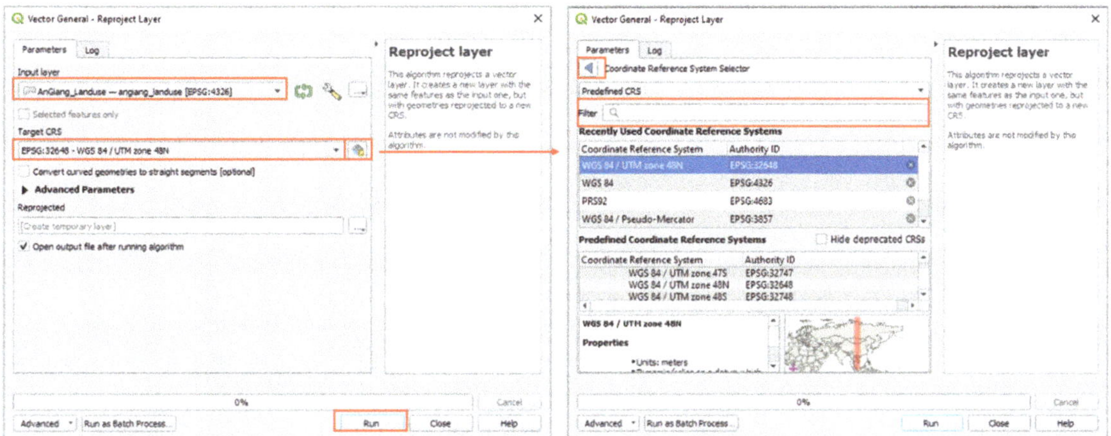

Source: Screenshots generated by the Asian Development Bank using QGIS 3.34.14. 2025.

(iii) After reprojecting the layer, open the attribute table of the new layer by right-clicking on the layer, and select **Open Attribute Table**.

(iv) Click to open the field calculator. Click **Create a new field**. Under **Output field name**, type "area_geo." Set **Output field type** to "Decimal Number (real)," **Output field length** to "10," and **Precision** to "3." In the Expression box, type the expression "area($geometry)." Select **OK** to calculate the area (Figure 4.21). The calculated area will be displayed in the attribute table.

Figure 4.21: Field Calculator

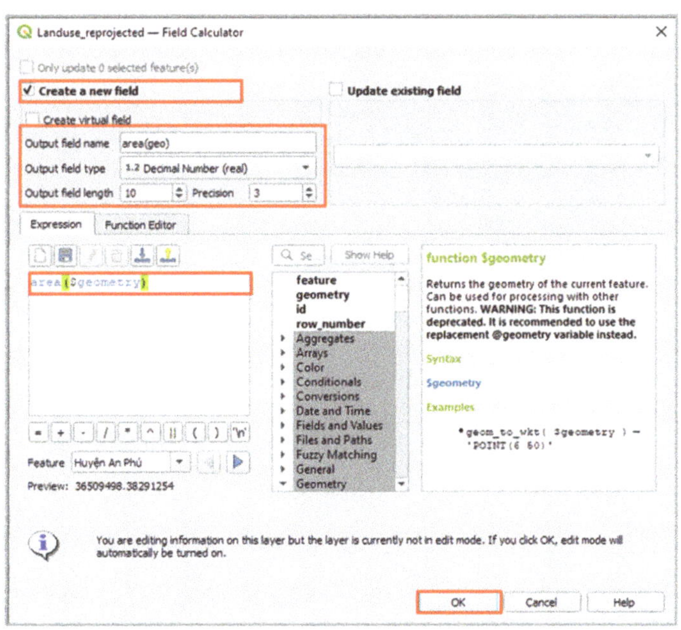

Source: Screenshot generated by the Asian Development Bank using QGIS 3.34.14. 2025.

Repeat similar steps to calculate the area using the **$area** expression. Compare the results in the attribute table (Figure 4.22). *Did you get similar results?*

Figure 4.22: Comparing the Calculated Areas

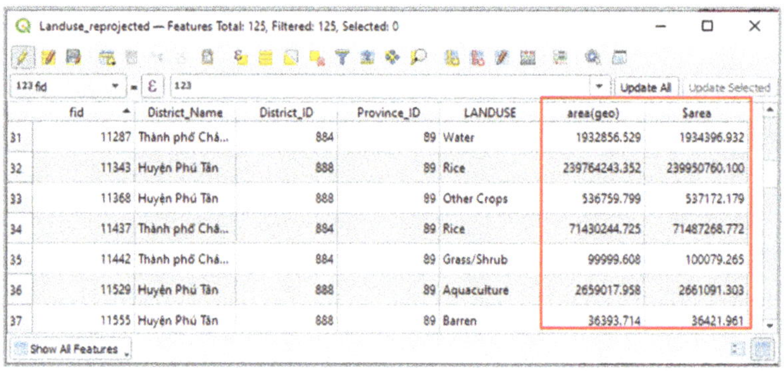

Source: Screenshot generated by the Asian Development Bank using QGIS 3.34.14. 2025.

Alternatively, to avoid duplicating the layer, it is possible to transform the Geographic CRS into a Projected CRS by using the transform function in combination with the area calculation expression.

To do this, go to the attribute table of the AnGiang_Landuse layer, which has a Geographic CRS. Open the field calculator and use the expression, **area(transform($geometry, 'EPSG:4326', 'EPSG:32648'))** to transform the CRS into EPSG:32648, and calculate the area as illustrated in Figure 4.23.

Figure 4.23: Changing a Vector Layer from a Geographic to a Projected Coordinate Reference System

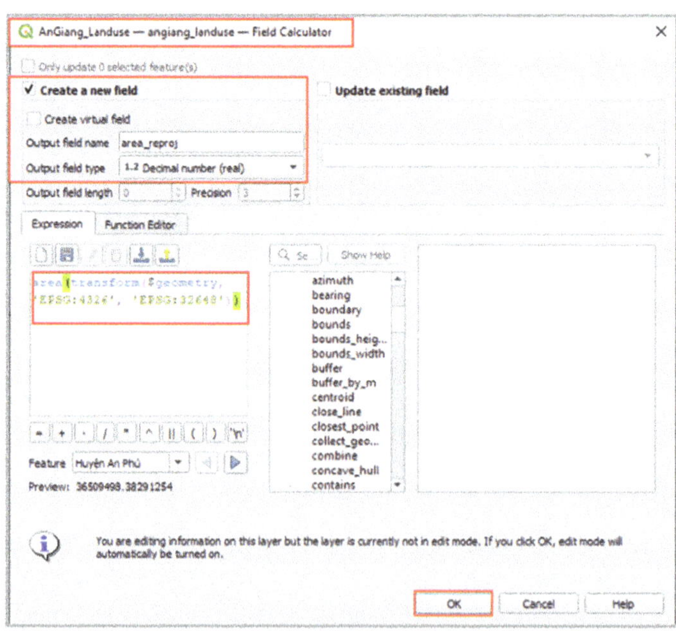

Source: Screenshot generated by the Asian Development Bank using QGIS 3.34.14. 2025.

4.5 Merging Tabular Data with Administrative Boundaries

We often have data in different formats, and information distributed over multiple files. Thus, it is important to learn how to join the attribute data from different data layers. The join functionality in QGIS allows us to combine data from multiple tables based on common values, e.g., identification (IDs) and categories.

In this exercise, we will use data from the *Statistical Yearbook of Viet Nam 2023* to merge tabular data with administrative boundaries (.shp file). The exercise file, which can be downloaded from the shared folder,[a] is MekongDelta_RiceStatistics.csv. This file contains data for paddy rice, including the areas planted and levels of production during major Mekong Delta region and the boundary shapefile for the Mekong Delta region (Mekong River Delta.gpkg).

[a] For access to shapefiles and other open-access datasets referenced in the handbook, go to the GitHub repository: https://github.com/ADB-Data-Division/QGIS-Handbook-2025

(i) To import MekongDelta_RiceStatistics.csv, click (**Add Delimited Text Layer**) from the **Manage Layers toolbar**.

In the **Manage Layers toolbar** dialog box, click on **Browse files**. In the **File Format Tab**, select **CSV (comma separated values)**. In the **Geometry Definition** tab, select **No geometry (attribute only table)**. Then click **Add** (Figure 4.24).

Figure 4.24: Importing Tabular Data in Comma-Separated Values Format

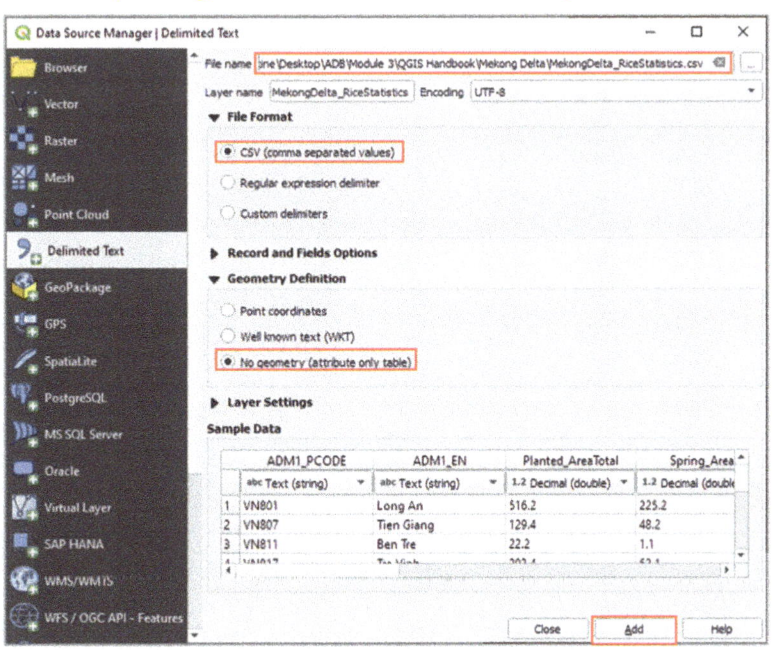

Source: Screenshot generated by the Asian Development Bank using QGIS 3.34.14. 2025.

(ii) Import the Mekong River Delta.gpkg file by dragging and dropping the file from the **Browser panel** into the map area.

(iii) Examine the attribute fields of both files. To join the two attribute tables, identify common field attributes of the two files (Figure 4.25). The common field attributes in this case are **ADM1_EN** (province names) and **ADM1_PCODE** (province codes) (Figure 4.26).

Figure 4.25: Tabular Layer

MekongDelta_RiceStatistics — Features Total: 13, Filtered: 13, Selected: 0

	ADM1_PCODE	ADM1_EN	Planted_AreaTotal	Spring_Area	Autumn_Area	Winter_Area	Spring_Prod
1	VN801	Long An	516.2	225.2	289.8	1.3	1526.6
2	VN807	Tien Giang	129.4	48.2	81.2	NULL	336.3
3	VN811	Ben Tre	22.2	1.1	16	5.2	5.7
4	VN817	Tra Vinh	203.4	62.1	140.2	1.1	378.7
5	VN809	Vinh Long	112.7	39.8	72.9	NULL	274
6	VN803	Dong Thap	498.3	190.2	308.3	NULL	1390.1
7	VN805	An Giang	616.2	227.7	385	3.5	1719.4
8	VN813	Kien Giang	712.9	281	363.3	68.5	2159.6
9	VN815	Can Tho	216.2	75	141.2	NULL	559.8
10	VN816	Hau Giang	177.8	75.5	102.3	NULL	585.4
11	VN819	Soc Trang	330.4	174.5	145	10.9	1187.2
12	VN821	Bac Lieu	191.6	43.3	101.8	46.5	335.7
13	VN823	Ca Mau	111.3	35.2	35.2	40.9	218.7

Show All Features

Source: Screenshot generated by the Asian Development Bank using QGIS 3.34.14. 2025.

Figure 4.26: Attribute Table of the Administrative Boundary Layer

Mekong River Delta — Features Total: 13, Filtered: 13, Selected: 0

	ADM1_EN	ADM1_VI	ADM1_PCODE	ADM0_EN	ADM0_VI	ADM0_PCODE
1	An Giang	An Giang	VN805	Viet Nam	Việt Nam	VN
2	Bac Lieu	Bạc Liêu	VN821	Viet Nam	Việt Nam	VN
3	Ben Tre	Bến Tre	VN811	Viet Nam	Việt Nam	VN
4	Ca Mau	Cà Mau	VN823	Viet Nam	Việt Nam	VN
5	Can Tho city	TP. Cần Thơ	VN815	Viet Nam	Việt Nam	VN
6	Dong Thap	Đồng Tháp	VN803	Viet Nam	Việt Nam	VN
7	Hau Giang	Hậu Giang	VN816	Viet Nam	Việt Nam	VN
8	Kien Giang	Kiên Giang	VN813	Viet Nam	Việt Nam	VN
9	Long An	Long An	VN801	Viet Nam	Việt Nam	VN
10	Soc Trang	Sóc Trăng	VN819	Viet Nam	Việt Nam	VN
11	Tien Giang	Tiền Giang	VN807	Viet Nam	Việt Nam	VN
12	Tra Vinh	Trà Vinh	VN817	Viet Nam	Việt Nam	VN
13	Vinh Long	Vĩnh Long	VN809	Viet Nam	Việt Nam	VN

Show All Features

Source: Screenshot generated by the Asian Development Bank using QGIS 3.34.14. 2025.

(iv) To merge the tabular data with the attribute table of the administrative boundaries, right-click **Mekong River Delta** boundaries in the **Layers panel**, and then select **Properties**.

In the **Layer Properties** dialog box, click Joins and then Select to add a new join. The **Add Vector Join** dialog box will pop up (Figure 4.27).

Figure 4.27: Join Function in Layer Properties

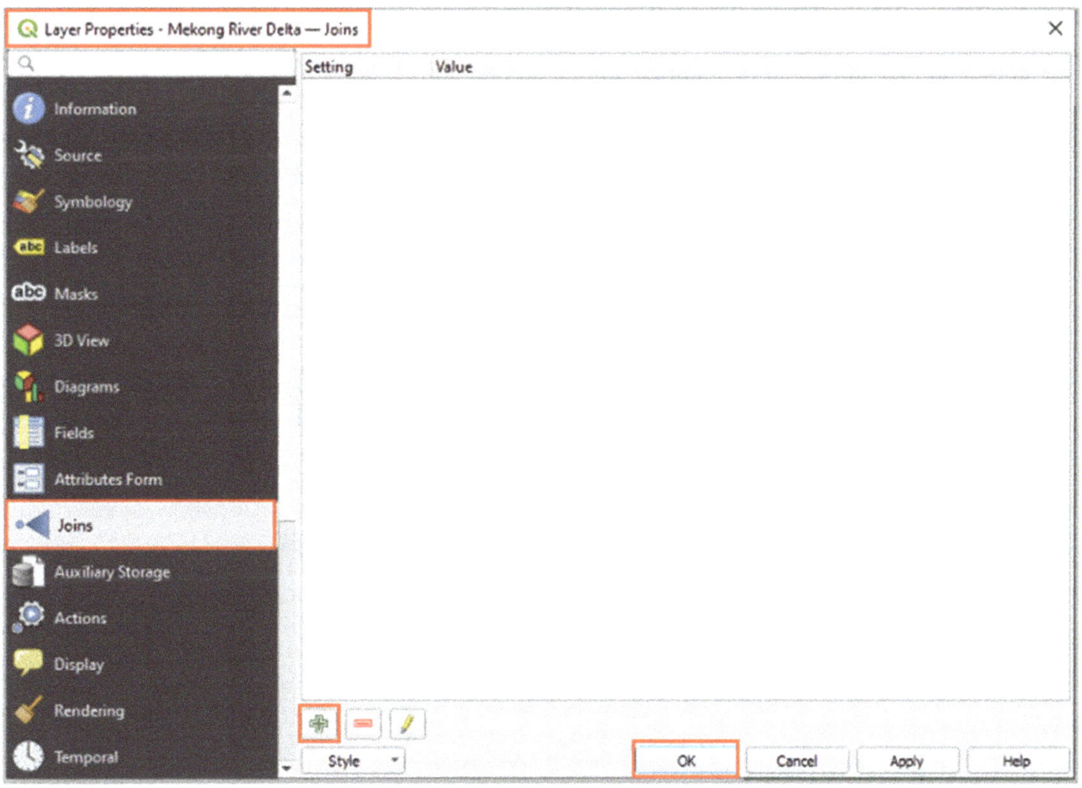

Source: Screenshot generated by the Asian Development Bank using QGIS 3.34.14. 2025.

(v) In the **Add Vector Join dialog box**, select **MekongDelta_RiceStatistics** under the **Join Layer tab**. In the Join field and Target field, select the common attribute field between the two layers. For this exercise, choose **ADM1_PCODE**. Select **Cache join layer in memory**. In the Custom field name prefix, delete "MekongDelta_RiceStatistics_," if you do not want to add this to each column name of the added attribute fields. Then click **OK** (Figure 4.28).

Figure 4.28: The Add Vector Join Dialog Box

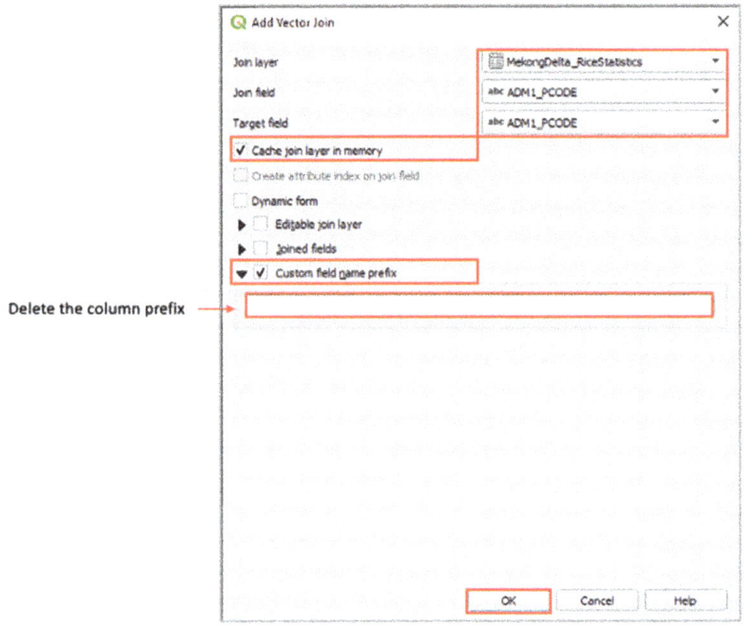

Source: Screenshot generated by the Asian Development Bank using QGIS 3.34.14. 2025.

(vi) In the **Join tab** of the **Layer Properties dialog box** (Figure 4.29), you can view your added joins. You can add more joins using the button; and you can delete joins using . Select **Apply** and **OK** to complete the process.

Figure 4.29: Adding Joins

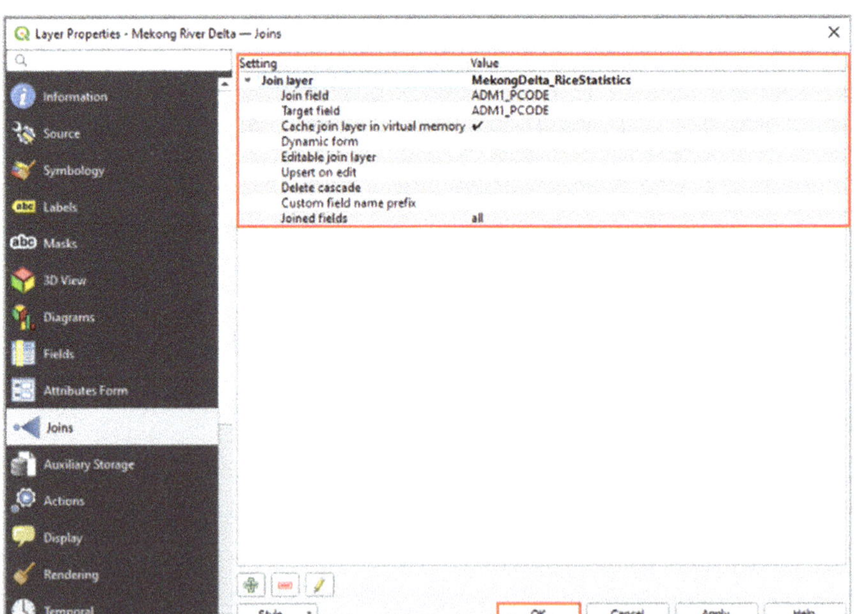

Source: Screenshot generated by the Asian Development Bank using QGIS 3.34.14. 2025.

(vii) Finally, open the **attribute table** of the **Mekong River Delta boundaries layer** and check the added columns. The newly joined columns in the attribute table on the Mekong River Delta boundaries layer are dependent on the MekongDelta_RiceStatistics layer. If you delete the tabular layer in your project, the joined columns in the Mekong River Delta boundaries layer will be deleted as well, including other calculated attributes.

To enable the Mekong River Delta boundaries layer to stand alone, export it as a new shapefile. To export a layer, right-click on it in the **Layers panel**, select **Export**, then **Save Features**, and a dialog box will appear with further settings (Figure 4.30).

Figure 4.30: Exporting a Layer

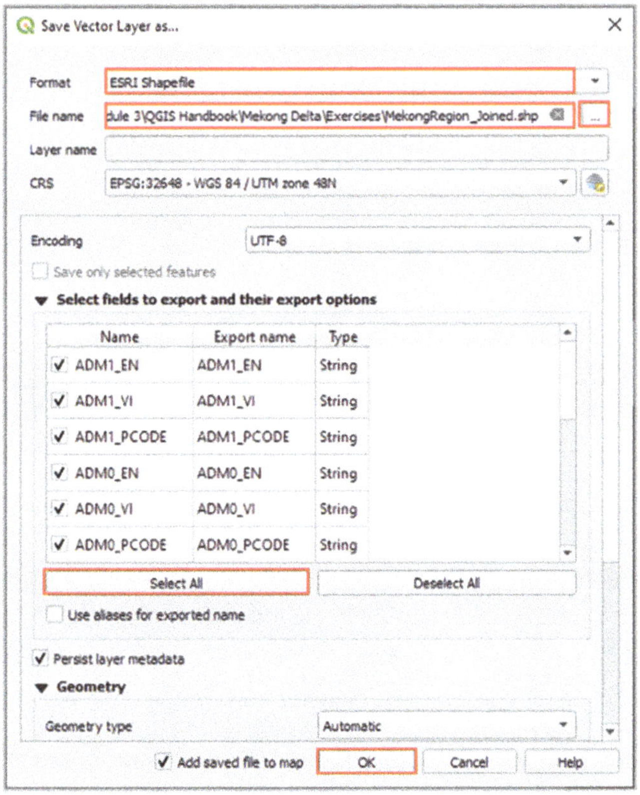

Source: Screenshot generated by the Asian Development Bank using QGIS 3.34.14. 2025.

4.6 Creating an Overlay Histogram on a Map

Integrating statistics with geospatial (location) data is valuable for data analysis, visualization, and dissemination. Depending on the context of your project, some data may become more intuitive when they are visualized on a map. For instance, viewing a histogram graph overlaid on a map provides us with more insights than just viewing the data in a table or a graph. For example, have a look at Table 4.1 and Figure 4.31.

Table 4.1: Rice Plantation Areas and Production Levels in the Mekong Delta Region, by Province, 2023

Province	Area ('000 hectares)				Production ('000 tons)			
	Annual Total	Spring	Autumn	Winter	Annual Total	Spring	Autumn	Winter
Long An	516.2	225.2	289.8	1.3	3,081.0	1,526.6	1,549.0	5.4
Tien Giang	129.4	48.2	81.2	...	791.0	336.3	454.7	...
Ben Tre	22.2	1.1	16.0	5.2	109.7	5.7	83.1	20.9
Tra Vinh	203.4	62.1	140.2	1.1	1,110.3	378.7	726.6	5.1
Vinh Long	112.7	39.8	72.9	...	676.2	274.0	402.2	...
Dong Thap	498.3	190.2	308.3	...	3,325.3	1,390.1	1,935.2	...
An Giang	616.2	227.7	385.0	3.5	4,067.7	1,719.4	2,334.6	13.7
Kien Giang	712.9	281.0	363.3	68.5	4,561.9	2,159.6	2,035.8	366.5
Can Tho	216.2	75.0	141.2	...	1,362.2	559.8	802.4	...
Hau Giang	177.8	75.5	102.3	...	1,186.7	585.4	601.3	...
Soc Trang	330.4	174.5	145.0	10.9	2,077.2	1,187.2	834.2	55.8
Bac Lieu	191.6	43.3	101.8	46.5	1,239.0	335.7	631.8	271.5
Ca Mau	111.3	35.2	35.2	40.9	568.4	218.7	174.5	175.2

... = data not available.

Source: Government of Viet Nam, General Statistics Office. 2024. Statistical Yearbook of Viet Nam 2023. https://www.nso.gov.vn/wp-content/uploads/2024/06/NIEN-GIAM-THONG-KE-2023_Ban-quyen.pdf.

Figure 4.31: Rice Plantation Areas and Production Levels in the Mekong Delta Region, 2023

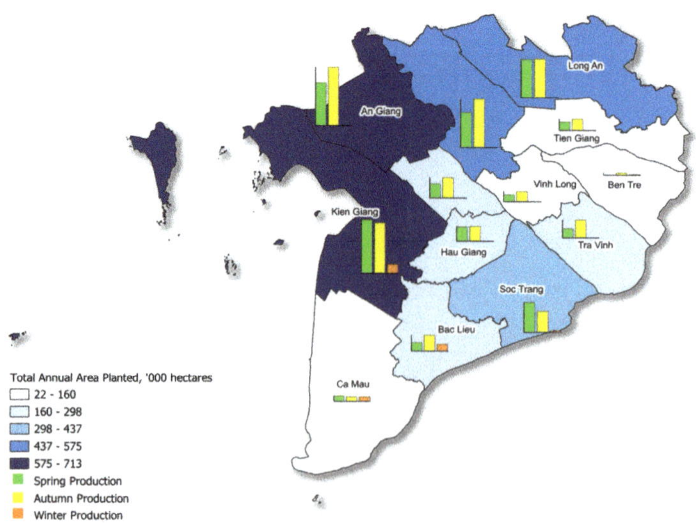

Sources: Government of Viet Nam, National Statistics Office. 2024. *Statistical Yearbook of Viet Nam 2023.* https://www.nso.gov.vn/wp-content/uploads/2024/06/NIEN-GIAM-THONG-KE-2023_Ban-quyen.pdf; Screenshot generated by the Asian Development Bank using QGIS 3.34.14. 2025.

In the following exercise, we will create a histogram based on the level of rice production and the proportion of land area planted with rice in each province of the Mekong Delta region. We will use the attribute table and the joined columns of the administrative layer from our previous exercise.

(i) Right-click on the joined layer in the **Layers panel**, then click **Open Attribute Table** (Figure 4.32).

Figure 4.32: Opening the Attribute Table

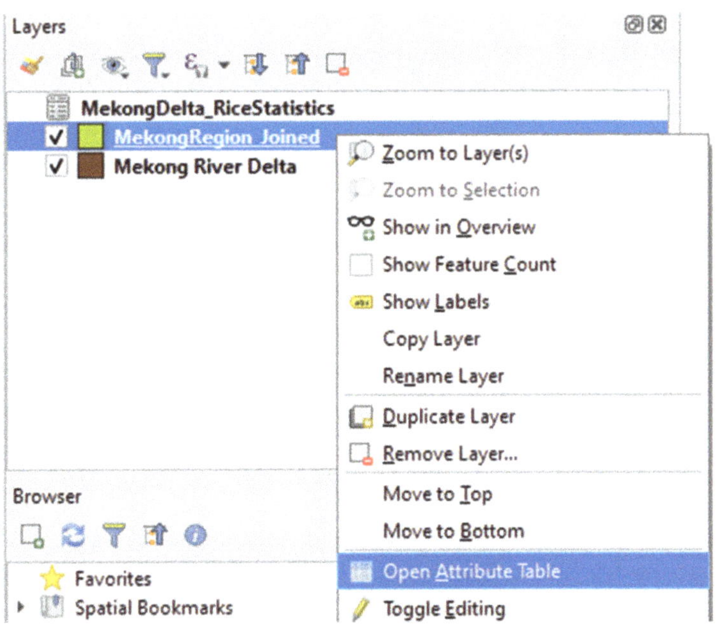

Source: Screenshot generated by the Asian Development Bank using QGIS 3.34.14. 2025.

(ii) To determine the proportion of area planted with rice in each province, start by calculating the rice-planted area for each province in the region. Click in the attribute table to open the **Field Calculator** (Figure 4.33). In the Expression box, type the expression, **$area**. Click **OK** and check the calculated area in the attribute table.

Figure 4.33: Opening the Field Calculator from the Attribute Table and Calculating the Proportion of Planted Areas

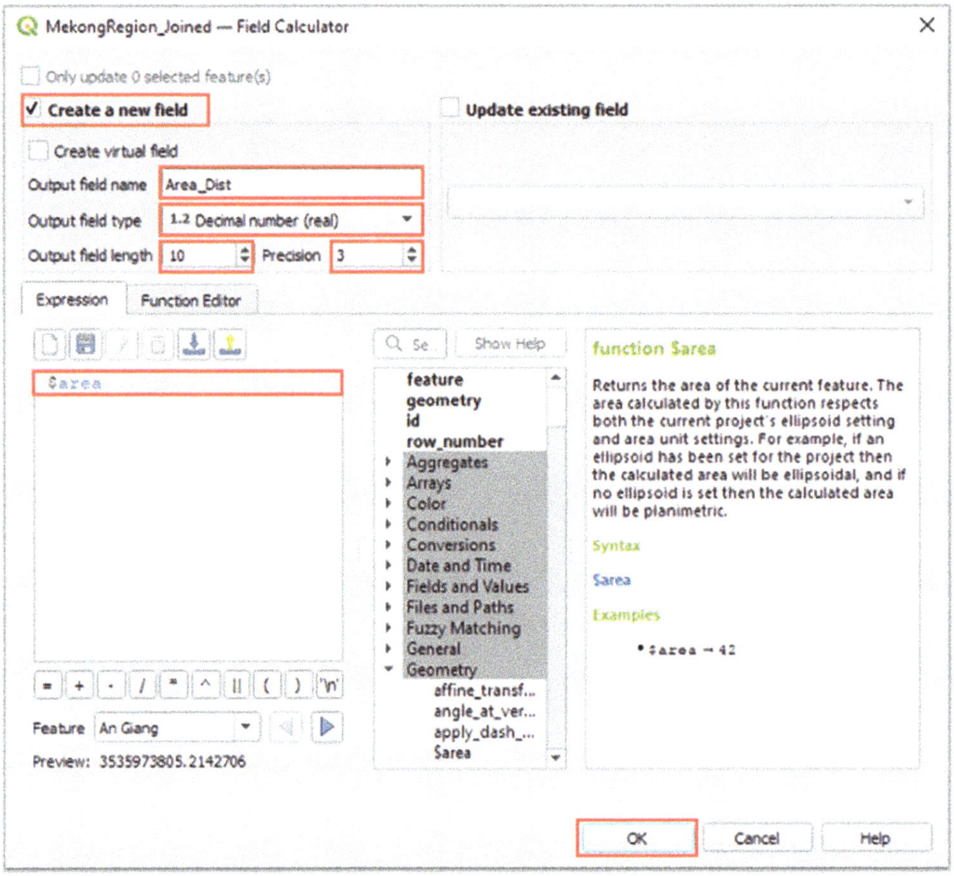

Source: Screenshot generated by the Asian Development Bank using QGIS 3.34.14. 2025.

(iii) Next, calculate the proportion of area under rice cultivation during the spring season in each province. Open the **Field Calculator** and click **Create a new field** (Figure 4.34). Type **%Area_Spri** in the **Output field name** box and select **Decimal number (real)** as the **Output field type**. In the **Expression box**, type the equation **"Spring_Are" / ("Area_Dist" / (10000 * 1000))** to calculate the proportion. In QGIS, the default unit of area measurement is square meters. The **integer (10000 * 1000)** is a conversion factor from square meters to thousand hectares, as the published statistics on rice are in thousand hectares. Click **OK** and check the calculated proportion in the attribute table.

To check and change the default units, go to the **main menu | Project | Properties | General | Measurements**.

Figure 4.34: Calculating the Proportion of Areas Planted with Rice

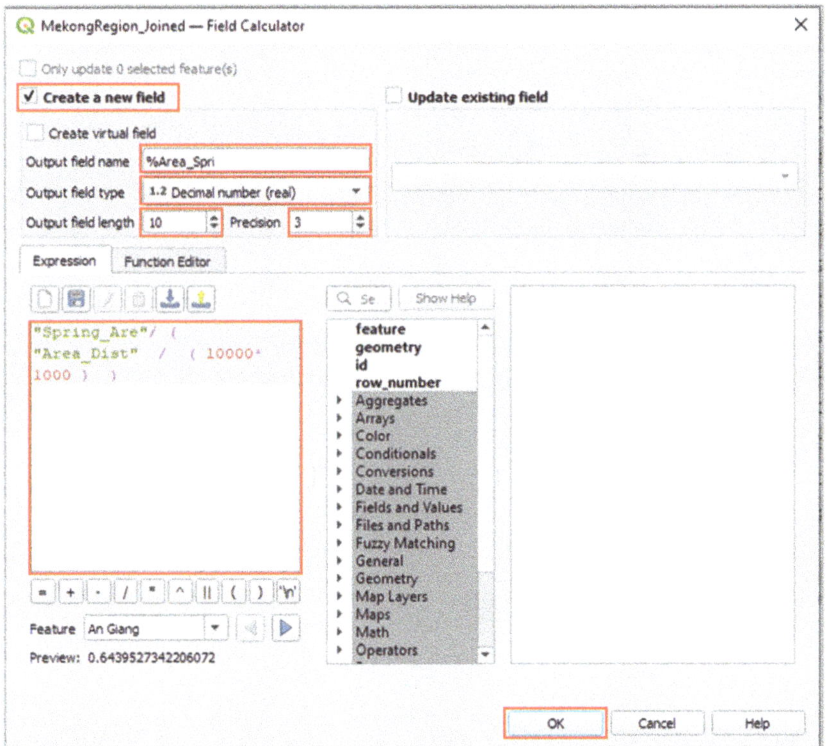

Source: Screenshot generated by the Asian Development Bank using QGIS 3.34.14. 2025.

Next, classify the proportion of area planted with rice for the season. This procedure is similar to that depicted in the styling exercises in Module 2. There are several layer-styling options available in QGIS for effectively visualizing spatial data, each serving a specific purpose, depending on the nature of the dataset and the intended analysis. These include **Single Symbol, Categorized, Graduated, Rule-Based,** and **Point Displacement**; as well as more advanced styling options such as **Merged Features, Inverted Polygon, 2.5D Renderer,** and **Embedded Symbols**. Each method provides a unique approach to presenting data on a map. These styling options will be discussed in more detail in Module 5 of the handbook.

In the **Layers panel**, click to open the **Layer Styling panel** (Figure 4.35).

In the Layer Styling panel, select **Graduated**. The graduated classification is used to visualize numerical values.

In the **Value tab**, select **%Area_Spri** and Blues in the Color ramp.
In the **Mode tab**, select **Equal Interval** to classify the values in equal intervals. Click **Classify**.

Notice that the map is now classified into five classes at equal intervals. The algorithm automatically creates the five classes, according to the optimal intervals calculated from the data. You can set the number of classes if you prefer more or fewer classifications in your projects.

Figure 4.35: Applying Graduated Symbology to Visualize Finite Attribute Data

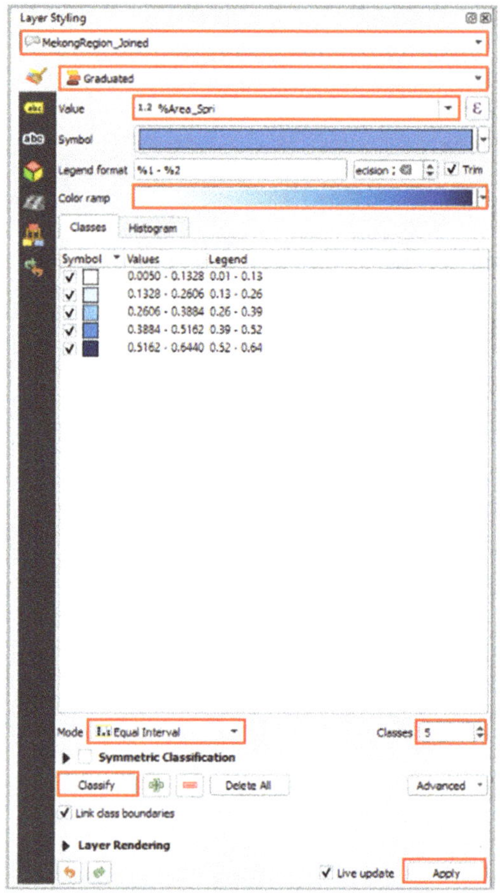

Source: Screenshot generated by the Asian Development Bank staff using QGIS 3.34.14. 2025.

(iv) To add more visual effects to the map, click the **Layer Rendering** tab. Select **Draw effects** and click [icon] to customize the visual effects of the map (Figure 4.36). The **Effects Properties** part of the **Layer Styling panel** will appear (Figure 4.37).

Figure 4.36: Enabling Draw Effects and Effects Properties

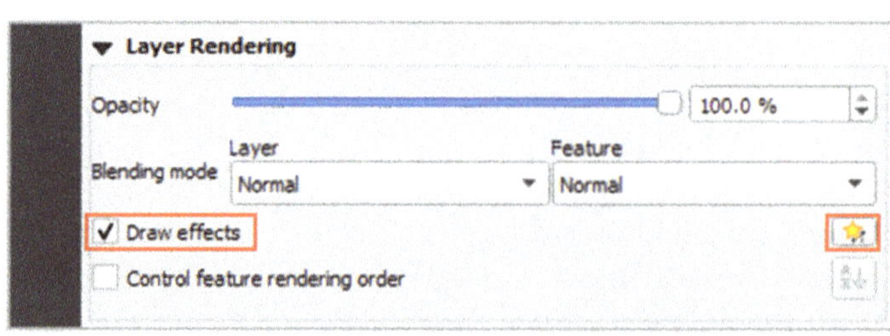

Source: Screenshot generated by the Asian Development Bank using QGIS 3.34.14. 2025. Manila.

(v) In the **Effects Properties**, select **Outer Glow** and **Drop Shadow** to provide more emphasis on the boundary lines of the map. Type **1 (mm)** for both **Spread** and **Blur radius**. Click **Apply** (Figure 4.37).

Figure 4.37: Settings for Effects Properties

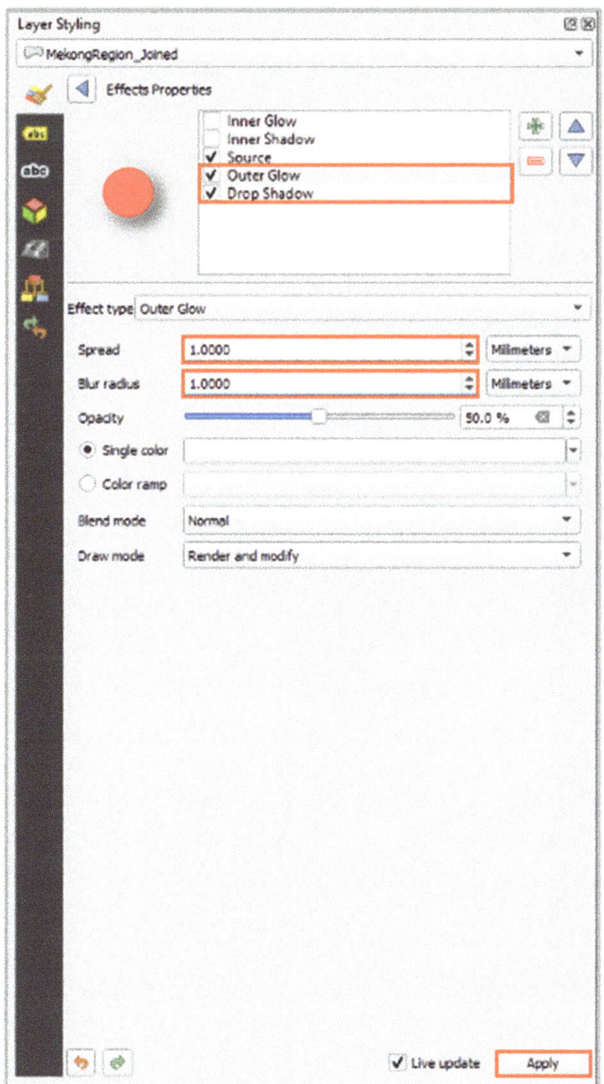

Source: Screenshot generated by the Asian Development Bank using QGIS 3.34.14. 2025.

(vi) To add the names of the provinces to the map, click ; or click on **Labels** to set up the labels on the map. Select **Single Labels** and set the **Value** to **ADM1_EN**.

Click (**Text** tab) to change the font, style, size, color, etc. of the text.

Click (**Buffer** tab) to create a buffer in the text and add highlights (Figure 4.38).

Figure 4.38: Rendering the Map Labels

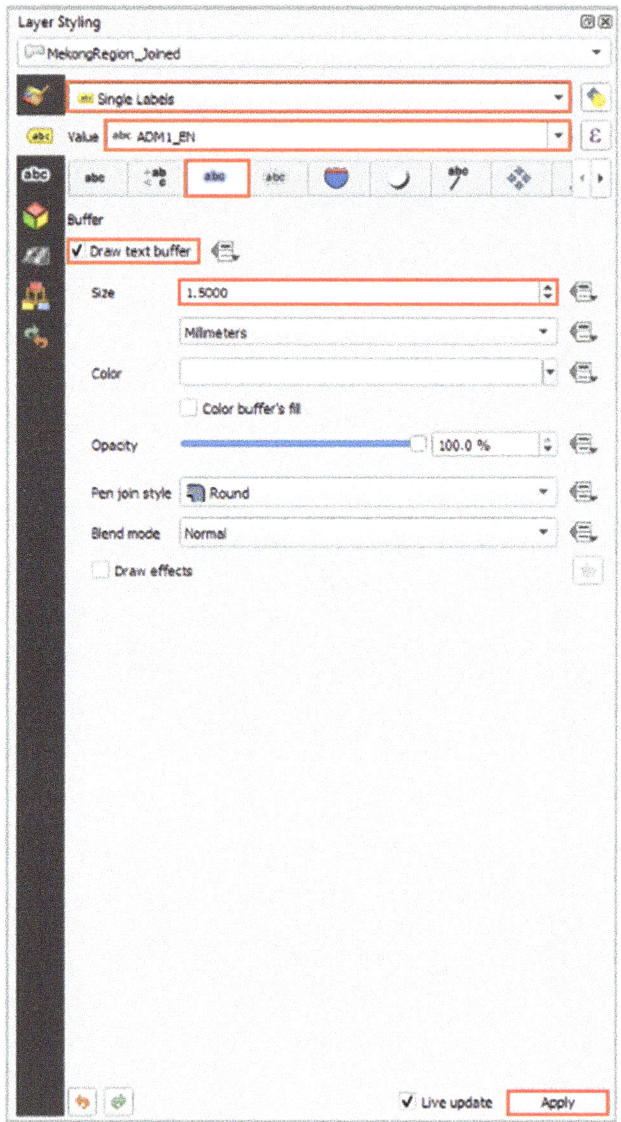

Source: Screenshot generated by the Asian Development Bank using QGIS 3.34.14. 2025.

(vii) The categorized map showing the proportion of area planted with rice during the spring season will look like Figure 4.39.

Figure 4.39: Map Showing the Proportion of Provinces Planted with Rice in the Mekong Delta Region, Spring 2023

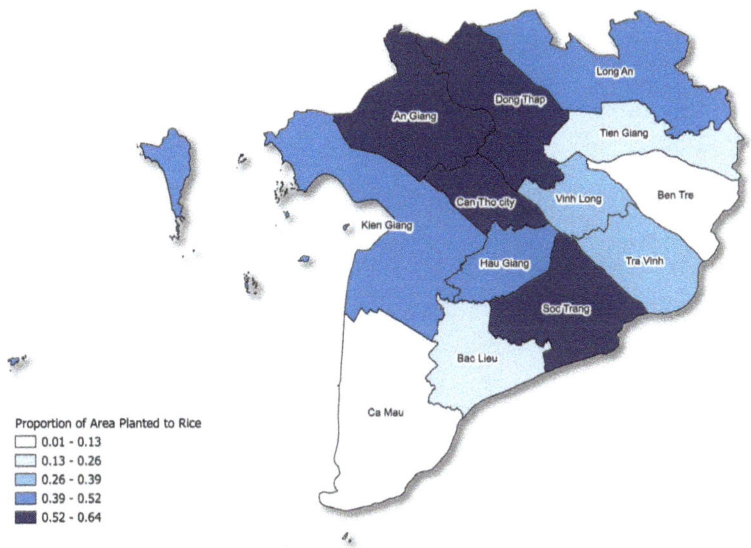

Source: Screenshot generated by the Asian Development Bank using QGIS 3.34.14. 2025.

(viii) Next, add histograms to the map to visualize rice productivity in the region. Right-click on the **MekongRegion_Joined** layer, select **Properties | Diagrams** (Figure 4.40).

Figure 4.40: Adding Attributes to the Histogram

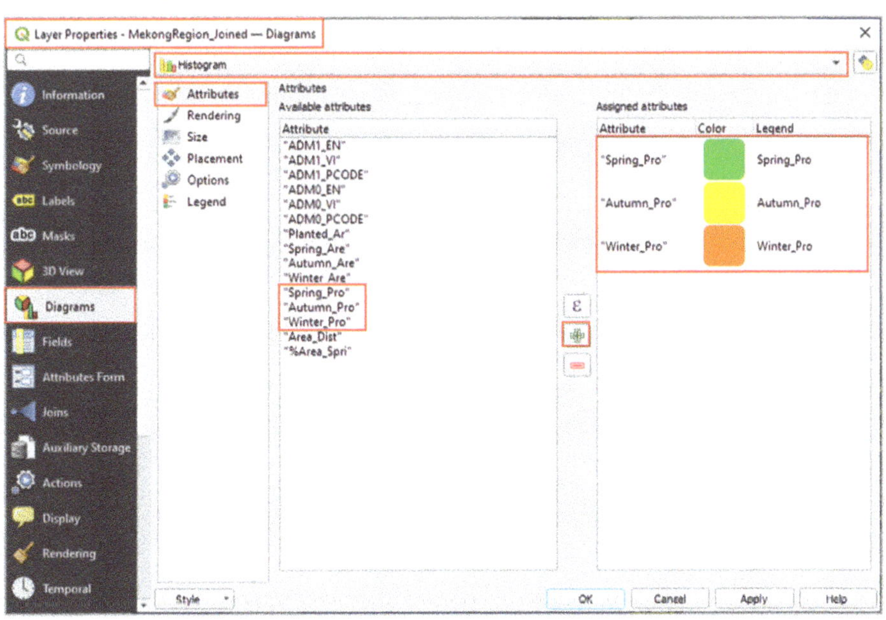

Source: Screenshot generated by the Asian Development Bank using QGIS 3.34.14. 2025.

In the drop-down menu, select **Histogram**. Click on the **Attributes tab** and assign field attributes ("Spring_Pro," "Autumn_Pro," and "Winter_Pro") using the histogram. In the **Assigned attributes box**, change the color and the legend of the histogram by clicking accordingly. Then click **Apply**.

(x) In the Rendering tab, set the **Bar width** and **Bar spacing** at levels proportionate to the map. Be sure to enable **Show all diagrams** to depict all the histograms in each administrative boundary (Figure 4.41).

Figure 4.41: Rendering the Bar Width and Spacing of Histograms

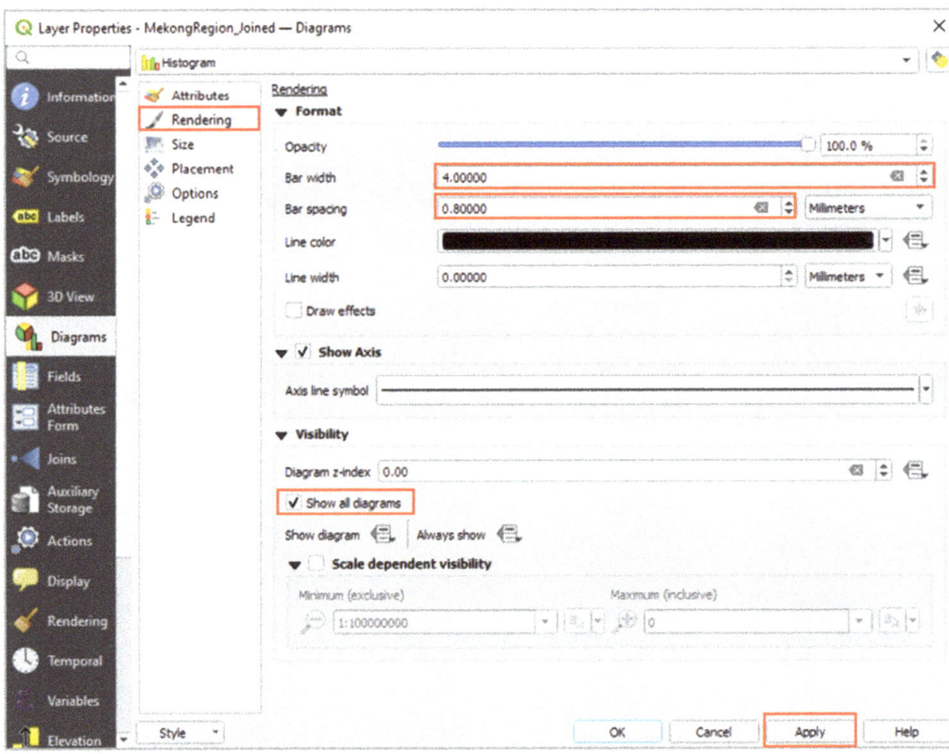

Source: Screenshot generated by the Asian Development Bank using QGIS 3.34.14. 2025.

(xi) In the **Size tab**, select the attribute with the highest value to configure the size of the bars. In this example, the attribute that contains the highest value is Autumn_Pro, which represents one of the provinces with the highest rice production during the year. To find the highest value, click the **Find** button, which will automatically identify the maximum value in the attribute table data. Set the bar length to 22. Click **Apply** (Figure 4.42).

Figure 4.42: Setting the Sizes of Bars in the Histogram

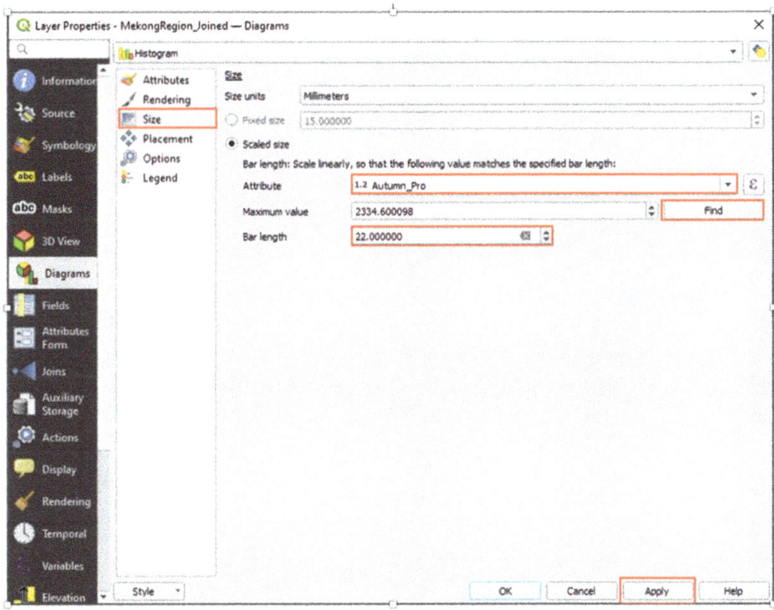

Source: Screenshot generated by the Asian Development Bank using QGIS 3.34.14. 2025.

(xii) In the **Placement tab**, select **Inside Polygon** to place the histogram inside the boundaries, where it belongs (Figure 4.43).

Figure 4.43: Setting the Bar Placement in the Histogram

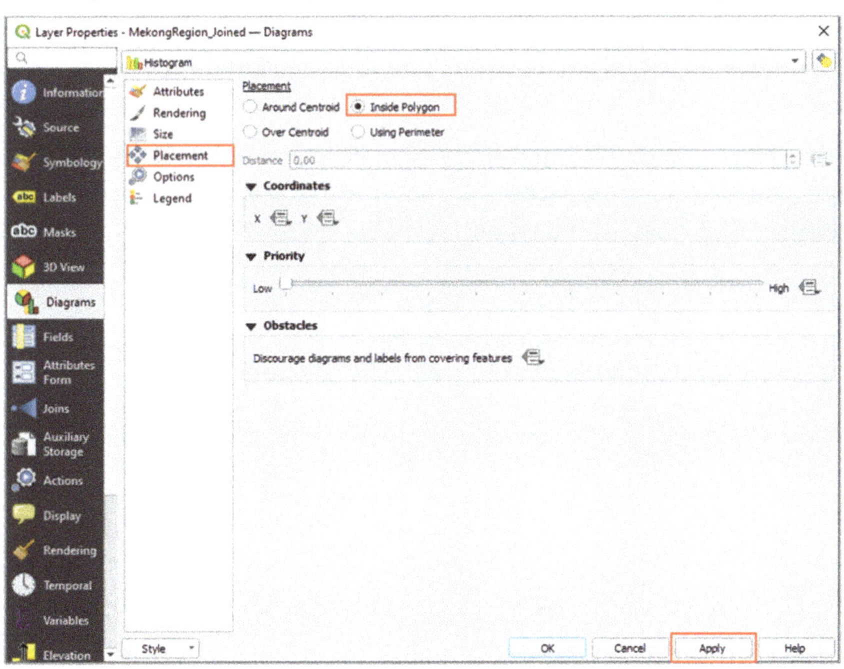

Source: Screenshot generated by the Asian Development Bank using QGIS 3.34.14. 2025.

(xiii) In the **Options tab**, set the **Bar Orientation** on **Up**. Click **Apply**, then **OK** (Figure 4.44).

Figure 4.44: Setting the Bar Orientation of the Histogram

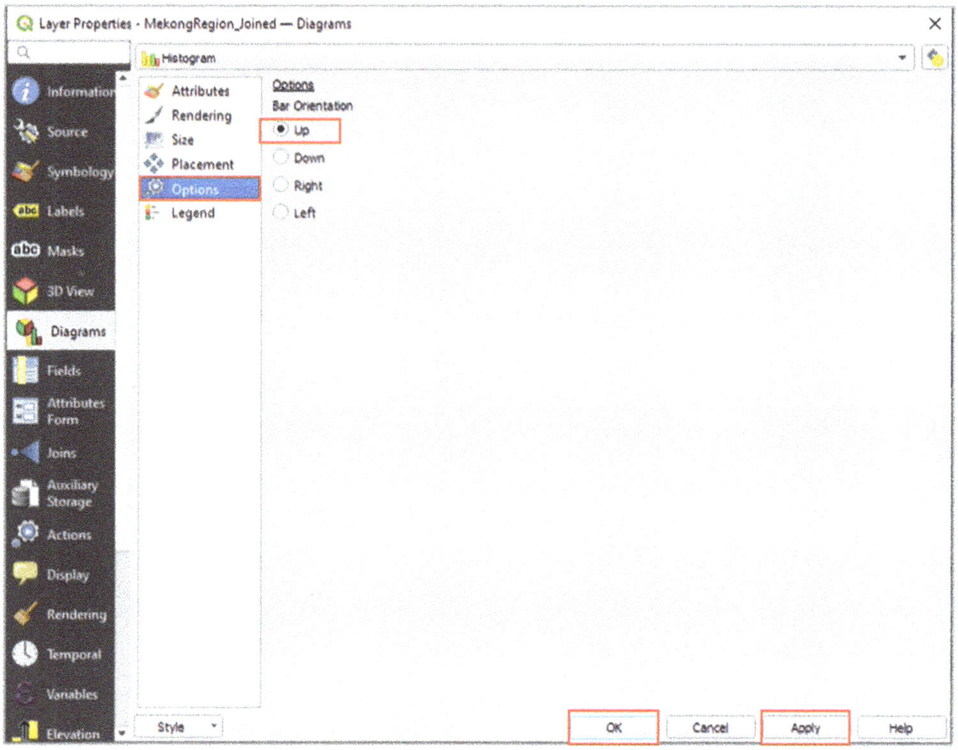

Source: Screenshot generated by the Asian Development Bank using QGIS 3.34.14 2025.

(xiv) Finally, the histogram of rice production for the year, visualized for each province in the Mekong Delta region, should resemble Figure 4.45. Note that the annual planted areas are also shown, following the procedure for categorizing the data on the map.

Figure 4.45: Histogram Overlaid on the Map of the Mekong Delta Region, Shown in QGIS

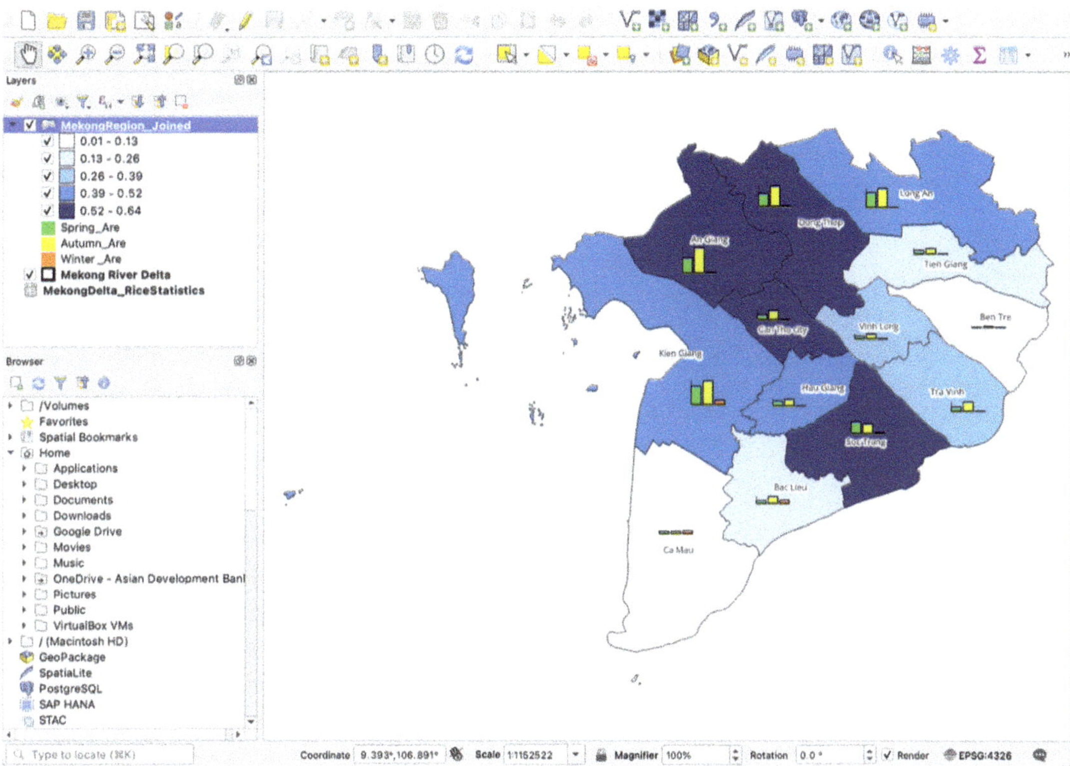

Sources: Government of Viet Nam, National Statistics Office. 2024. *Statistical Yearbook of Viet Nam 2023.* https://www.nso.gov.vn/wp-content/uploads/2024/06/NIEN-GIAM-THONG-KE-2023_Ban-quyen.pdf; Screenshot generated by the Asian Development Bank using QGIS 3.34.14. 2025.

Module 5: Basic Geographic Information System Mapping

This module provides a practical guide for creating basic maps using QGIS, focusing on the essential elements for effective map development. The ability to create and export maps is a fundamental skill in agricultural statistics, as it enhances the presentation and interpretation of spatial data.

The topics covered in this module are critical for several reasons. Accurate and visually compelling maps facilitate the understanding of spatial patterns, trends, and relationships within agricultural data. Mastery of these tools and techniques ensures that data is presented in a clear and impactful manner, thereby supporting better insights and informed decisions.

This module builds on the foundational knowledge provided in modules 1 to 4, which cover basic QGIS functionalities and data handling. It guides users through the process of creating thematic maps, incorporating essential map elements, and exporting maps for presentation or publication. The skills acquired in this module are essential for effectively leveraging QGIS in agricultural statistics.

The topics covered in this module are as follows:

5.1 Creating a Thematic Map
- 5.1.1 Categorized Symbology
- 5.1.2 Graduated Symbology

5.2 Creating a Basic Map with Scale, a Legend, and Text

5.3 Exporting a Basic Map as a PDF File

By the end of this module, users will be able to produce a basic thematic map using QGIS, incorporating key elements such as scale, a legend, and labels. The output map shown in Figure 5.1 shows the areas planted with rice in the Mekong Delta region in 2023. This output serves as a demonstration of the mapping process in QGIS using the tools and techniques covered in this module.

Figure 5.1: Extent of Rice Plantation in the Mekong Delta Region, by Province, 2023

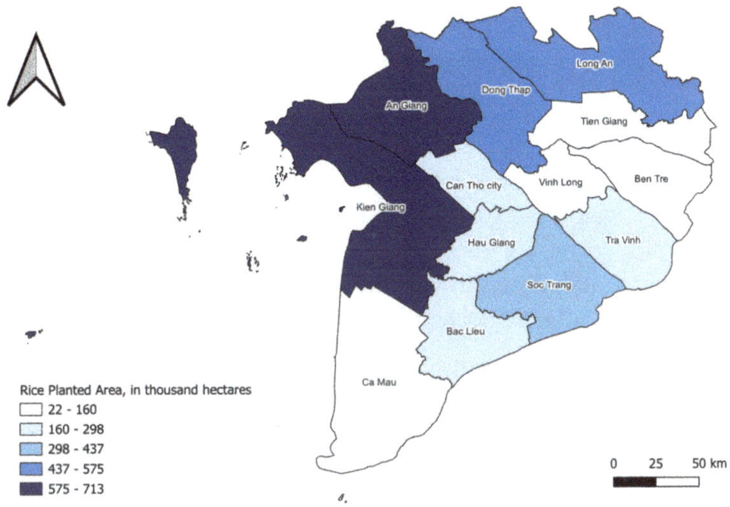

Source: Screenshot generated by the Asian Development Bank using QGIS 3.34.14. 2025.

5.1 Creating a Thematic Map

A map serves as a powerful tool for communicating geographic information. In the context of mapping, **symbology** refers to the use of symbols, colors, and patterns to represent data on a map, making it easily interpretable for the audience. Symbology is crucial for conveying the underlying data in a clear and meaningful way.

There are several types of symbology for vector layers available in QGIS, each tailored to different types of analysis and presentation needs. These include **Single Symbol** (uniform styling for all features), **Categorized** (distinct styles based on attribute values), **Graduated** (symbol variation based on numeric ranges), **Rule-Based** (custom styling using logical expressions), and **Point Displacement** (overlapping points separated for clarity). Additional advanced options include **Merged Features** (multiple features combined into a single symbol), Inverted Polygons (areas styled outside the polygon geometry), **2.5D Renderer** (pseudo-3D effect added to polygons for enhanced visualization), and Embedded Symbols (symbols nested within other symbols for more complex styling). For this module, the focus will be on two commonly used types of symbology: **Categorized** and **Graduated**.

To apply symbology in QGIS, a shapefile containing numeric or categorical data is required. The symbology can be configured using the **Symbology tab** under **Layer Properties**. This tab provides various options for customizing the appearance of the data, ensuring that the resulting map effectively communicates the intended information.

In the following exercise, we will work with geospatial data previously used in earlier modules. Specifically, this module utilizes the land use and land cover (LULC) map of An Giang Province, Viet Nam, derived from the Japan Aerospace Exploration Agency (JAXA) Earth Observation Research Center (EORC) LULC Map of Viet Nam for 2020 (version 23.09), along with administrative boundary data for An Giang Province.

To ensure continuity, it is recommended that you complete the raster data preprocessing steps outlined in **Module 3** prior to beginning this module. These steps include downloading, importing, clipping, and saving the raster data, followed by the data's conversion into vector format and initial classification. If you have completed Module 3, the resulting vectorized file, **AnGiang_Landuse.gpkg**, should be available to serve as the primary input for the activities in this module.

You will also need the **MekongRegion_Joined** file, which contains the combined boundary layer and the paddy rice statistics from an exercise in Module 4. If you are starting with this module, but have not completed the earlier activity, the **MekongRegion_Joined** file can be downloaded directly from the shared folder.[a]

[a] For access to shapefiles and other open-access datasets referenced in the handbook, go to the GitHub repository: https://github.com/ADB-Data-Division/QGIS-Handbook-2025

(xv) Add the vector land use layer **AnGiang_Landuse.gpkg** in QGIS by dragging and dropping it from the **Browser panel** into the **map area**.
(xvi) Right-click on the layer and select **Properties**. Alternatively, you can access a layer's properties by double-clicking on the layer itself.
(xvii) Click on the **Symbology** tab (Figure 5.2).

Figure 5.2: Changing the Symbology of a Layer

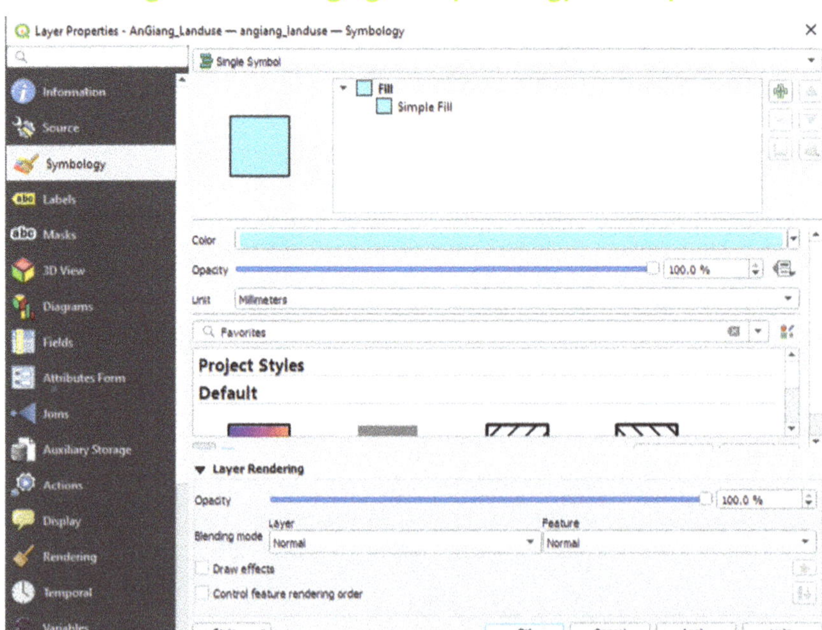

Source: Screenshot generated by the Asian Development Bank using QGIS 3.34.14. 2025.

5.1.1 Categorized Symbology

Categorized symbology represents categorical data, assigning each category a unique color or symbol. This type of symbology is particularly useful for visualizing discrete data, such as land use types or soil classifications. For example, categorized symbology can be used to differentiate among various types of land cover, such as forests, wetlands, and agricultural areas. It can also be used to represent different crop types or farming practices, providing a clear visual distinction among categories based on shared characteristics.

(i) Navigate to the **Symbology tab** under **Layer Properties**. Click on the drop-down menu. It will show the selection of **Single Symbol**. You should change that to **Categorized** (Figure 5.3).

Figure 5.3: Selecting Categorized Symbology

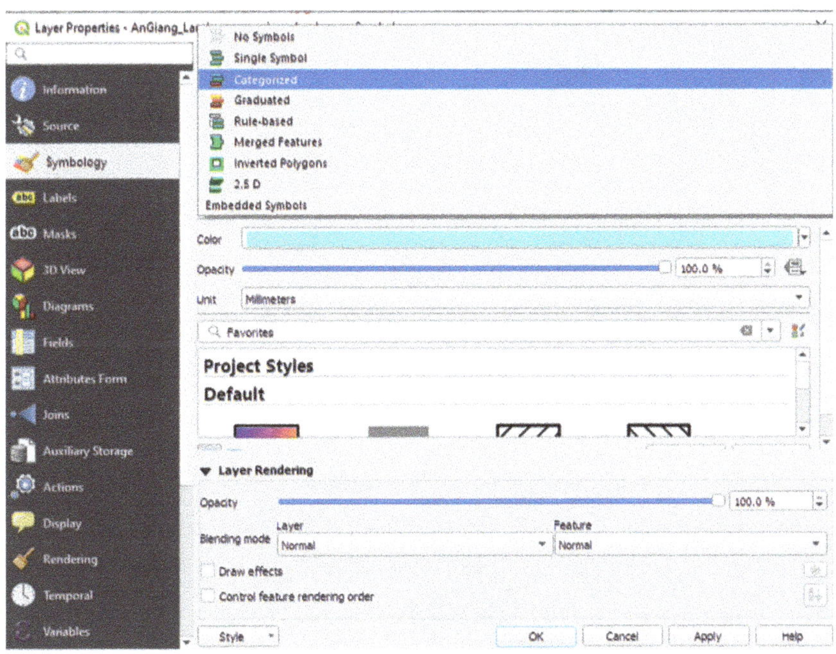

Source: Screenshot generated by the Asian Development Bank using QGIS 3.34.14. 2025.

(ii) Set the **Value** field to **LANDUSE** to categorize the data based on land use. Select **Random colors** for the **Color ramp** to visually differentiate each category. Click **Classify** to populate the Value field with the different land use categories (Figure 5.4).

Figure 5.4: Classifying Data into Different Categories

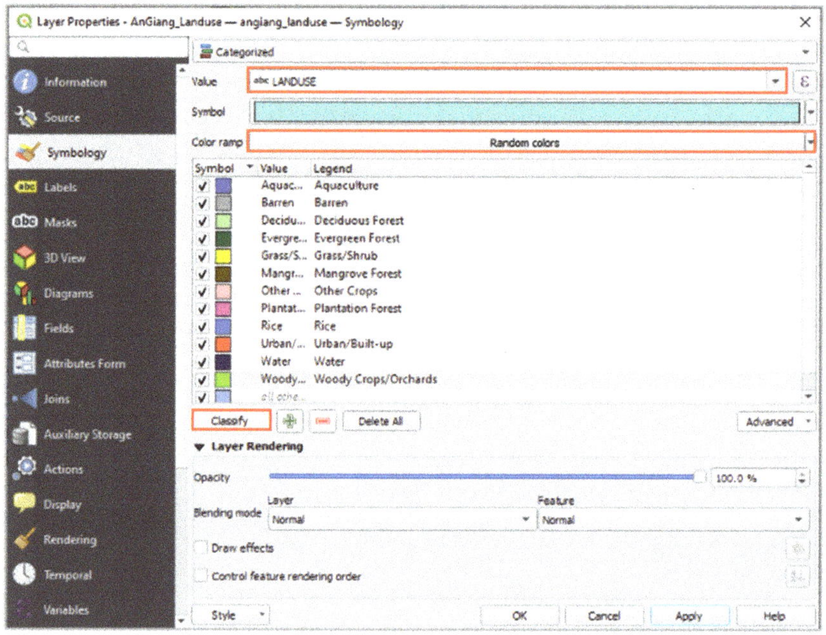

Source: Screenshot generated by the Asian Development Bank using QGIS 3.34.14. 2025.

(iii) Click **OK**. The result will resemble Figure 5.5. Since random colors are being used, the map's appearance may differ among users.

Figure 5.5: Output of Land Use Classification for An Giang Province, Viet Nam, Using Random Colors

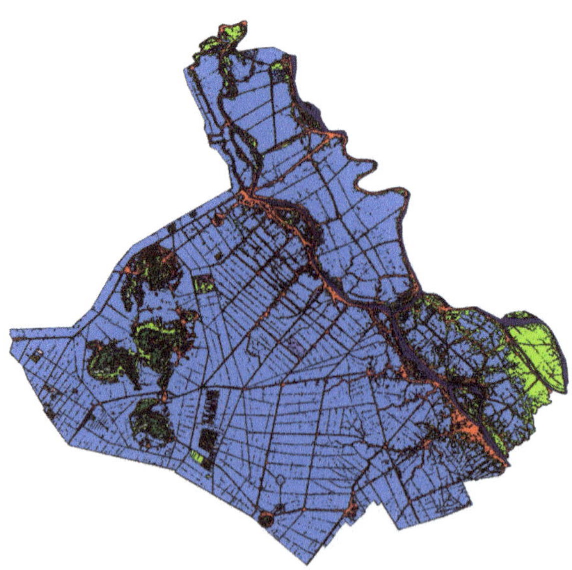

Source: Screenshot generated by the Asian Development Bank using QGIS 3.34.14. 2025.

(iv) Click on the arrow next to **AnGiang_Landuse** in the **Layers panel** to view the categories (Figure 5.6).

Figure 5.6: Land Use Categories in the Layers Panel

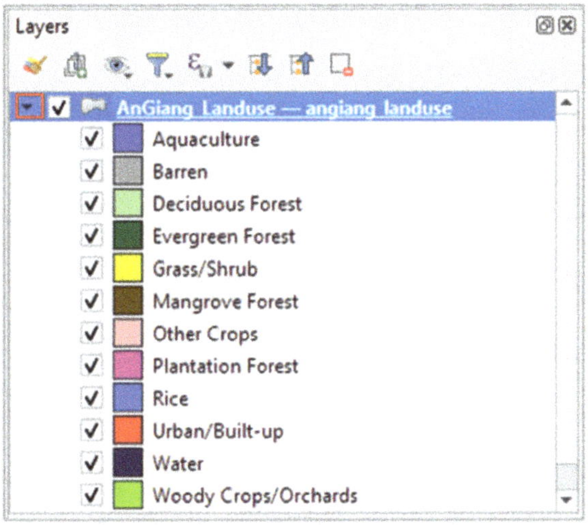

Source: Screenshot generated by the Asian Development Bank using QGIS 3.34.14. 2025.

5.1.2 Graduated Symbology

Graduated symbology represents numeric data by dividing it into classes, with each class assigned a different color or symbol based on its value. This approach is effective for depicting continuous, numerical, and quantitative data, such as crop yields or soil fertility. In the following exercise, each province in the Mekong Delta region will be classified according to the annual total paddy rice area planted. Using the attribute table and the joined columns of the administrative boundary layer from the previous module (Table 4.1), we can plot these values with unique symbols or color gradients for each category, making it easier to interpret and analyze the data.

(i) Add the vector land use layer (**MekongRegion_Joined**) in QGIS by dragging and dropping it from the **Browser panel** into the **map area**.
(ii) Right-click on the layer and select **Properties**. Navigate to the **Symbology** tab, click on the drop-down menu, which will show **Single Symbol** as the selected option. Switch from Single Symbol to **Graduated** (Figure 5.7).

Figure 5.7: Selecting Graduated Symbology

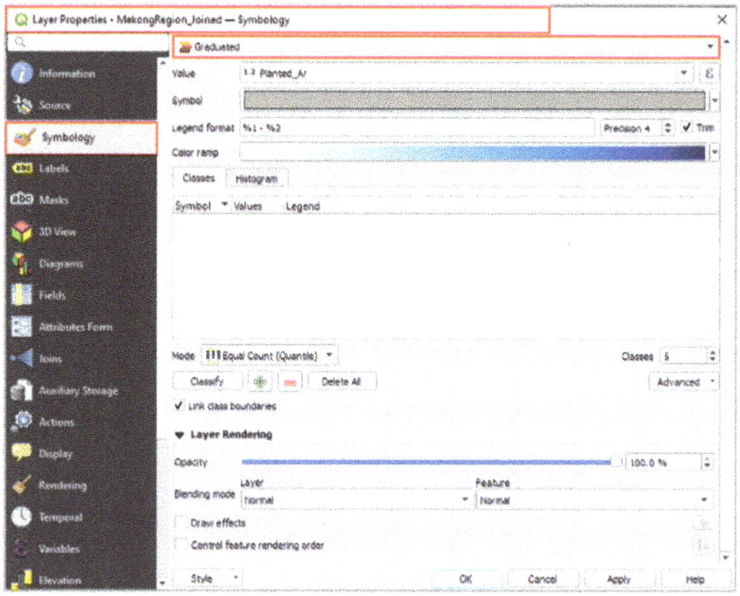

Source: Screenshot generated by the Asian Development Bank using QGIS 3.34.14. 2025.

(iii) In the **Value tab**, select **Planted_Ar** and the **Blues** color scheme in the Color ramp. You may also use other color schemes of your preference. In the Mode tab, select **Equal Interval** to classify the values into equal intervals (Figure 5.8).

(iv) Click **Classify** to populate the value field with different categories. The map will now be divided into five classes at equal intervals. The algorithm automatically creates the five classes according to the optimal intervals calculated from the data. You can also set the number of classes using the **Classes** button if you prefer more or fewer classifications.

Note that, in QGIS as commonly implemented, range-based symbology methods—such as graduated and categorized styles—handle class breaks in a way where the lower bound of each interval is treated as inclusive, meaning values equal to the lower limit are included in that class. The upper bound tends to be exclusive, meaning values equal to the upper limit are assigned to the next class, with the exception of the final class, which includes both bounds to ensure full coverage of the data range.

Figure 5.8: Selecting "Equal Interval" to Classify the Values

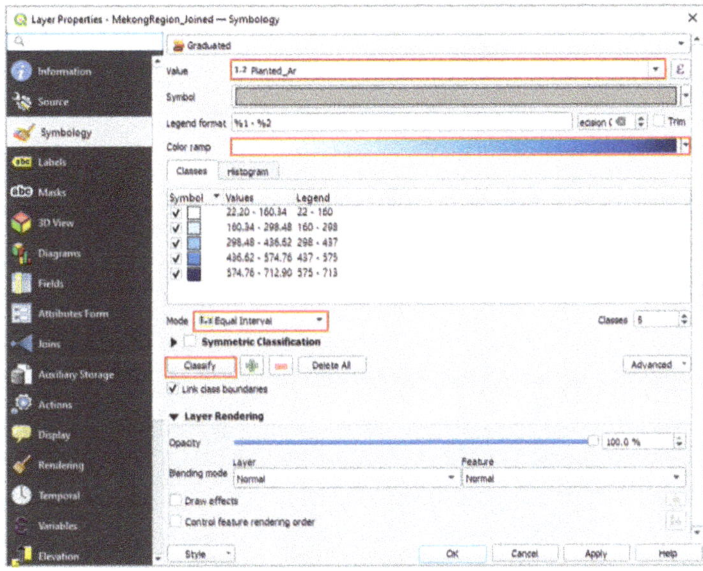

Source: Screenshot generated by the Asian Development Bank using QGIS 3.16. 2022.

(v) Click **OK**. The result will resemble Figure 5.9.

Figure 5.9: Graduated Map of the Extent of Rice Plantation in the Mekong Delta Region, 2023

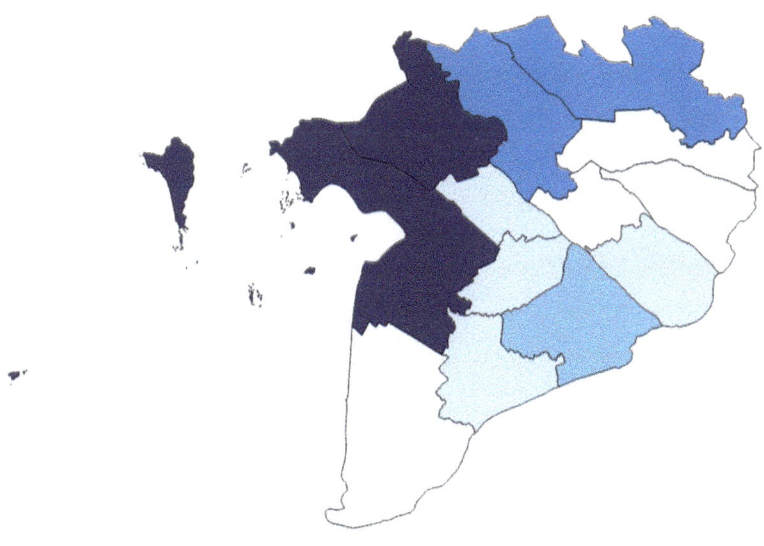

Source: Screenshot generated by the Asian Development Bank using QGIS 3.34.14. 2025.

(vi) To add the names of the provinces to the map, click the tab to set up the labels on the map, then select **Single Labels** and set the **Value tab** to **ADM1_EN**. To customize the text, click the tab to change the font, style, size, and color. Finally, click the tab to create a buffer around the text and add highlights. Click **Apply** (Figure 5.10). The result will resemble Figure 5.11.

Figure 5.10: Adding Labels to the Map

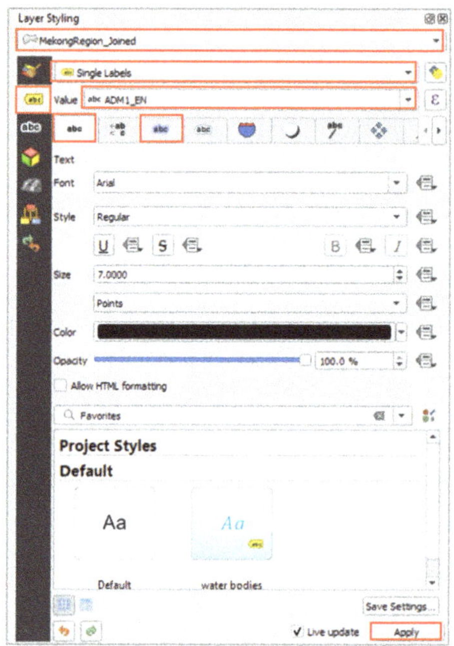

Source: Screenshot generated by the Asian Development Bank using QGIS 3.34.14. 2025.

Figure 5.11: Graduated Map of Rice Plantation Rates with the Provincial Names Added

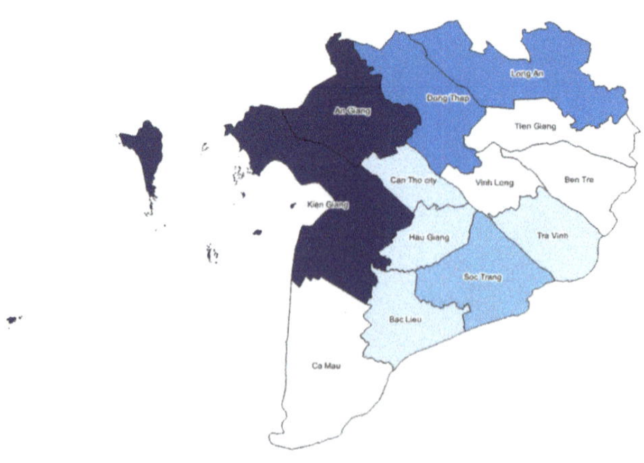

Source: Screenshot generated by the Asian Development Bank using QGIS 3.34.14. 2025.

5.2 Creating a Basic Map with Scale, a Legend, and Text

In agricultural statistics, presenting data effectively is crucial for analysis and decision-making. Adding essential elements such as a scale, a legend, and text provides valuable context for stakeholders. After generating the initial map, you can enhance its clarity and usability by adding map elements. The print layout feature allows you to incorporate commonly used components such as map titles, legends, scale bars, and more, making the data more accessible and easier to interpret.

(i) To create a new print layout, click on Project in the main menu, then select New Print Layout. You will be prompted to choose a title for the new layout. Alternatively, you could click on the icon for New Print Layout, , in the Project toolbar to open the tool (Figure 5.12).

Figure 5.12: Creating a Print Layout

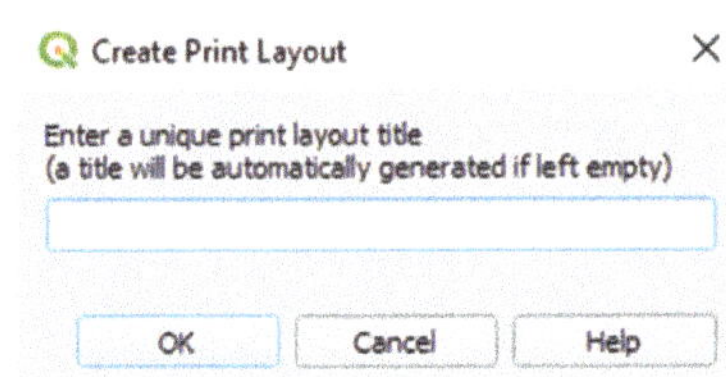

Source: Screenshot generated by the Asian Development Bank using QGIS 3.34.14. 2025.

(ii) To insert your map into the print layout, select the **Add Map toolbar button**, , in the left-hand panel. Then hold down the left mouse button and draw a rectangle on the canvas to define the map area (Figure 5.13).

Figure 5.13: Inserting a Map into a Print Layout

Source: Screenshot generated by the Asian Development Bank using QGIS 3.34.14. 2025.

The map will look like Figure 5.14 in the print layout interface. You can modify various aspects of your map and layout elements in the **Item Properties window**. This includes **map orientation, map scale, size**, and **position**.

Figure 5.14: Map Display in the Print Layout Interface

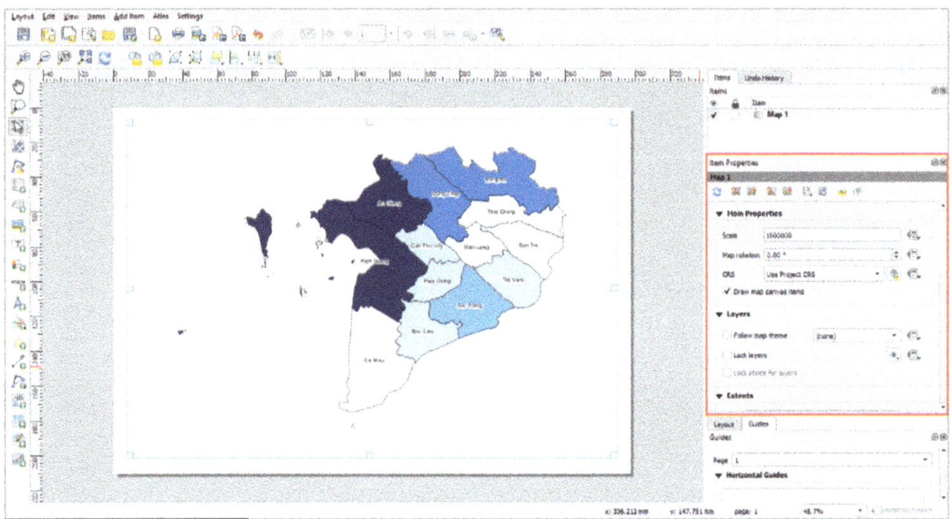

Source: Screenshot generated by the Asian Development Bank using QGIS 3.34.14. 2025.

(iii) To add a legend, select the **Add Legend toolbar button**, , and draw a rectangle on the canvas while holding down the left mouse button. Check the **Only show items inside the linked map** option to display only the visible items in the legend. You can also customize the **font size, style**, and color of the legend by clicking the **Font** and **Text Formatting** options (Figure 5.15).

Figure 5.15: Adding a Legend to a Map in Print Layout

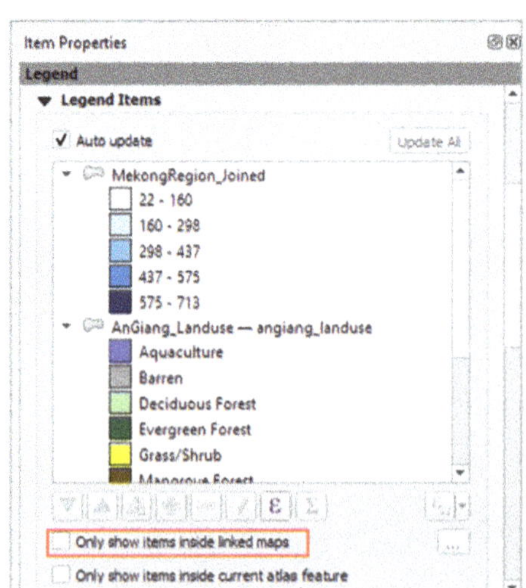

Source: Screenshot generated by the Asian Development Bank using QGIS 3.34.14. 2025.

Unchecking the **Auto update** option will enable you to manually refresh or update the legend as needed. For instance, clicking the **Edit selected item properties** button will allow you to manually edit the labels in the legend (Figure 5.16).

Figure 5.16: Updating Legend Items Manually

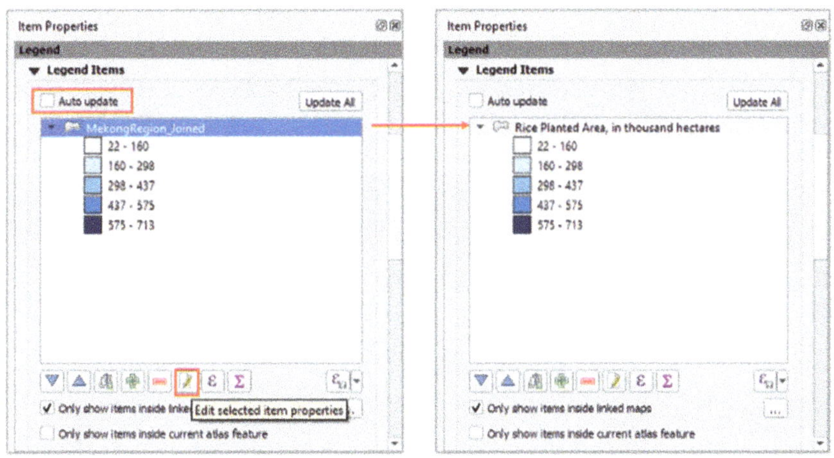

Source: Screenshot generated by the Asian Development Bank using QGIS 3.34.14. 2025.

(iv) To add a north arrow in the print layout, select the **Add North Arrow tool**, , place it on the layout, and adjust its properties as needed.

Similarly, to add a scale bar, select the **Add Scale Bar tool**, , place it on the layout, and customize its settings accordingly, as illustrated in Figure 5.17.

Figure 5.17: Map Layout with Key Elements

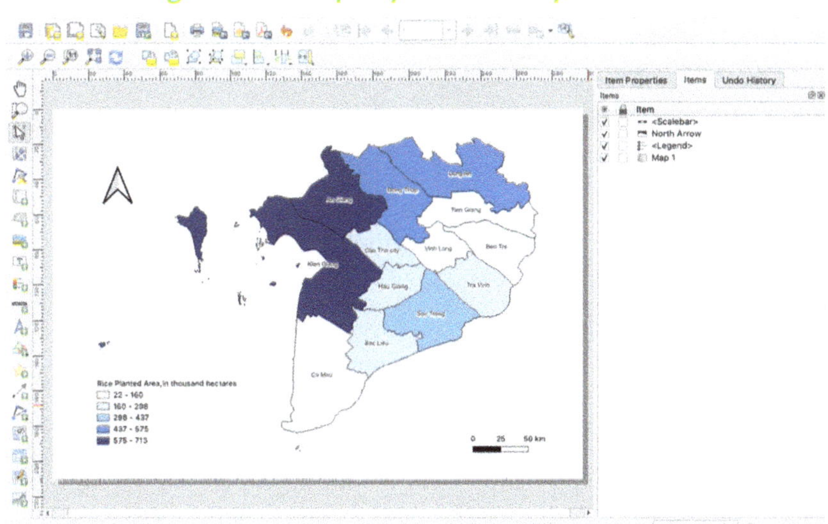

Source: Screenshot generated by the Asian Development Bank using QGIS 3.34.14. 2025.

(v) To add a title, select the **Add Label tool button**, . Click and drag the button to the place where you want to insert the title (Figure 5.18). Note that the tool can be used to add text anywhere within the map layout area.

Figure 5.18: Adding a Title or Text to a Map

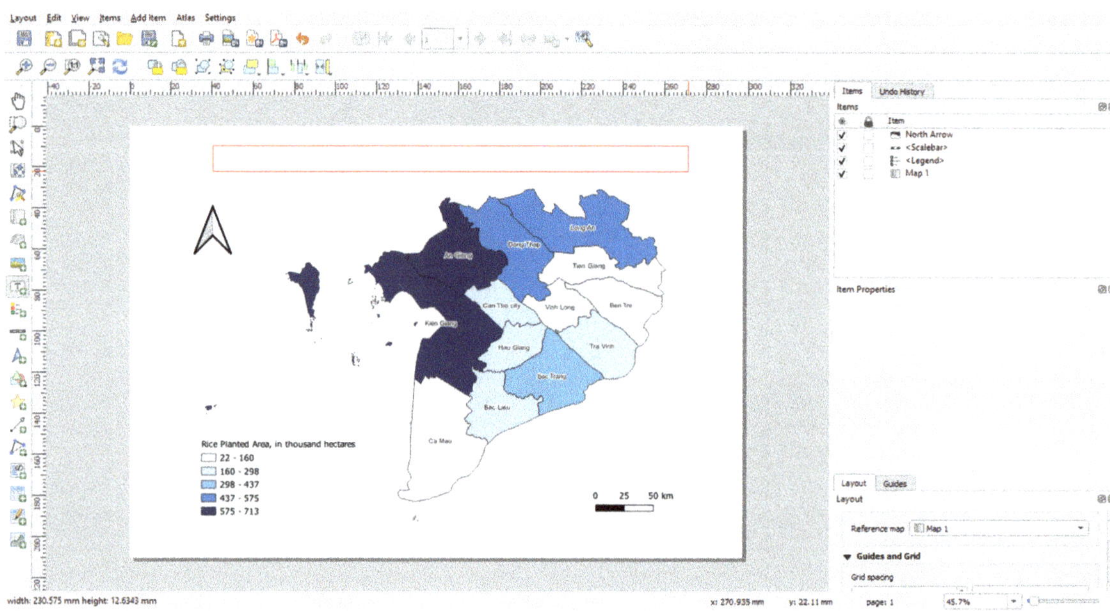

Source: Screenshot generated by the Asian Development Bank using QGIS 3.34.14. 2025.

Note that all the map elements in the layout are listed in the item window box, to the right. Select an element you wish to edit, and it will appear in the item property window, below.

5.3 Exporting a Basic Map as a PDF File

QGIS offers various options for saving your files in different formats. Once your map is finalized, you can export it as an image (e.g., JPEG, PNG, GIF, GeoTIFF), as a PDF, or SVG. The **export buttons**, , are located near the top-left corner of the **layout window**.

This exercise will focus on exporting the map as an image (JPEG).

(i) Click , which is the **Export as Image** button.
(ii) Navigate to the desired folder and enter a file name for your image. Click Save.

A dialog box with image-export options will appear. This dialog box will allow you to configure various settings, including resolution and other export parameters, to ensure that the image meets specific requirements. After setting the parameters, click **Save** (Figure 5.19).

Figure 5.19: Image Export Options

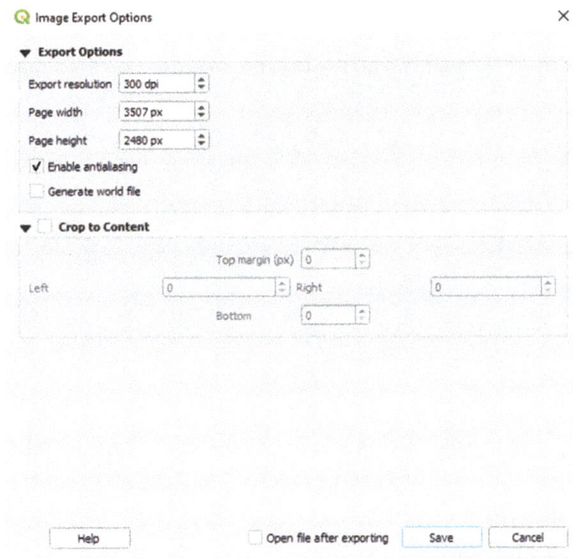

Source: Screenshot generated by the Asian Development Bank using QGIS 3.34.14. 2025.

(iii) Access the folder where you saved the image to check your map. The final output will resemble Figure 5.20.

Figure 5.20: Final Output Map

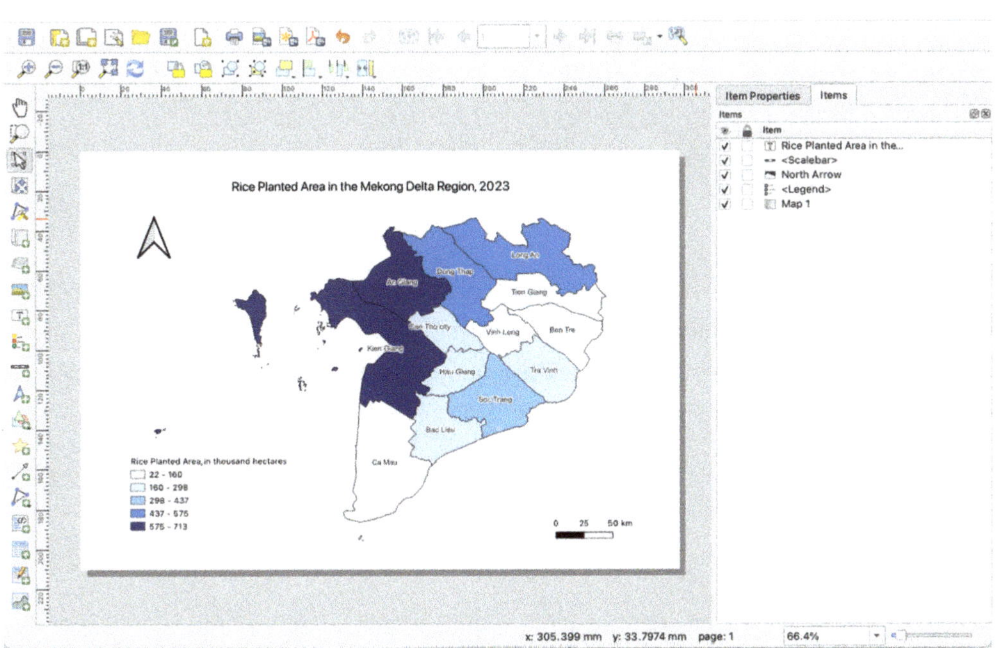

Source: Screenshot generated by the Asian Development Bank using QGIS 3.34.14. 2025.

Module 6: Basic Geographic Information System Vector Analysis Tools—Working with Global Positioning System Data

This module focuses on analyzing Global Positioning System (GPS) data in QGIS to strengthen agricultural data collection and analysis. Accurate GPS data and effective geographic information system (GIS) workflows enable precise field mapping, crop monitoring, and resource management, thereby improving the quality of agricultural data.

This module builds on the foundational knowledge provided in the previous chapters, which covered basic QGIS functionalities and data handling. It will guide users through the process of importing GPS data, utilizing vector-analysis tools, and applying digitizing techniques.

In this module, you will go through the following topics:

6.1 Importing Global Positioning System Point Coordinates into QGIS
- 6.1.1. What is a Global Positioning System?
- 6.1.2. Marking a Waypoint Using a Global Positioning System Device
- 6.1.3. Importing Global Positioning System Point Coordinate Data into QGIS

6.2 Digitizing Tools to Draw a Point, Line, or Polygon
- 6.2.1. Drawing a Point
- 6.2.2. Drawing a Line
- 6.2.3. Drawing a Polygon

6.4 Creating Polygons from Global Positioning System Point Corners

6.5 Calculating the Distance from Global Positioning System Points to Other Points of Interest

By the end of this module, you will understand how to work with GPS data in QGIS and will be able to apply these skills to agricultural statistics. This knowledge will help optimize field operations, improve crop management, and enhance overall agricultural productivity.

6.1. Importing Global Positioning System Point Coordinates into QGIS

6.1.1. What is a Global Positioning System?

GPS is a space-based radio-navigation system consisting of a network of satellites. It broadcasts navigation signals and includes ground stations and satellite control stations that help triangulate the observed location. Currently, 31 GPS satellites orbit the Earth at an altitude of approximately 11,000 miles, providing accurate position, velocity, and time information to users worldwide. Operated and maintained by the Government of the United States, GPS is just one of several Global Navigation Satellite Systems (GNSSs) in use around the world, alongside Japan's Quasi-Zenith Satellite System (QZSS); the BeiDou Navigation Satellite System (BDS), of the People's Republic of China; the European Union's Galileo; and Russia's GLONASS (FAA, n.d.).

A GPS receiver determines its location by analyzing signals from multiple satellites. By calculating the time it takes for each signal to reach the receiver, it can triangulate its position in terms of latitude, longitude, and sometimes elevation (Esri, n.d.-b). Most GPS receivers also support the storage and management of spatial data in three primary forms:

- **waypoints:** specific geographic coordinates saved by the user;
- **routes:** ordered sequences of waypoints that form a path; and
- **tracks:** recorded paths based on the receiver's movement over time (Garmin, n.d.-c).

Waypoints, routes, and tracks are the three basic feature types of GPS data. In QGIS, waypoints are displayed in point layers, while routes and tracks are shown in line string layers, enabling spatial analysis and mapping workflows (Esri, n.d.-b).

Figure 6.1 shows the Garmin eTrex 20x, a handheld GPS device commonly used in field-based data collection and navigation. Garmin is one of several manufacturers of GPS equipment, and the eTrex 20x model was selected for this module due to its accessibility, ease of use, and support for both the GPS and GLONASS satellite systems (Garmin 2020). Its features make it suitable for illustrating basic GPS data concepts such as waypoints, routes, and tracks (Garmin, n.d.-b).

Figure 6.1 also illustrates the key physical components of the Garmin eTrex 20x. While the layout and terminology may vary slightly across different brands and models, the following features are commonly found in most handheld GPS devices:

- **Display screen.** The screen provides a visual interface for maps, coordinates, menus, and navigational data, allowing users to view waypoints, routes, and tracks in real time.
- **Thumb Stick (joystick controller).** This multidirectional input device allows users to scroll through menus, pan across maps, and make selections. It replaces the touch-screen interface found in some newer models, and is preferred in rugged environments where gloves are often worn.
- **Zoom keys (+ / -).** These buttons allow users to zoom in and out of the map display, helping to adjust the level of detail shown. This function is essential for both route planning and real-time navigation.

- **Menu key.** This key provides access to contextual options depending on the current screen. A double press typically returns the user to the main menu, offering quick navigation across device functions.
- **Back key.** This key is used to return to the previous screen or to cancel an action. It is a standard feature across most GPS interfaces for supporting intuitive navigation.
- **Power or backlight key.** This key serves a dual function: turning the device on and off and adjusting screen brightness. It is especially useful in low-light or high-glare conditions.
- **Battery compartment.** Located at the back of the device, the compartment houses two AA batteries. Its design allows for easy replacement in the field, a feature common to many handheld GPS units.

Figure 6.1: Garmin eTrex 20x Global Positioning System Device

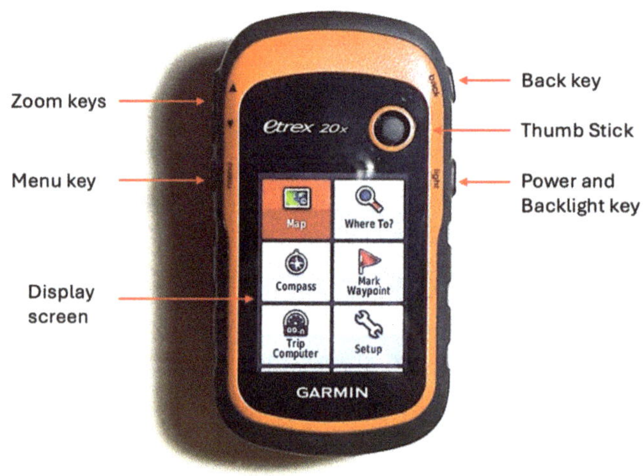

Source: Image taken by the Asian Development Bank. 2025.

6.1.2. Marking a Waypoint Using a Global Positioning System Device

A waypoint is a marked location stored on a GPS device. It is created by pressing the device's **Mark button**, which pins a waypoint at the current GPS location or at the map cursor if a different location is being viewed. To mark a specific location, follow these steps:

(iv) Turn on the device. Press and hold the power button until the screen lights up.
(v) Wait for a satellite connection. Ensure that the device is connected to satellites, so it may obtain accurate location data. To check if the device is connected to satellites, navigate to the **Satellite page** using the Thumb Stick. Figure 6.2 displays the status of a satellite connection.

Figure 6.2: Satellite Connectivity Status

Source: Image taken by the Asian Development Bank. 2025.

(vi) Access the main menu. Use the Thumb Stick to navigate to **Mark Waypoint** and press the Thumb Stick to select it (Figure 6.3).

Figure 6.3: Marking a Waypoint

Source: Image taken by the Asian Development Bank. 2025.

(vii) Edit the waypoint details. You can revise the name, symbol, and other details of the waypoint by selecting the respective fields and making changes. Use the Thumb Stick to move the cursor and edit the waypoint details (Figure 6.4).

Figure 6.4: Editing Waypoint Details

Source: Image taken by the Asian Development Bank. 2025.

(viii) After making the desired changes, select **Done** to save the waypoint.

6.1.3. Importing Global Positioning System Point Coordinate Data into QGIS

There are several file formats used for storing GPS data. In QGIS, the format used is the GPS Exchange Format (GPX). GPX is the standard interchange format that can store multiple waypoints, tracks, and routes within a single file.

(i) To import a GPX file into QGIS, follow the same steps as those for importing a vector file: through the Browser panel, through the main menu, or by simply dragging and dropping the file into the map area. In the dialog box that appears (Figure 6.5), select the feature type (waypoints, tracks, routes) you want to import. Click **Add Layers** to load the selected data into QGIS.

Figure 6.5: Importing a Global Positioning System Exchange Format File into QGIS

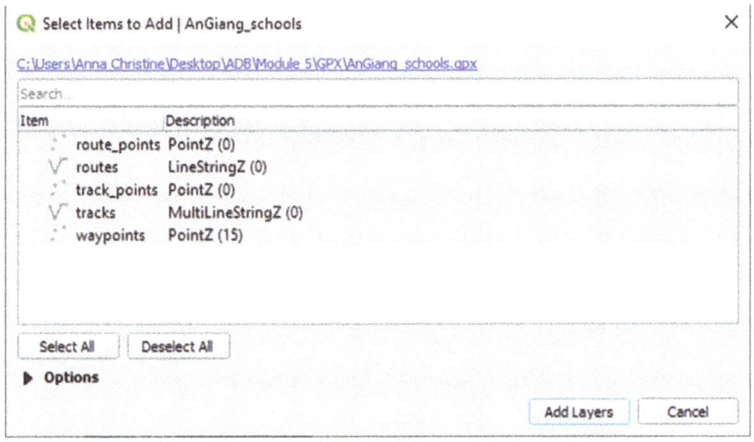

Source: Screenshot generated by the Asian Development Bank using QGIS 3.34.14. 2025.

In most agricultural surveys using GPS, multiple GPX files need to be imported in a batch. In QGIS, a plugin called "**Batch GPX Importer**" can be used for this. If the plugin is not yet installed, refer to Module 1, section 1.3.5 for the steps of installing the plugin.

In the following exercise, the GPX files extracted from OpenStreetMap, which contains locations of schools in Viet Nam's An Giang Province (An Giang Schools), will be used. The administrative boundaries of An Giang Province at the provincial level will also be used in the exercise. Download this file from the shared folder[a] and load it into QGIS as part of the workflow. Change the layer's symbology to display the boundary using a simple line style.

[a] For access to shapefiles and other open-access datasets referenced in the handbook, go to the GitHub repository: https://github.com/ADB-Data-Division/QGIS-Handbook-2025

(ii) To use the **Batch GPS Importer** plugin, go to the **Menu**, select **Vector**, then **Batch GPS Importer** (Figure 6.6). Alternatively, you can enable the plugins toolbar and click the icon.

Figure 6.6: Using the Batch Global Positioning System Importer Plugin

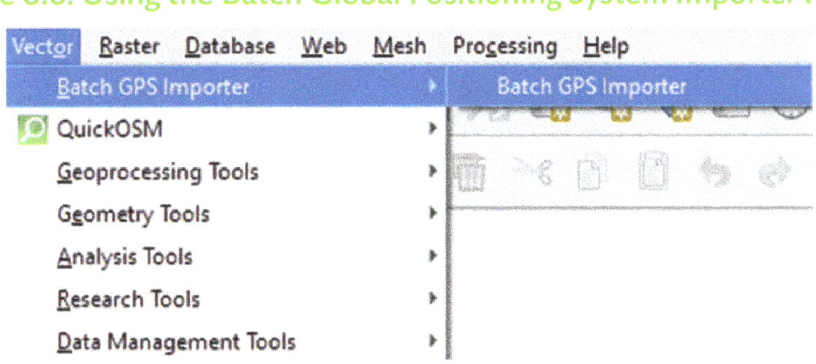

Source: Screenshot generated by the Asian Development Bank using QGIS 3.34.14. 2025.

(iii) In the plugin interface, navigate to the directory containing your GPS files and upload the entire folder. Configure any additional settings as needed (e.g., selecting specific feature types such as waypoints, tracks, or routes). Click **Import** to load all the selected GPS files into QGIS (Figure 6.7).

Figure 6.7: Navigating to the Global Positioning System Exchange Format File

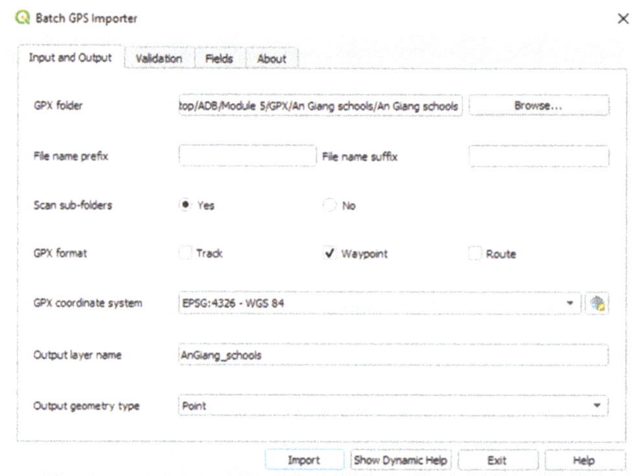

Source: Screenshot generated by the Asian Development Bank using QGIS 3.34.14. 2025.

(iv) A prompt will appear indicating that the import process has started. It will also notify you of the number of features successfully imported. Click **Exit** to close the log. You will now see the locations of waypoints on the map. In this example, they are the schools in the province (Figure 6.8).

Figure 6.8: Waypoints in the Map Display

Source: Screenshot generated by the Asian Development Bank using QGIS 3.34.14. 2025.

6.2. Digitizing Tools to Draw a Point, Line, or Polygon

The digitization of spatial features such as points, lines, and polygons is a fundamental process in most agricultural statistical analyses. It involves converting geographic data into digital formats that can be manipulated and analyzed using a GIS. By digitizing elements such as farm boundaries, irrigation systems, crop locations, and infrastructure, agricultural statisticians can create detailed and accurate maps that are critical for effective planning, monitoring, and decision-making.

This section highlights QGIS tools used to digitize vectors, demonstrating their application in agricultural statistics. The integration of digitized features with data from sources such as remote sensing and field surveys will create a comprehensive spatial database, enhancing resource management and environmental response.

6.2.1. Drawing a Point

(i) In the main menu, click **Layer**, then select **Create Layer** and choose **New Shapefile Layer** (Figure 6.9).

Figure 6.9: Creating a New Shapefile Layer

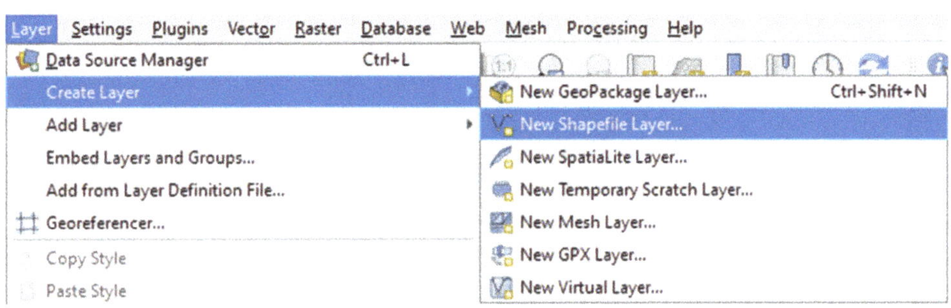

Source: Screenshot generated by the Asian Development Bank using QGIS 3.34.14. 2025.

(ii) In the dialog box that appears, enter the file name and select the folder where you want to save your file. Then set the **Geometry type** to **Point**.

In the **New Field** section, enter the name of the field you want to add and select the data type (e.g., text, integer, decimal). Click the **Add to Fields List** button to add the field to your shapefile (Figure 6.10). Repeat this step for any additional fields you want to include.

Adding attributes is important, as demonstrated in agricultural surveys, where it enables more effective operations management through the tagging of specific data points.

Figure 6.10: Filling in Attributes for a New Point Layer

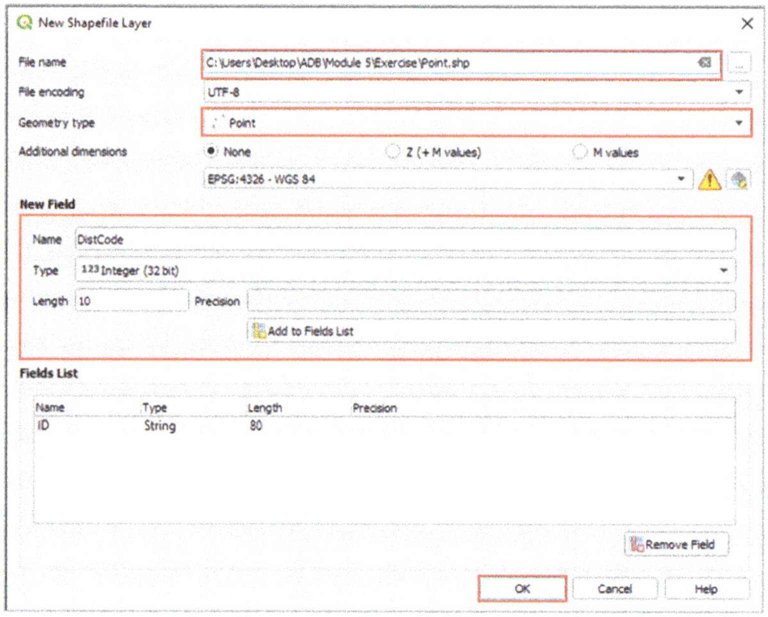

Source: Screenshot generated by the Asian Development Bank using QGIS 3.34.14. 2025.

Click **OK** to create the new layer.

(iii) The new point layer will appear in the **Layers panel**. Right-click on it and select **Toggle Editing** to enable editing mode, or simply click to enable the toggle editing mode (Figure 6.11).

Figure 6.11: Enabling Toggle Editing for Layer Modification

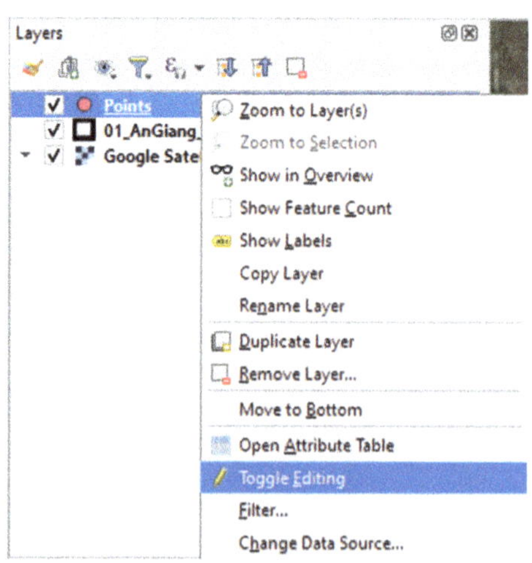

Source: Screenshot generated by the Asian Development Bank using QGIS 3.34.14. 2025.

(iv) In the **Digitizing Toolbar,** click on the **Add Point Feature** tool or click on the icon. Click on the map where you want to place the point. A dialog box will appear prompting you to enter the attributes for the point. In this example, fill out the point identification (ID) and district code (Figure 6.12).

Figure 6.12: Adding Attributes to the New Point Layer

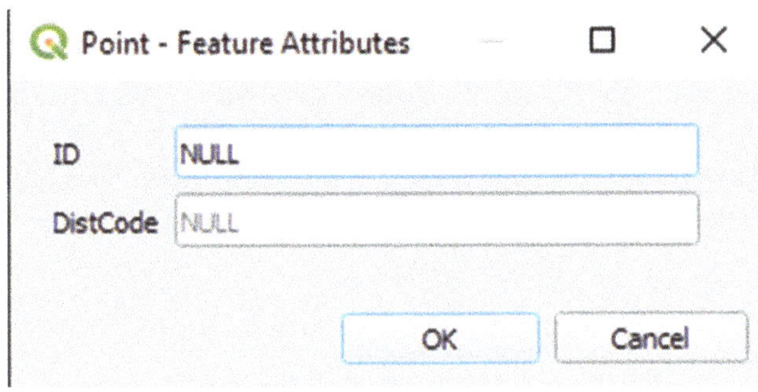

Source: Screenshot generated by the Asian Development Bank using QGIS 3.34.14. 2025.

If you need to place the point at the exact coordinates, you can use the **Advanced Digitizing Panel**. To enable it, click **View** in the main menu, select **Panels**, then **Advanced Digitizing Panel**. Enter the coordinates of the points manually (Figure 6.13).

Figure 6.13: Placing Exact Point Coordinates Using the Advanced Digitizing Toolbar

Source: Screenshot generated by the Asian Development Bank using QGIS 3.34.14. 2025.

(v) After adding the points, click the **Save Layer Edits** button or the icon in the toolbar to save your changes. To stop editing, right-click on the layer again and select **Toggle Editing**.

(vi) Ensure that the points are correctly placed by zooming in and checking the locations on the map. To change the layer properties, refer to Module 2, section 2.4, "Visualizing and Styling Data Layers." For adding additional fields or attributes, see Module 4, section 4.3, "Adding a New Field, Renaming, and Organizing Fields."

6.2.2. Drawing a Line

(i) To draw a line, follow the same procedure used for creating a shapefile for a point layer, as described in the previous section, i.e., click on **Layer | Create Layer | New Shapefile Layer**. Then set the **Geometry type** to **LineString**. In the **New Field** section, enter the name of the field you want to add and select the data type (e.g., text, integer, decimal). Click **OK** to create the new layer (Figure 6.14).

Figure 6.14: Filling in Attributes for a New Line Layer

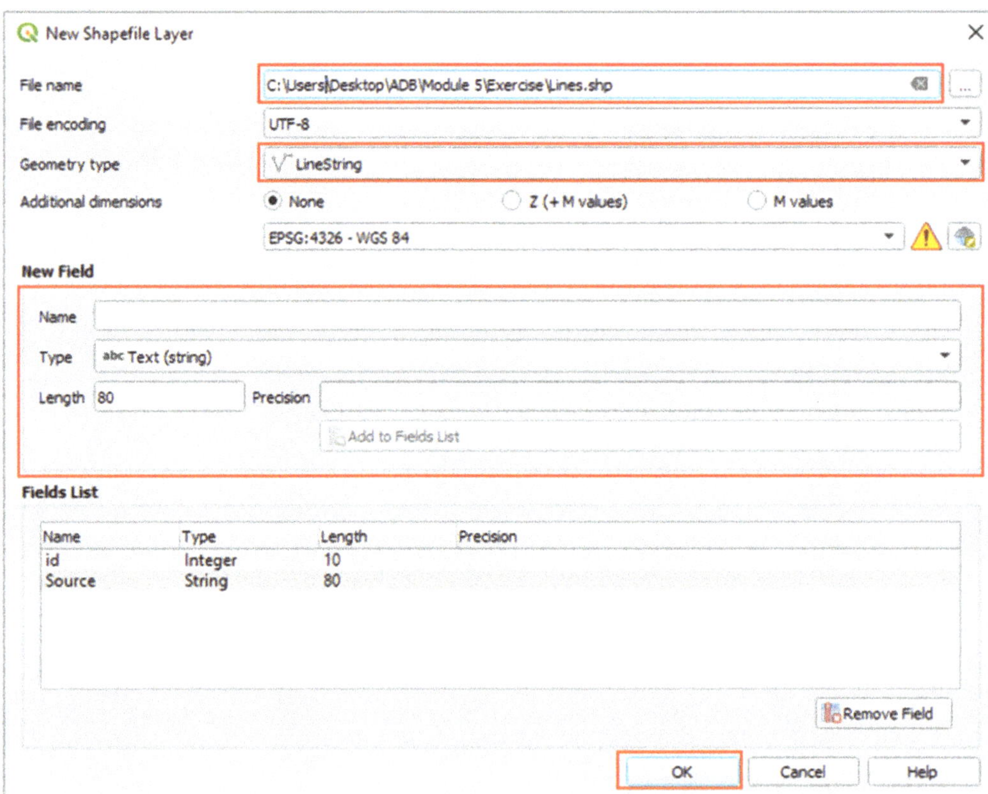

Source: Screenshot generated by the Asian Development Bank using QGIS 3.34.14. 2025.

(ii) The new line layer will appear in the **Layers panel**. Right-click on it and select **Toggle Editing** to enable editing mode, or simply click to enable the toggle editing mode.

(iii) In the **Digitizing Toolbar**, click on the **Add Line Feature** tool or click on the icon. Then click on the map to start drawing your line. Continue clicking to add vertices, then right-click to finish. A dialog box will appear where you must enter the attributes for the line. In this example, fill out the line ID and the irrigation source (Figure 6.15).

Figure 6.15: Adding Attributes to the New Line Layer

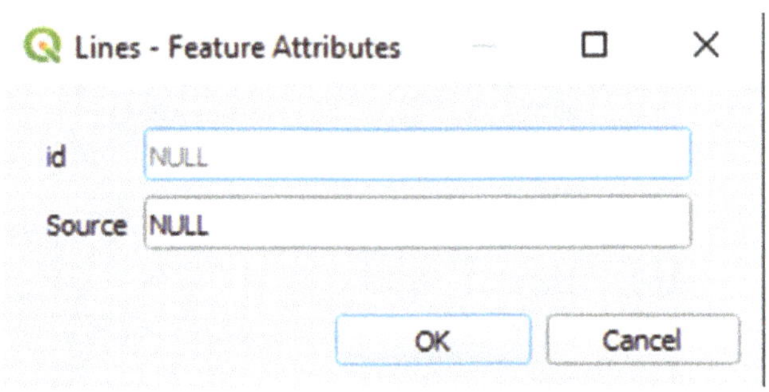

Source: Screenshot generated by the Asian Development Bank using QGIS 3.34.14. 2025.

(iv) In the **Advanced Digitizing Panel**, you can manually enter the exact coordinates for each vertex of your line. To enable it, click **View** in the main menu, select **Panels**, then **Advanced Digitizing Panel**. Click to place the first vertex, then enter the exact coordinates for subsequent vertices manually. Right-click to finish drawing the line.

(v) Click the **Save Layer Edits** button or the icon in the toolbar to save your changes after drawing all the lines in the layer. To stop editing, right-click on the layer again and select **Toggle Editing**.

(vi) To change layer properties, refer to Module 2, section 2.4, “Visualizing and Styling Data Layers.” For adding additional fields or attributes, see Module 4, section 4.3, “Adding a New Field, Renaming, and Organizing Fields.”

The output example will resemble the one shown in Figure 6.16.

Figure 6.16: Line Feature in the Map Display

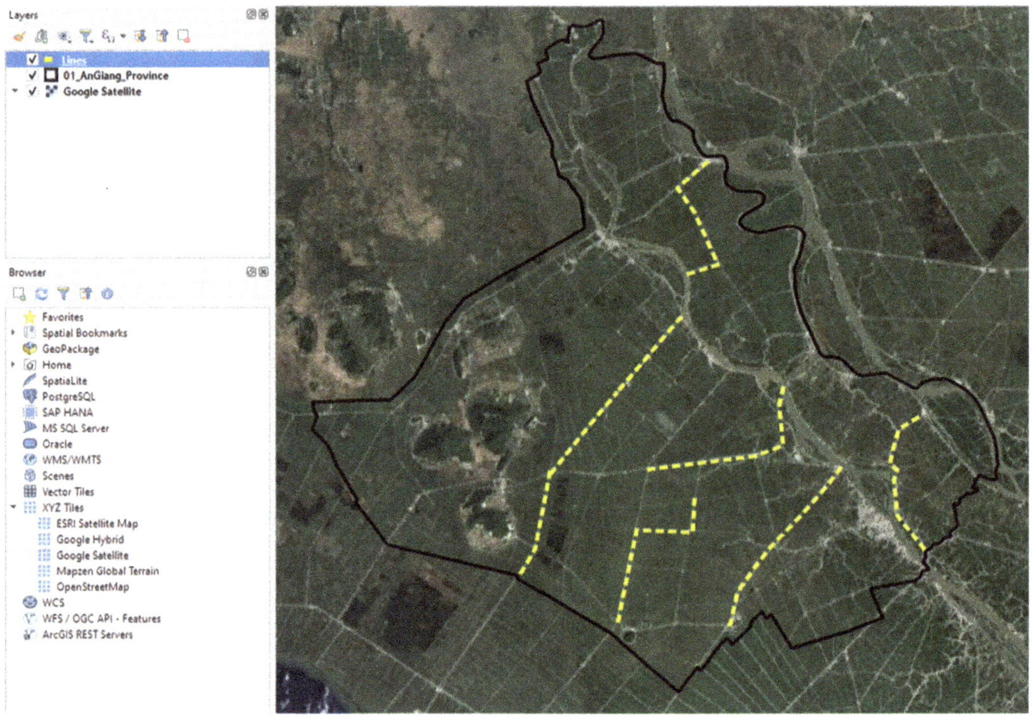

Source: Screenshot generated by the Asian Development Bank using QGIS 3.34.14. 2025.

6.2.3. Drawing a Polygon

(i) To draw a polygon, follow the same procedure used for creating a shapefile for a point or a line layer, as described in the previous two sections, i.e., click on **Layer | Create Layer | New Shapefile Layer**. Then set the **Geometry type** to **Polygon** (Figure 6.17).

In the **New Field** section, enter the name of the field you want to add and select the data type (e.g., text, integer, decimal). Filling in this section will allow you to tag polygons with specific attributes, such as land use, which is essential for agricultural surveys and detailed analysis. For instance, identifying the land use of a farm parcel helps in understanding crop distribution and soil types. Adding fields ensures that each polygon contains sufficient information for later analysis, making data collection and interpretation more efficient.

Figure 6.17: Filling in Attributes for a New Polygon Layer

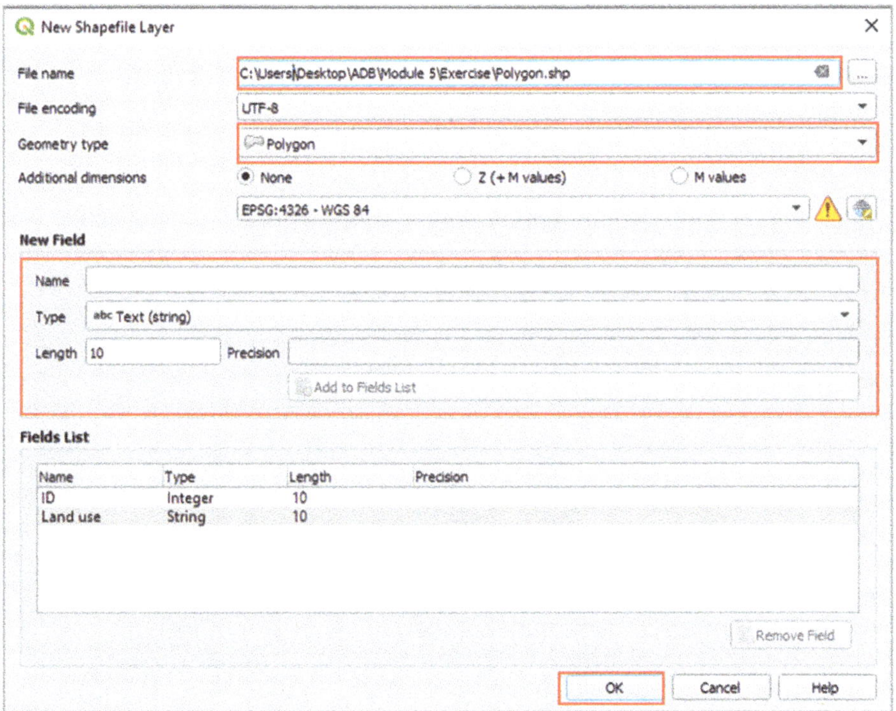

Source: Screenshot generated by the Asian Development Bank using QGIS 3.34.14. 2025.

Click **OK** to create the new layer.

(ii) The new polygon layer will appear in the **Layers panel**. Right-click on it and select **Toggle Editing** to enable editing mode, or simply click to enable the toggle editing mode.

(iii) In the **Digitizing Toolbar**, click on the **Add Polygon Feature** tool or click on the icon. Click on the map to start drawing your polygon. Continue clicking to add vertices, then right-click to finish. A dialog box will appear, and you must enter the polygon's attributes in the box. In this example, enter the parcel ID and the land use of the polygon (Figure 6.18).

Figure 6.18: Adding Attributes to the New Polygon Layer

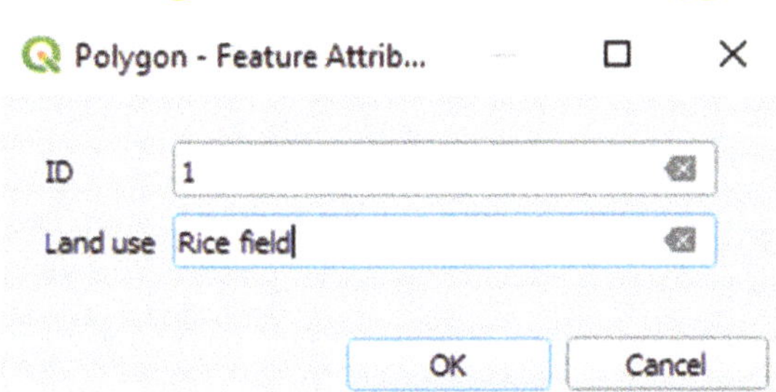

Source: Screenshot generated by the Asian Development Bank using QGIS 3.34.14. 2025.

(iv) In the **Advanced Digitizing Panel**, you can manually enter the exact coordinates for each vertex of your polygon. To enable it, click **View** in the main menu, select **Panels**, then **Advanced Digitizing Panel**. Click to place the first vertex, then enter the exact coordinates for subsequent vertices manually. Right-click to finish drawing the polygon.

(v) Click the **Save Layer Edits** button or the icon in the toolbar to save your changes after all the lines in the layer have been drawn. To stop editing, right-click on the layer again and select **Toggle Editing**.

(vi) To change layer properties, refer to Module 2, section 2.4, "Visualizing and Styling Data Layers." For adding additional fields or attributes, see Module 4, section 4.3, "Adding a New Field, Renaming, and Organizing Fields."

The output example will resemble the one shown in Figure 6.19.

Figure 6.19: Polygon Features on a Map Display

Source: Screenshot generated by the Asian Development Bank using QGIS 3.34.14. 2025.

6.3 Creating Polygons from Global Positioning System Point Corners

For the following exercise, the **GPS_Waypoints file** will be used. Download it from the shared folder.[a]

[a] For access to shapefiles and other open-access datasets referenced in the handbook, go to the GitHub repository: https://github.com/ADB-Data-Division/QGIS-Handbook-2025

(i) Add the vector layer **GPS_Waypoints.gpkg** to QGIS.

Note that when you import a GPS Exchange Format (GPX) file into QGIS, it will usually be in read-only format, which means that you cannot modify its attributes or geometry directly. To enable editing, you should export the GPX layer into an editable format such as Esri shapefile (.shp), GeoJSON (.geojson), or GeoPackage (.gpkg). These formats support full editing capabilities and are better suited for data manipulation and analysis.

To export a GPX file into an editable format in QGIS, right-click on the layer (e.g., waypoints or tracks) in the **Layers panel** and select **Export | Save Features As**. In the dialog box that appears, choose your preferred format, such as **Esri shapefile (.shp), GeoJSON (.geojson)**, or **GeoPackage (.gpkg)**. Specify the file name and location. Click **OK** to complete the export. The new layer will be added to your map and will be editable.

(ii) Right-click on the **GPS_Waypoints** layer and select **Open Attribute Table**. Click the **Toggle Editing** button to enable the editing mode. Create a new attribute column for the order of points by clicking the **New Field** button or the icon, naming the column (e.g., "Order"), and setting the type to **Integer** (Figure 6.20).

To check which feature corresponds to which point on the map, click on the attribute table; then note the highlighted point on the map and identify it (Module 4, section 4.2).

(iii) Manually enter the order for each point in the new column (Figure 6.21). Put the points in chronological order and ensure that they do not cross.

Figure 6.20: Adding a New Field to the Attribute Table

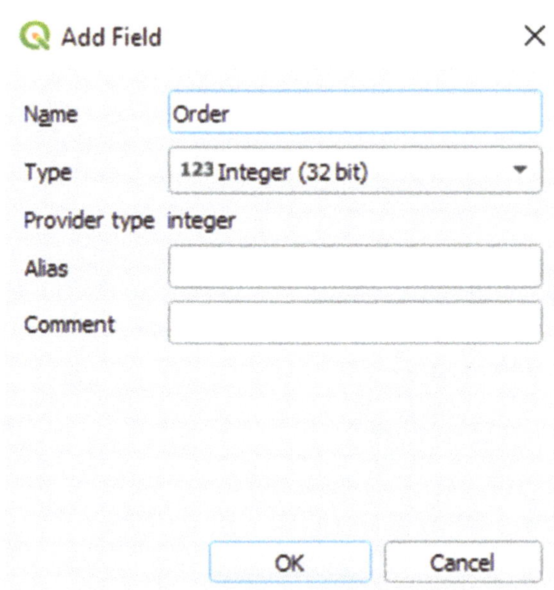

Source: Screenshot generated by the Asian Development Bank using QGIS 3.34.14. 2025.

Figure 6.21: Manual Ordering of Features in the Attribute Table

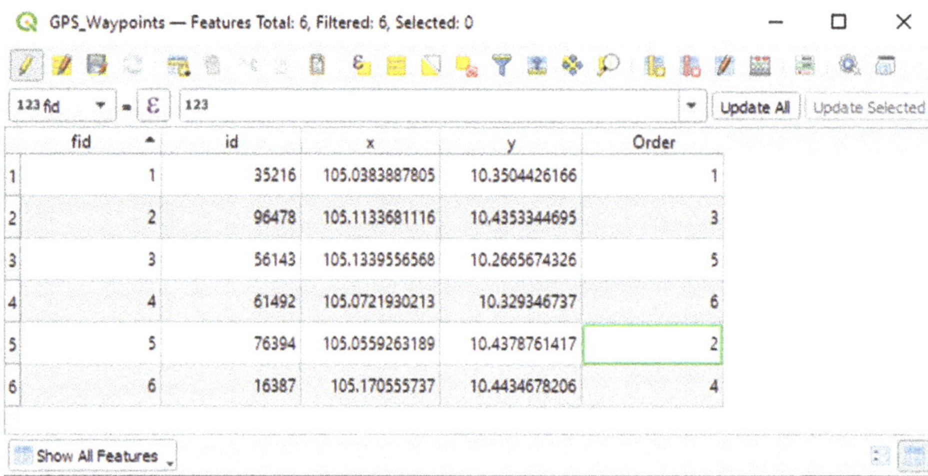

	fid	id	x	y	Order
1	1	35216	105.0383887805	10.3504426166	1
2	2	96478	105.1133681116	10.4353344695	3
3	3	56143	105.1339556568	10.2665674326	5
4	4	61492	105.0721930213	10.329346737	6
5	5	76394	105.0559263189	10.4378761417	2
6	6	16387	105.170555737	10.4434678206	4

Source: Screenshot generated by the Asian Development Bank using QGIS 3.34.14. 2025.

Save the edits by clicking the **Save Edits** button, and then click **Toggle Editing** again to stop editing.

(iv) Open the **Processing Toolbox** by selecting **Processing**, then **Toolbox**. In the search bar, type **Points to Path** and select the **Points to path** tool (Figure 6.22). In the **Points to Path** window, select your layer as the input layer and choose the order field you created earlier. Specify a group field, if necessary; then choose an output file location, and click **Run** to create a line layer based on the points (Figure 6.23).

Figure 6.22: Opening the Processing Toolbox Menu

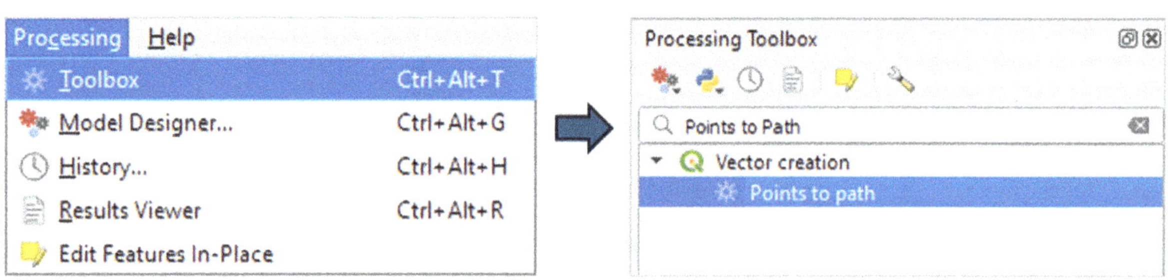

Source: Screenshot generated by the Asian Development Bank using QGIS 3.34.14. 2025.

Figure 6.23: Using the Points to Path Tool in the Processing Toolbox

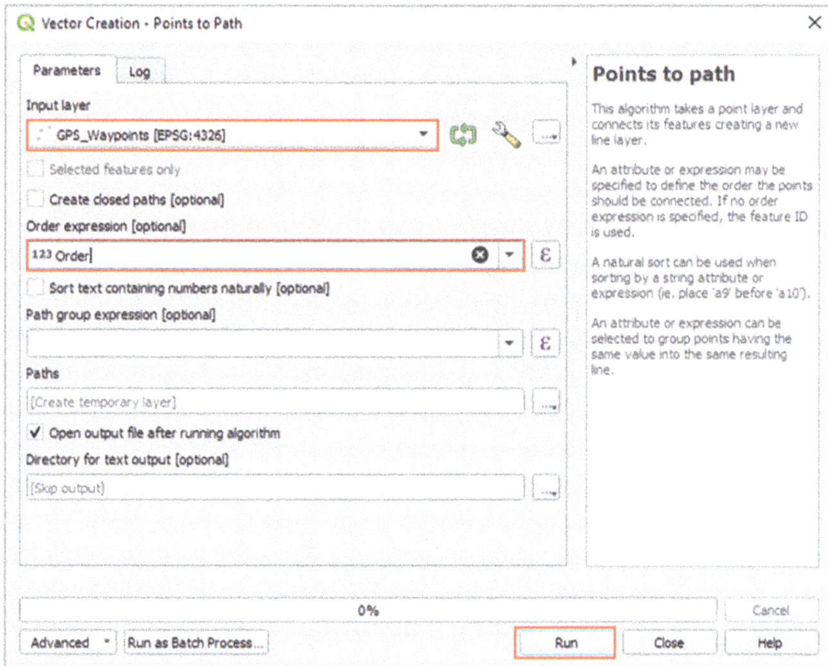

Source: Screenshot generated by the Asian Development Bank using QGIS 3.34.14. 2025.

Depending on how you order the GPS waypoints, the resulting output should resemble what is shown in Figure 6.24.

Figure 6.24: Path Visualization on Map Display

Source: Screenshot generated by the Asian Development Bank using QGIS 3.34.14. 2025.

(v) In the Processing Toolbox, search for **Lines to Polygons** and select the **Lines to Polygons** tool. Alternatively, you can also go to **Vector**, select **Geometry Tools**, then select **Lines to Polygons**. In the **Lines to Polygons** window, select the line layer created in the previous step as the input layer. Choose an output file location and click **Run** to create a polygon layer from the line (Figure 6.25).

Figure 6.25: Using the "Lines to Polygons" Tool in the Processing Toolbox

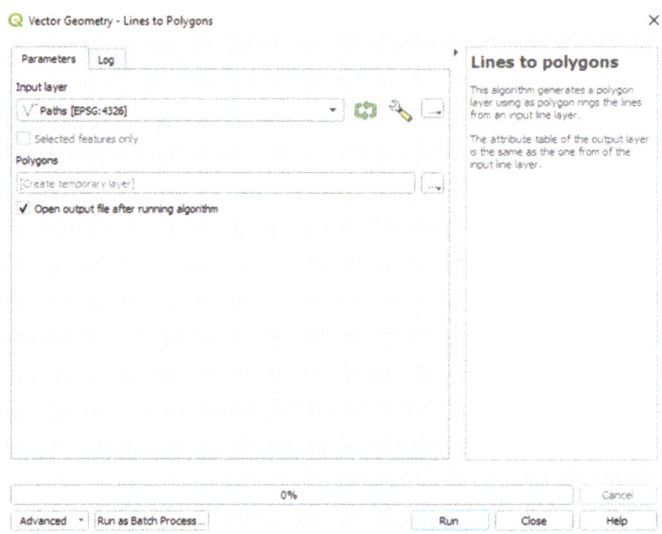

Source: Screenshot generated by the Asian Development Bank using QGIS 3.34.14. 2025.

The output polygon should resemble the polygon shown in Figure 6.26.

Figure 6.26: Polygon Output Generated from Lines

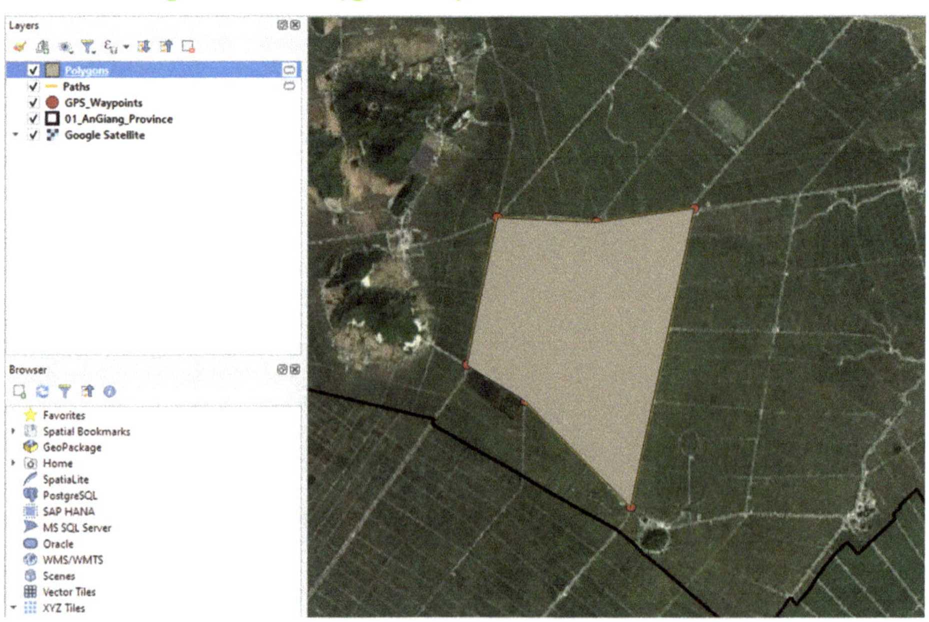

Source: Screenshot generated by the Asian Development Bank using QGIS 3.34.14. 2025.

(vi) You can also use the **Points2One** plugin to create polygons. If the plugin is not yet installed, refer to Module 1, section 1.3.5, for the steps to install it. Search for the NextGIS Points2One plugin among the list of "Not Installed" plugins.

To use the plugin, go to **Vector**, select **Points2One**, or simply click the P2 icon from the Vector toolbar (Figure 6.27).

Figure 6.27: Opening the Points2One Plugin

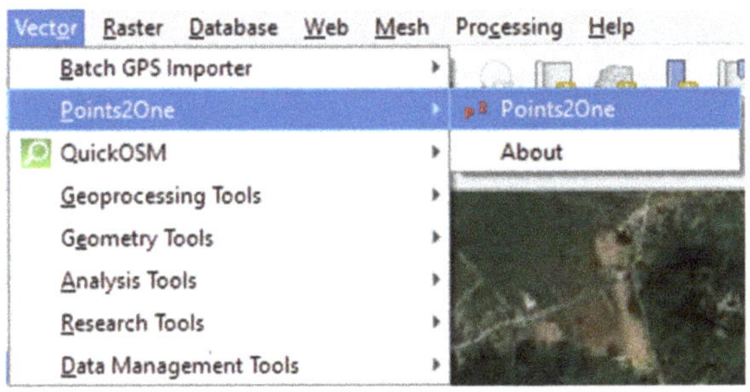

Source: Screenshot generated by the Asian Development Bank using QGIS 3.34.14. 2025.

In the **Points2One dialog box**, select your point layer and designate it as the **Input Layer**, choose **Create polygons**, sort the vertices using the "Order" option to ensure that the points connect correctly, specify the **Output shapefile** name and location, check **Add result to canvas**, and click **OK to create the polygon** (Figure 6.28).

Figure 6.28: Creating a Polygon Using the Points2One Plugin

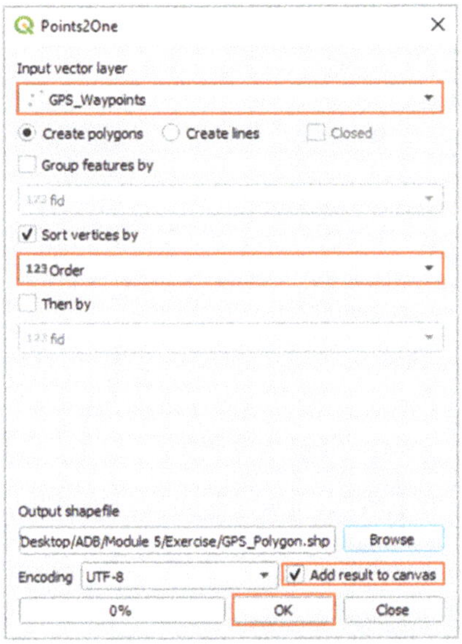

Source: Screenshot generated by the Asian Development Bank using QGIS 3.34.14. 2025.

Check the output polygon after running the algorithm.

6.4 Calculating the Distance from Global Positioning System Points to Other Points of Interest

(i) Load the GPX files into the An Giang Province schools folder using the **Batch GPS Importer** plugin, and load the **GPS_Waypoints layer** into QGIS. To learn more about using the **Batch GPS Importer** plugin, refer to section 6.1.3 of this module.
(ii) **Reproject** both of these layers onto **UTM zone 48N**. To learn more about reprojecting a layer, refer to Module 4, section 4.4, "Calculating a Variable (with Area Calculation as an Example)".
(iii) Go to **Vector | Analysis Tools**, then select **Distance Matrix** (Figure 6.29).

Figure 6.29: Opening the Distance Matrix Tool

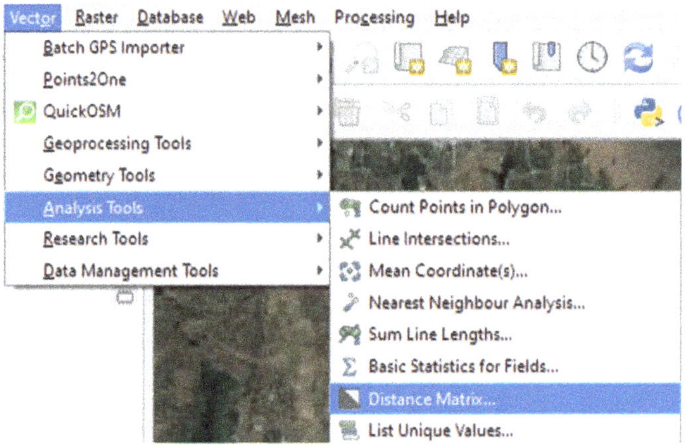

Source: Screenshot generated by the Asian Development Bank using QGIS 3.34.14. 2025.

(iv) In the **Distance Matrix dialog box**, select the An Giang_schools layer as the **Input Point Layer**. Select the GPS_Waypoints layer as the **Target Point Layer**. Identify the unique IDs of both layers in their attributes.

The **Distance Matrix tool** offers different output matrix types, each providing unique insights, depending on the context of the analysis. These matrix types include:

a. **Linear distance matrix.** This matrix type calculates the straight-line (Euclidean) distance between points, useful for simple distance measurements where obstacles or terrain are not considered. For instance, calculating the straight-line distance between various fields or plots of land can help in planning irrigation systems or determining the proximity of different crops to each other for optimal growth conditions.
b. **Network distance matrix.** This matrix type calculates distances based on a network, such as roads or pathways. It considers the actual travel distance along the network, making it more accurate for real-world scenarios in which travel paths are constrained by infrastructure. For example, measuring the travel distance between farms and distribution centers using road networks is useful for logistics planning, ensuring efficient transportation routes for delivering produce to markets or processing facilities.

c. **Great circle distance matrix.** This matrix type calculates the shortest distance between two points on the surface of a sphere, which is useful for global-scale distance measurements, such as those between cities on different continents. For example, determining the shortest distance between agricultural research stations located in different countries can be useful for international collaboration and for coordinating research on global agricultural challenges.

For this exercise, select **Linear distance matrix**. Choose **Only the nearest (k) target points** and set **k** to 1 if you want the nearest point, or set **k** to 0 if you want to know the distance from each input point layer to the target point layer. Specify the **Output matrix type** (e.g., Linear distance matrix). Choose the **Output file** location and name. Click **Run** to execute the analysis (Figure 6.30).

Figure 6.30: Configuring the Parameters for Distance Matrix Analysis

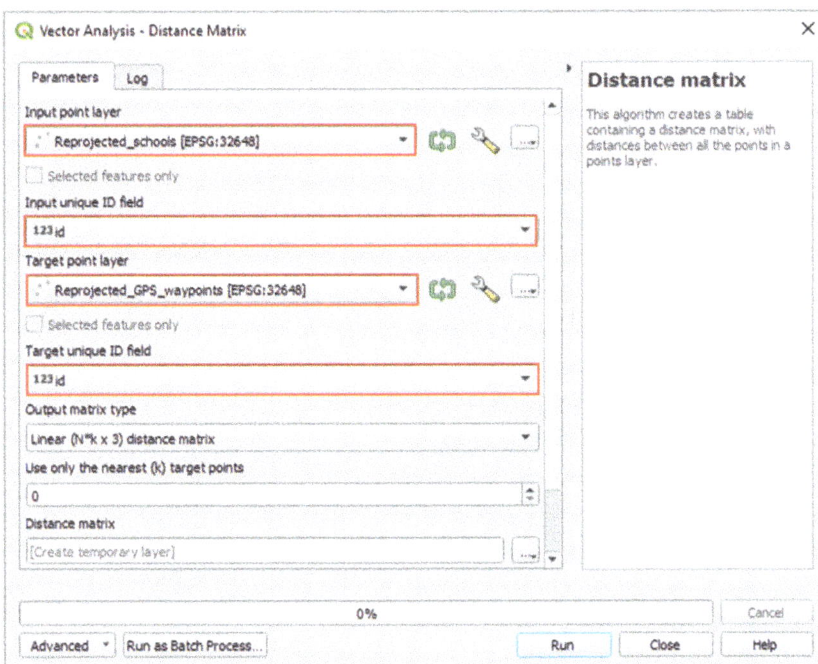

Source: Screenshot generated by the Asian Development Bank using QGIS 3.34.14. 2025.

(v) Open the **attribute table** to view the distances between each GPS point and the points of interest (Figure 6.31).

Figure 6.31: Attribute Table Outputs Showing Distance Calculations to Points of Interest

a

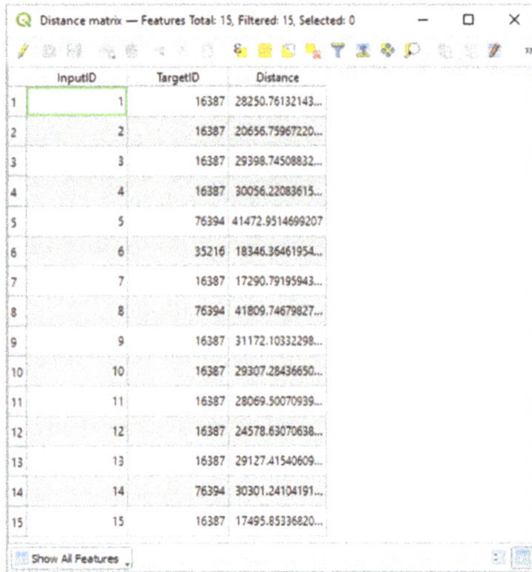

Distance matrix — Features Total: 15, Filtered: 15, Selected: 0

	InputID	TargetID	Distance
1	1	16387	28250.76132143...
2	2	16387	20656.75967220...
3	3	16387	29398.74508832...
4	4	16387	30056.22083615...
5	5	76394	41472.9514699207
6	6	35216	18346.36461954...
7	7	16387	17290.79195943...
8	8	76394	41809.74679827...
9	9	16387	31172.10332298...
10	10	16387	29307.28436650...
11	11	16387	28069.50070939...
12	12	16387	24578.63070638...
13	13	16387	29127.41540609...
14	14	76394	30301.24104191...
15	15	16387	17495.85336820...

Show All Features

b

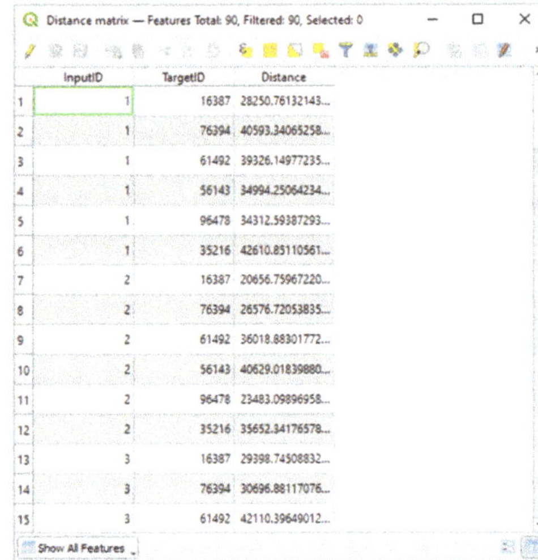

Distance matrix — Features Total: 90, Filtered: 90, Selected: 0

	InputID	TargetID	Distance
1	1	16387	28250.76132143...
2	1	76394	40593.34065258...
3	1	61492	39326.14977235...
4	1	56143	34994.25064234...
5	1	96478	34312.59387293...
6	1	35216	42610.85110561...
7	2	16387	20656.75967220...
8	2	76394	26576.72053835...
9	2	61492	36018.88301772...
10	2	56143	40629.01839880...
11	2	96478	23483.09896958...
12	2	35216	35652.34176578...
13	3	16387	29398.74508832...
14	3	76394	30696.88117076...
15	3	61492	42110.39649012...

Show All Features

Note: (a) represents output when k = 1, indicating the calculated distance to the nearest point of interest; (b) represents the output when the distances to all points of interest are calculated.

Source: Screenshots generated by the Asian Development Bank using QGIS 3.34.14. 2025.

(vi) To verify, use the **Measure Line tool** or the icon in the **Attributes toolbar** to manually measure the distances (Figure 6.32).

Figure 6.32: Measuring Distances with the Measure Line Tool

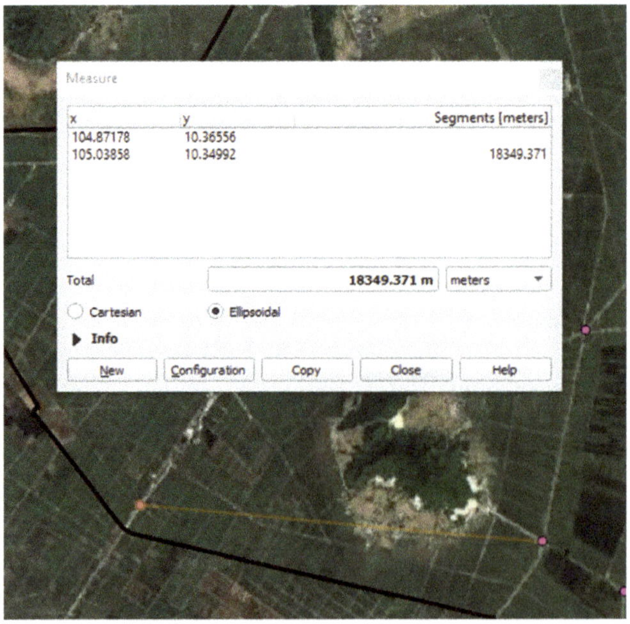

Source: Screenshot generated by the Asian Development Bank using QGIS 3.34.14. 2025.

Module 7: Basic Geographic Information System Vector Analysis Tools

This module focuses on the basic geographic information system (GIS) vector analysis tools essential for agricultural statistics. These tools are fundamental for improving the accuracy and efficiency of data collection and analysis in agricultural surveys, particularly when working with enumeration areas.

Enumeration areas are operational geographic units used for the collection of census data. They are subdivisions of the country and are frequently utilized as a national sampling frame for a range of follow-up surveys, enabling accurate mapping, resource allocation, and data analysis. Enumeration areas also facilitate the sampling and weighting of data, and for the presentation and dissemination of results. The process of identifying enumeration areas requires significant resources. It can be done manually by digitizing high-resolution satellite images or physically by walking around area boundaries (World Bank 2019). Whether created digitally or manually, enumeration areas are designed with clear, non-overlapping borders; and each has a population that is both sizable enough to provide privacy and anonymity and small enough for an enumerator to cover during a census period. Hence, counting the number of residents within an enumeration area is a crucial task.

The topics covered in this module are critical for several reasons. Accurate vector analysis and effective GIS management enhance the quality of agricultural data, supporting better insights and informed decisions. Mastery of these tools and techniques ensures that agricultural data are collected and analyzed in a detailed and impactful manner.

In this module, you will go through the following topics:

- 7.1 Splitting and Merging Polygons
  - 7.1.1 Selecting Features
  - 7.1.2 Splitting Polygons
  - 7.1.3 Merging Polygons
- 7.2 Fixing Gaps, Overlaps, and Invalid Geometries
  - 7.2.1 Identifying Invalid Geometries
  - 7.2.2 Digitizing Features
- 7.3 Editing the Attribute Table
- 7.4 Calculating Areas and the Number of Points Within Them

By the end of this module, you will be able to use these GIS vector analysis tools to optimize field operations, improve data accuracy, and support effective agricultural surveys and analyses.

7.1 Splitting and Merging Polygons

For this module, the district boundaries (ADM2) of An Giang Province, Viet Nam, will be used since enumeration-area boundaries are not available.

Additionally, sample data will be used for the exercises in this module. Download these from the shared folder.[a]

[a] For access to shapefiles and other open-access datasets referenced in the handbook, go to the GitHub repository: https://github.com/ADB-Data-Division/QGIS-Handbook-2025

7.1.1 Selecting Features

(vii) Add the vector layer **AnGiang_Districts in QGIS** by dragging and dropping it or **Browser panel** into the **map area**.

(viii) Click the **Select Features** icon, , to select any district in An Giang Province. The **Select Features** icon can be found in the **Selection toolbar** (Figure 7.1). To add or view other toolbars, go to **Menu | View | Toolbars** or right-click on any empty space in the toolbar area. Click on the check box to activate or deactivate a toolbar in the interface. You may also refer to Module 1, section 1.3.4, "QGIS Graphical User Interface," for more details.

Figure 7.1: Toolbars in the QGIS Graphical User Interface

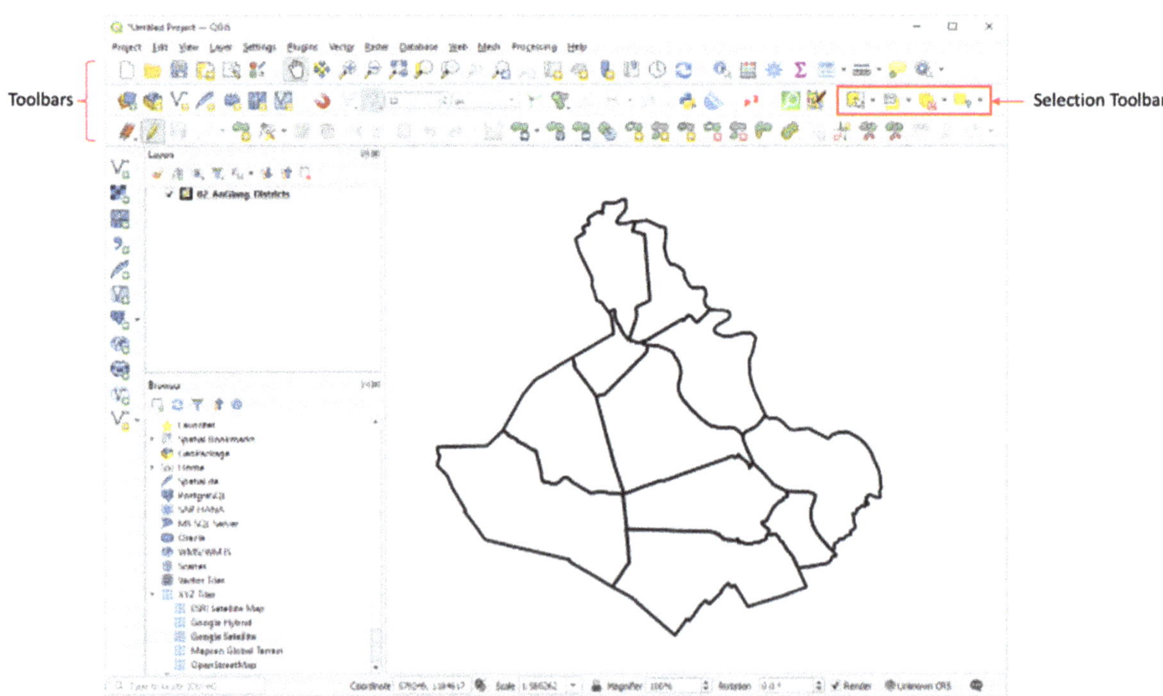

Source: Screenshot generated by the Asian Development Bank using QGIS 3.34.14. 2025.

In the example shown in Figure 7.2, the selected district is Cho Moi District (Huyen Cho Moi). To verify the selected district, open the **attribute table** and check the highlighted feature. For more information about the **Selection toolbar**, refer to Module 4: Attribute Tables.

Figure 7.2: Selecting Features in the Map View

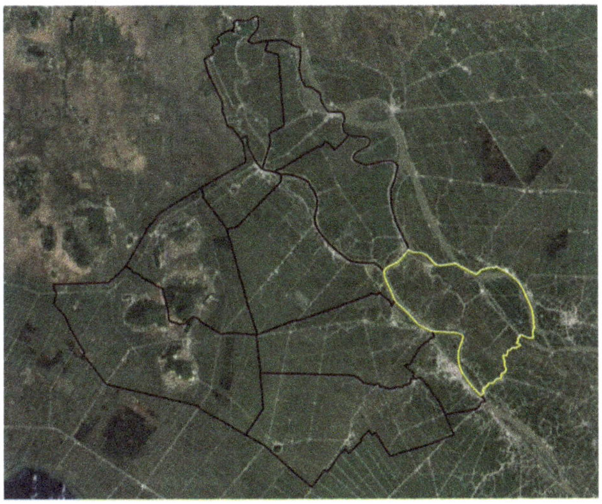

Source: Screenshot generated by the Asian Development Bank using QGIS 3.34.14. 2025.

7.1.2 Splitting Polygons

Enable the **advanced digitizing toolbar** (Figure 7.3). To activate it, go to **Menu | View | Toolbars | Advanced Digitizing Toolbar**. To split the polygon (the district in this example), ensure that the **Toggle editing** mode is enabled by clicking the icon in the **digitizing toolbar**.

Figure 7.3: Advanced Digitizing Tools

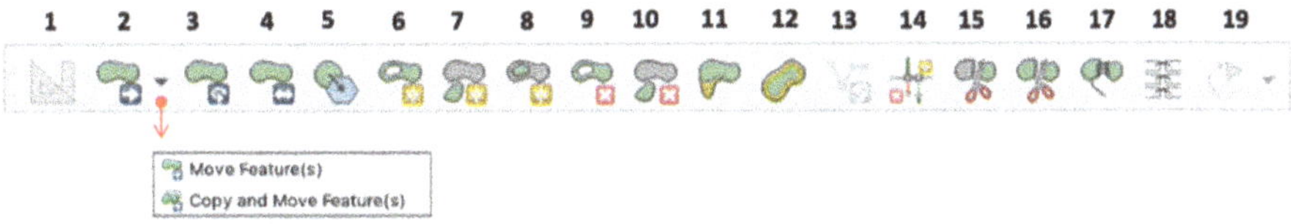

1. **Enable Advanced Digitizing Tools**: Provides access to precision editing tools.
2. **Move Feature**: Moves selected features to a new location.
3. **Rotate Feature**: Rotates selected features around a specified point.
4. **Scale Feature**: Resizes selected features proportionally by a specified scaling factor
5. **Simplify Feature**: Reduces the number of vertices in a feature while preserving its overall shape.
6. **Add Ring**: Adds a ring (hole) inside a polygon feature.
7. **Add Part**: Adds a new part to a multipart feature.
8. **Fill Ring**: Fills a ring (hole) within a polygon feature.
9. **Delete Ring**: Removes a ring (hole) from a polygon feature.
10. **Delete Part**: Removes a part from a multipart feature.
11. **Reshape Features**: Modifies the shape of selected features by drawing a new shape.
12. **Offset Curve**: Creates a parallel copy of selected features at a specified distance.
13. **Reverse Line**: Reverses the direction of a line feature.
14. **Trim/Extend Feature**: Trims or extends a feature to meet another feature.
15. **Split Feature**: Splits a feature into two or more parts.
16. **Split Part**: Splits a part of a multipart feature.
17. **Merge Selected Features**: Combines multiple selected features into one.
18. **Merge Attributes of Selected Features**: Merges the attributes of selected features into one.
19. **Rotate Point Symbols**: Rotates point symbols to a specified angle.

Source: Screenshot generated by the Asian Development Bank using QGIS 3.34.14. 2025.

(ix) Click the **Split Features** icon, , to divide the polygon and set new boundaries. Draw a line or segment by placing vertices. Ensure that the start and end vertices lie outside the polygon area to be split (Figure 7.4). Then, right-click to finish the process.

Note that **Enable Advanced Digitizing Tools**, accessed through the icon, is automatically enabled, allowing the vertices to be placed with exact coordinates. If you want the segment to have exact vertices with specific coordinates, refer to Module 6.2.2 on drawing lines.

Figure 7.4: Splitting Polygons

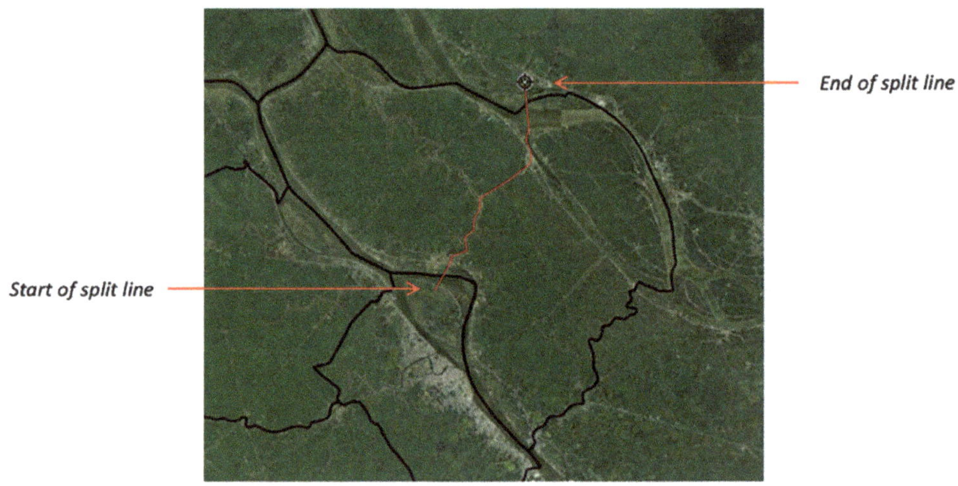

Source: Screenshot generated by the Asian Development Bank using QGIS 3.34.14. 2025.

Check the newly created polygons and save the changes in the layer (Figure 7.5).

Figure 7.5: Polygons Resulting from Splitting a Single Feature

Source: Screenshot generated by the Asian Development Bank using QGIS 3.34.14. 2025.

(x) Open the **attribute table**. Observe that there are now two entries for Cho Moi District after the splitting process (Figure 7.6).

Figure 7.6: Attributes of a Split Polygon

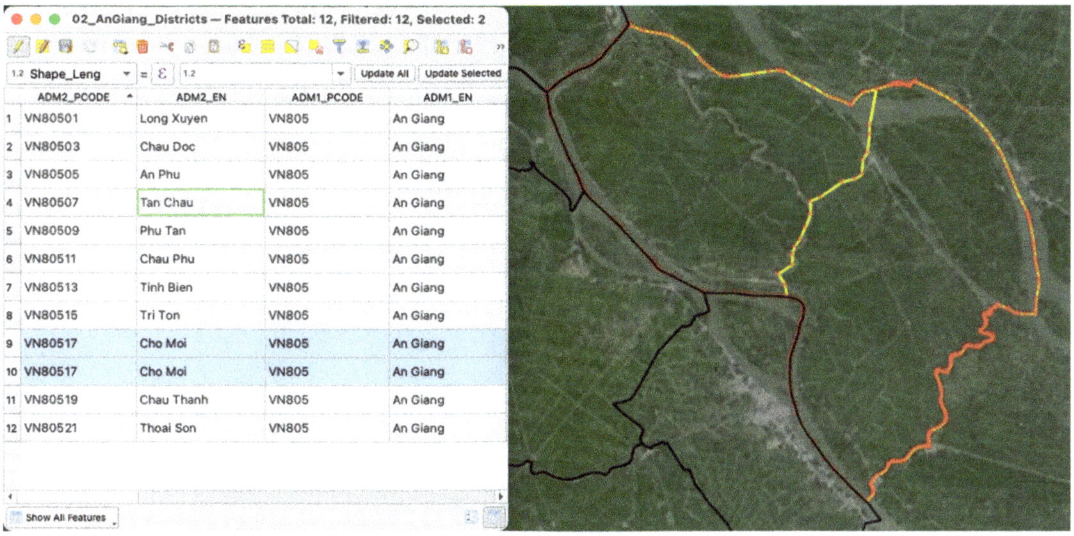

Source: Screenshot generated by the Asian Development Bank using QGIS 3.34.14. 2025.

7.1.3 Merging Polygons

There are different ways to merge polygons. Each method offers advantages, and the selection of method depends on your specific requirements and the level of detail needed for your project.

1. Using the Merge Selected Features Tool in the Advanced Digitizing Toolbar

This tool manually merges the selected polygons with precise control over the editing process. The detailed control enables you to edit and adjust the boundaries and attributes of the merged polygons. This is useful when you need to merge polygons and make specific adjustments to the boundaries, while ensuring accurate boundary alignment. For example, if you need to merge several small plots into a larger field for a new irrigation system, you can use the **Merge Selected Features tool** in the **Advanced Digitizing Toolbar** to adjust the boundaries to ensure that the new field is accurately depicted.

(i) To merge polygons, make sure that the **Toggle editing** mode is enabled. Select the polygons, or in this example, select the districts, that you want to merge. Click the **Select Features** icon, 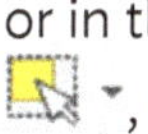, then click and drag across the districts you intend to merge. Alternatively, you can also hold the Shift button while selecting the districts (Figure 7.7).

Figure 7.7: Selecting the Polygons for Merging

Source: Screenshot generated by the Asian Development Bank using QGIS 3.34.14. 2025.

(ii) Click the **Merge Selected Features** icon, , in the **Advanced Digitizing Toolbar**. A dialog box will appear, allowing you to confirm the merge. The options listed below give you flexibility in handling attributes during the merging process.

a. **Taking attributes from selected features.** This option merges features and retains the attributes from the selected features. It allows you to choose which attributes to keep from the features being merged.
b. **Taking a feature from largest area.** Selecting this option ensures that the merged feature
c. inherits the attributes from the feature with the largest land area.
d. **Skipping all fields.** This option merges the features without retaining any attributes, creating a totally new feature without any attribute data from the original features.
e. **Removing a feature from selection.** This option allows you to remove a selected feature from the merge process, enabling the exclusion of certain features from being merged.

In the example in Figure 7.8, the merged polygon initially takes on the attributes of the first selected feature that was being merged: Chau Thanh District.

Figure 7.8: Merging Polygons with Attribute-Handling Options

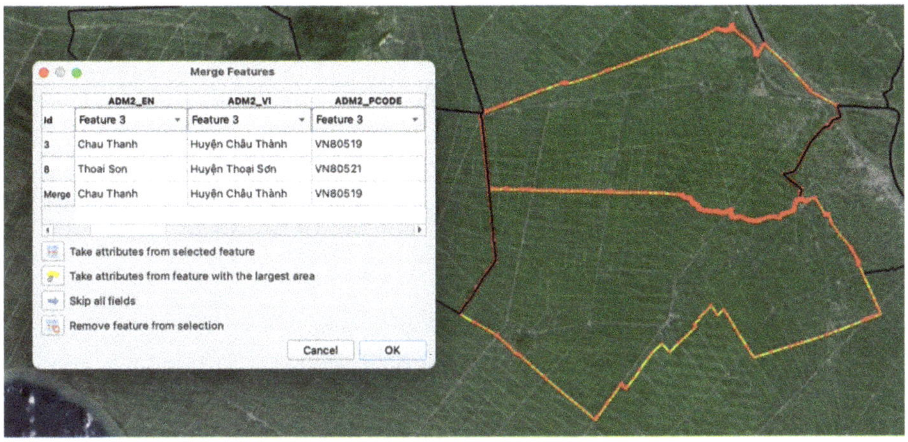

Source: Screenshot generated by the Asian Development Bank using QGIS 3.34.14. 2025.

(iii) In this case, the '**Take attributes from feature with the largest area**' option was applied, so the merged feature takes on the attributes of Thoai Son District. Click OK to confirm. Open the **attribute table** and note that there are now 10 features, reduced from the initial 11 features after the merging process had taken place, as shown in Figure 7.9.

Figure 7.9: Checking Merged Features in the Attribute Table

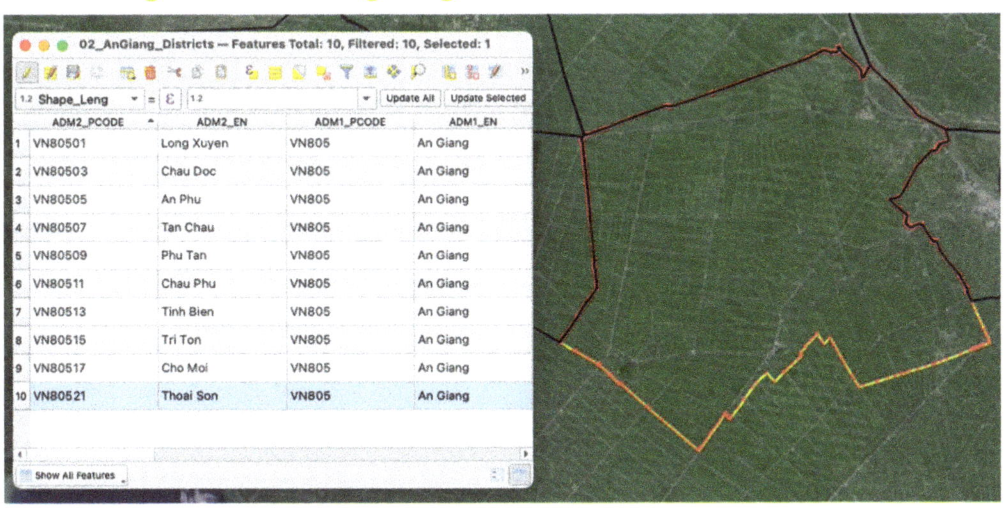

Source: Screenshot generated by the Asian Development Bank QGIS 3.34.14. 2025.

2. Using the Dissolve Tool

The Dissolve tool merges polygons based on a common attribute, removing internal boundaries to create a single, unified polygon. It is efficient for merging large datasets based on specific criteria, such as soil or crop types. For example, you can use this tool to create a single polygon representing all fields with the same crop type, thereby improving large-scale agricultural planning and analysis. If you want to analyze the total area of all fields planted with rice, you can use the Dissolve tool to select that crop type attribute. The Dissolve tool will merge all the rice fields into a single polygon, making it easier to calculate the total area and plan resources accordingly.

(i) Using the Dissolve tool, select the polygons or the districts that you want to merge. Click the **Select Features** icon, , and then click and drag over to the districts that you want to merge. You can also hold the Shift button while selecting the districts. To compare the Dissolve tool with the Merge Selected Features tool, let us select the same districts as those referred to in the previous example: Thoai Son and Chau Thanh (Figure 7.8).

(ii) Go to **Vector | Geoprocessing Tools | Dissolve**. Alternatively, you can go to the menu and click **Processing | Toolbox**, then go to **Dissolve** in the search bar (Figure 7.10).

Figure 7.10: Opening the Dissolve Tool

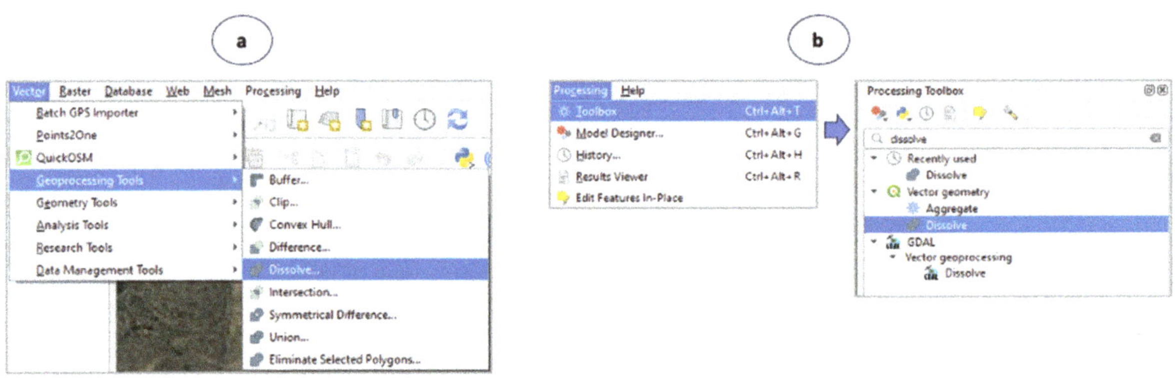

Note: Image "(a)" shows how to open the Dissolve tool through the Vector menu, while the two images designated as "(b)" show how to open the Dissolve tool through the Processing menu.

Source: Screenshot generated by the Asian Development Bank using QGIS 3.34.14. 2025.

(iii) Configure the tool. Do this by checking the **Select features only** option, since the selected polygons are part of a larger polygon in the layer. You may choose a field in which you want to dissolve the polygons (optional). This is referred to as the **Dissolve field**. When this is specified, QGIS will group and merge polygons that share a value in that field, and the resulting output will retain that value as an attribute. If no Dissolve field is selected, QGIS will merge all the selected polygons into a single feature. In such cases, the output polygon will inherit attributes from one of the input features, typically the first in the dataset order.

Specify the output file or create a temporary layer. Click the **Run** button to execute the dissolve process (Figure 7.11).

Figure 7.11: Configuring the Dissolve Tool

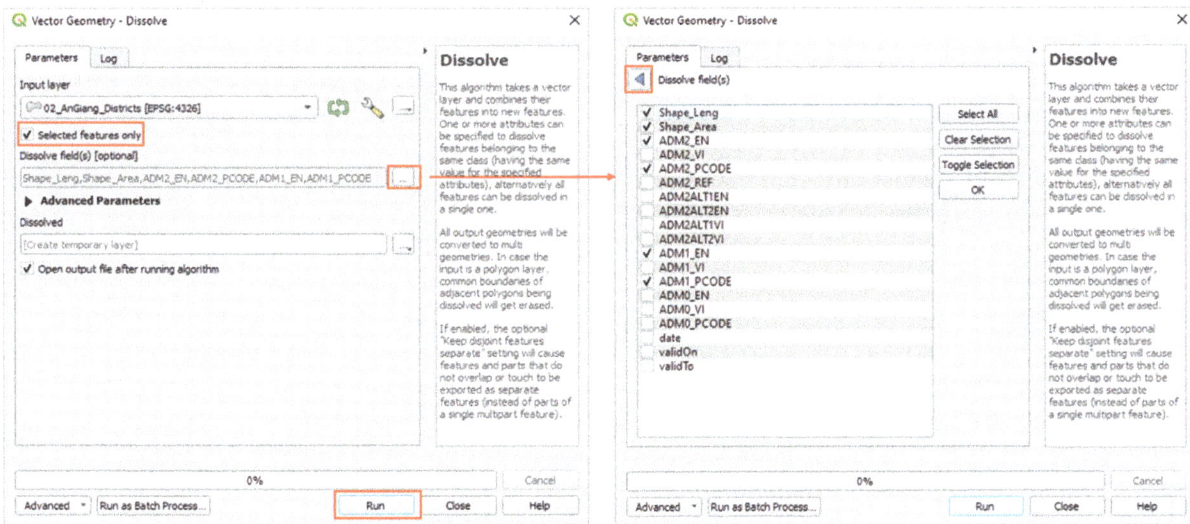

Source: Screenshot generated by the Asian Development Bank using QGIS 3.34.14. 2025.

(iv) The dissolved polygon layer will be added to the map canvas, as shown in Figure 7.12. Check the attributes to ensure that everything was merged correctly.

Figure 7.12: Checking the Attributes of a New Merged Layer

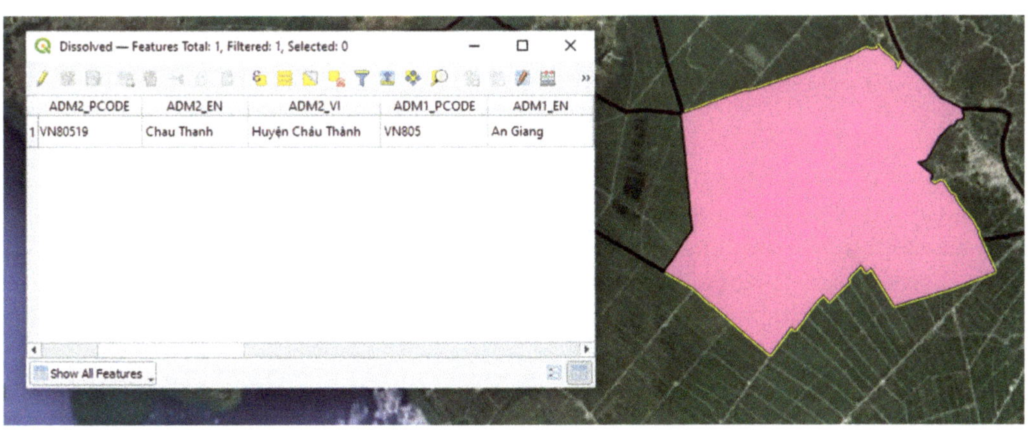

Source: Screenshot generated by the Asian Development Bank using QGIS 3.34.14. 2025.

7.2 Fixing Gaps, Overlaps, and Invalid Geometries

Gaps are empty spaces between polygons that should be adjacent. These can cause inaccurate area calculations, misrepresentation of spatial coverage, and errors in spatial analysis for such purposes as zoning or land use. In the context of multipolygon features, gaps within a single feature may be intentional (to represent dikes, irrigation paths, access roads through agricultural fields, etc.) and are acceptable when each polygon is logically separate, but grouped with others. However, gaps between features that are meant to be continuous are typically errors, often resulting from digitizing mistakes or misalignment, especially when topology rules require polygons to be adjacent (QGIS Project, n.d.-c).

Overlaps occur when polygons incorrectly share the same space. This can lead to double-counting in area or attribute analysis, and may violate spatial rules in cadastral or zoning maps (QGIS Project, n.d.-c).

Invalid geometries are spatial data errors that occur when polygons, lines, or points do not adhere to the rules of geometry. These errors, as illustrated in Figure 7.13, can stem from various sources, including inaccuracies in data collection, digitizing mistakes, or issues during geoprocessing operations. Common types of invalid geometries include self-intersecting polygons, which create loops by crossing over themselves; duplicate vertices (or nodes), where multiple vertices occupy the same location; unclosed rings, where polygons fail to form a closed loop; holes that are not contained within any polygon; and overlapping polygons. These errors can disrupt spatial operations, prevent proper rendering, and cause failures in geoprocessing tasks (Pebesma 2017; QGIS Project, n.d.-b).

Figure 7.13: Common Polygon Geometry Errors: Gaps, Overlaps, and Invalid Geometries

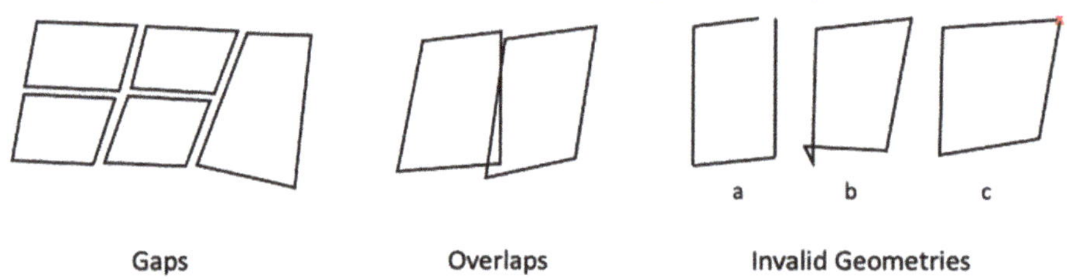

Note: The examples of invalid geometries here include (a) an unclosed polygon, (b) a self-intersecting polygon, and (c) a polygon with a duplicate node (marked “x” in red).

Source: Asian Development Bank visualization. 2025.

In agricultural mapping, invalid geometries can lead to significant problems. For example, when mapping fields using Global Positioning System (GPS) data, self-intersecting polygons might inaccurately represent fields, resulting in erroneous area calculations and flawed resource planning. Ensuring valid geometries is essential for accurate analysis and decision-making in agricultural applications, as it maintains the integrity of spatial data and supports effective resource management.

In the following exercise, we will use the dataset **AnGiang_polygons**, extracted from OpenStreetMap, which contains polygons representing buildings, including residential areas, business establishments, offices, and other structures in An Giang Province. Download these files from the shared folder.[a]

[a] For access to shapefiles and other open-access datasets referenced in the handbook, go to the GitHub repository: https://github.com/ADB-Data-Division/QGIS-Handbook-2025

7.2.1 Identifying Invalid Geometries

1. Using the Geometry Checker Plugin

(i) Add the vector layer **AnGiang_polygons** by dragging and dropping it from the **Browser panel** into the **map area**. Alternatively, you can add the layer by clicking the **Layer tab** in the **Menu Toolbar**, then selecting **Add Layer** and **Add Vector Layer**. The vast proportion of An Giang Province consists of rice plantation areas. To check most of the polygons, zoom into the districts, such as **Long Xuyen**, as shown in Figure 7.14, to see the polygons representing residential areas and other features.

Figure 7.14: Zoomed-In Views of Polygons in Long Xuyen District, An Giang Province

Source: Screenshot generated by the Asian Development Bank using QGIS 3.34.14. 2025.

(ii) To identify the gaps, overlaps, and invalid geometries, open the **Geometry Checker Plugin**. Note that this plugin is not installed by default in some QGIS versions (to install, click on **Plugins | Manage and install plugins | Check Geometries | Install)**. Open the tool by clicking on **Vector**, then **Check Geometries** (Figure 7.15).

Figure 7.15: Opening the Geometry Checker Plugin

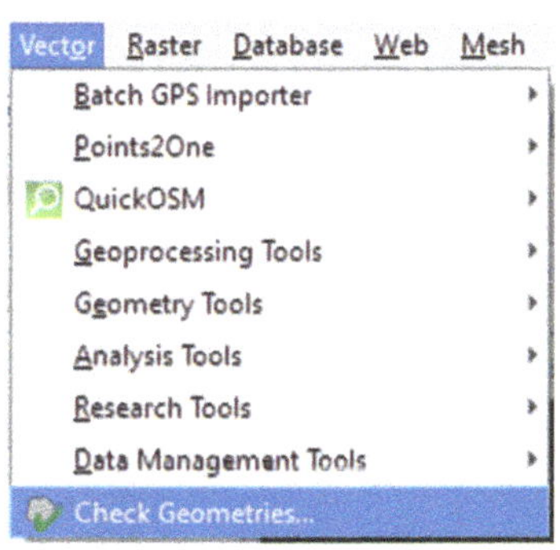

Source: Screenshot generated by the Asian Development Bank using QGIS 3.34.14. 2025.

(iii) A dialog box will appear with several options for configuring the check. Select the input layer—in this example, the **AnGiang_polygons** layer. Choose the types of geometry errors to check for, such as self-intersections, duplicate vertices, and overlapping polygons. Specify the output file for the results. Click **Run** to execute the check (Figure 7.16).

Figure 7.16: The Geometry Checker Plugin

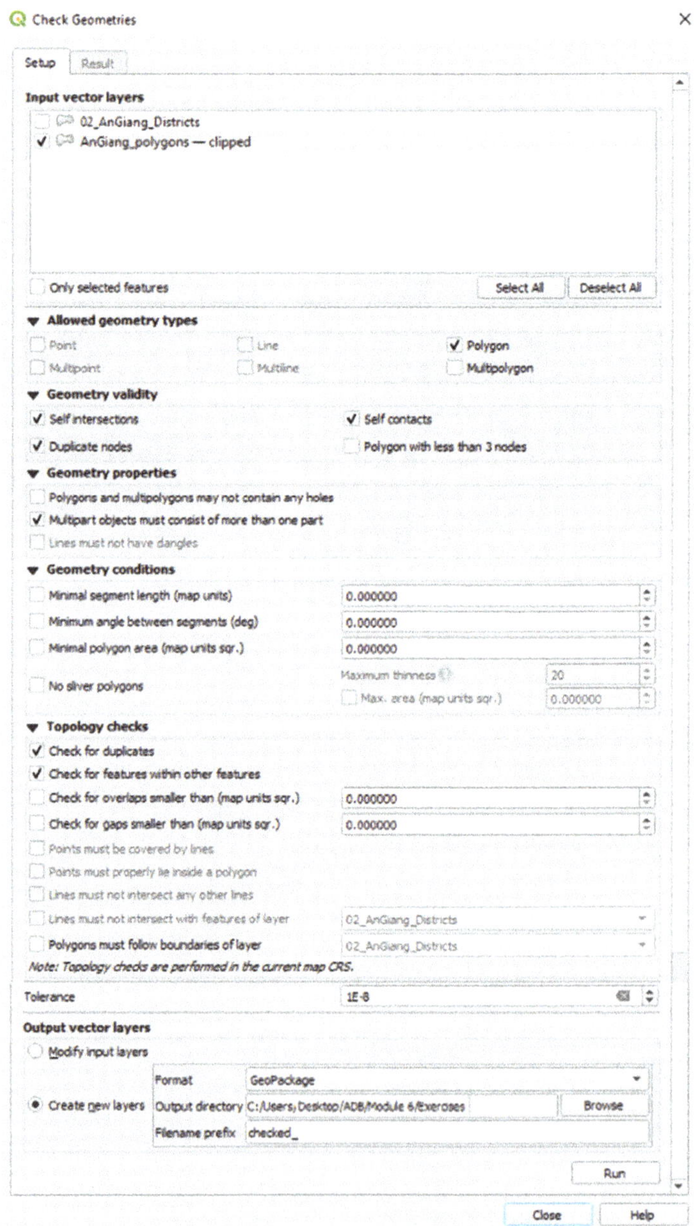

Source: Screenshot generated by the Asian Development Bank using QGIS 3.34.14. 2025.

(iv) The **Result** tab (Figure 7.17) displays a detailed table listing all the geometry errors found. Each row represents a single error and includes columns specifying Layer, Object ID, Error (e.g., self-intersection, duplicate nodes), Coordinates, Value, and Resolution. This table provides comprehensive information on the location and nature of each error, along with suggestions for resolution. An overview layer of the errors is added to the map canvas, prefixed with **checked_**.

Figure 7.17: Result Tab List of Geometry Errors

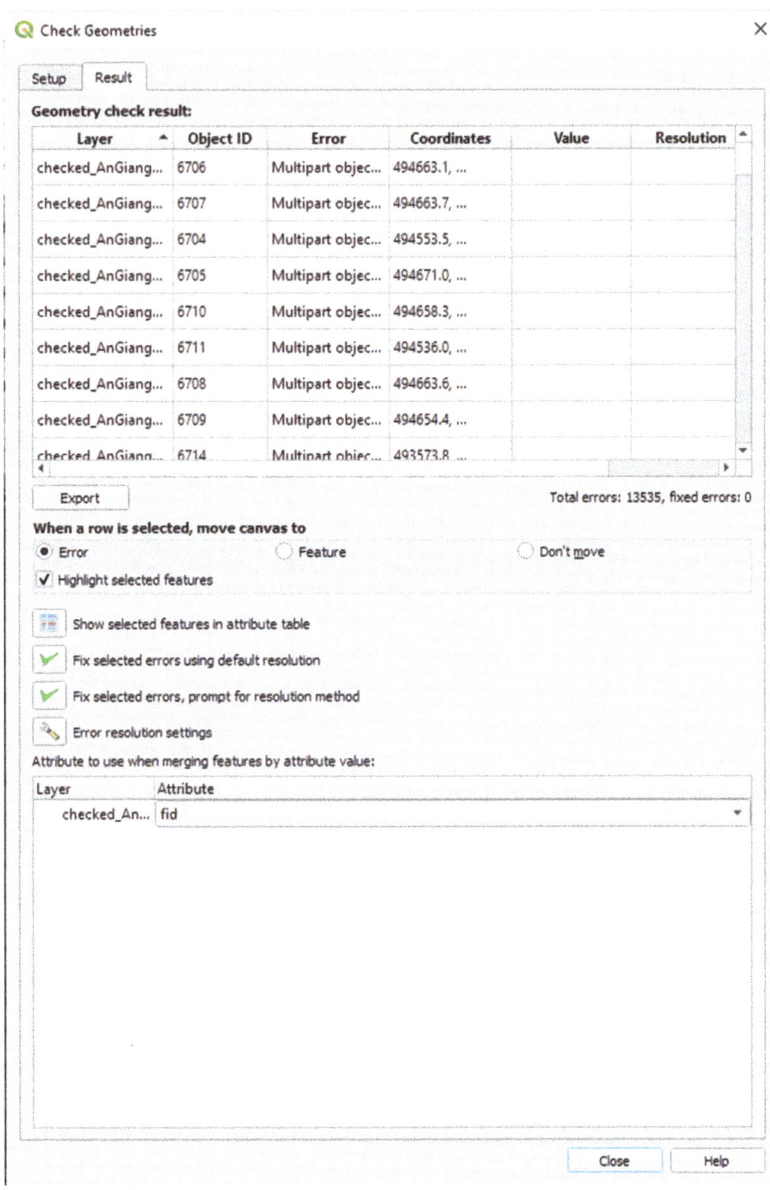

Source: Screenshot generated by the Asian Development Bank using QGIS 3.34.14. 2025.

(v) After running the **Geometry Checker Plugin**, several further options are available for efficiently identifying and correcting geometry issues in the dataset. These tools include options to zoom in on, select, fix, or delete errors directly on the map. Additionally, the list of errors can be exported for further analysis or record-keeping. The geometry-correction tools available are:

a. **Show selected features in attribute table.** This option allows you to view the selected features in the attribute table, filtering it to display only the chosen records. This helps you focus on specific errors and their attributes without the distraction of other data.
b. **Fix selected feature using default resolution.** This tool automatically resolves selected geometry errors using predefined methods, which are configured based on typical error types and their standard fixes. It is useful for quickly addressing common issues without your having to manually specify resolution parameters.
c. **Fix selected errors, prompt for resolution method.** For more control over error resolution, this option lets you choose how to resolve each error. When it is selected, QGIS has prompts asking for the specific resolution method for each error, allowing for tailored fixes based on the nature of the error. This is particularly useful for complex or less common errors that require individual handling.
d. **Error resolution settings.** These settings allow the configuration of error resolutions by default. Preferences can be set for various types of geometry errors, specifying the methods and parameters to be used for automatic resolution. This customization ensures that the error resolution process aligns with specific project requirements and standards.

(vi) Click **Set Error Resolutions** to configure the automatic resolution of geometry errors. In this example, the AnGiang_polygons layer was found to have several errors, including duplicate features, which can be resolved by automatically deleting duplicates (Figure 7.18).

Figure 7.18: Configuring the Error Resolution Settings

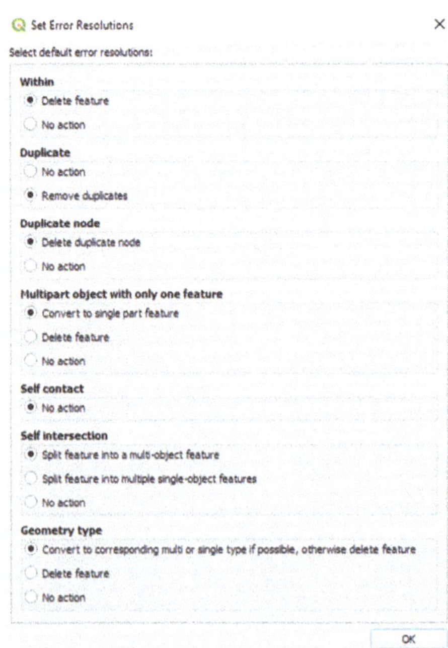

Source: Screenshot generated by the Asian Development Bank using QGIS 3.34.14. 2025.

Another common error involves the geometry type, when a feature's geometry does not match the expected type for its layer. For instance, if a layer is supposed to contain only polygon geometries, but a feature within that layer is a point or line, this would be flagged as a geometry type error. Such mismatches can cause problems with spatial analysis and visualization.

A **multi-part object with only one feature** error occurs when a geometric object is stored as a multi-part object, but contains only a single part (Figure 7.19). This can be resolved by converting the multi-part object to a **single-part object**.

Figure 7.19: Automatic Geometry Fix Confirmation

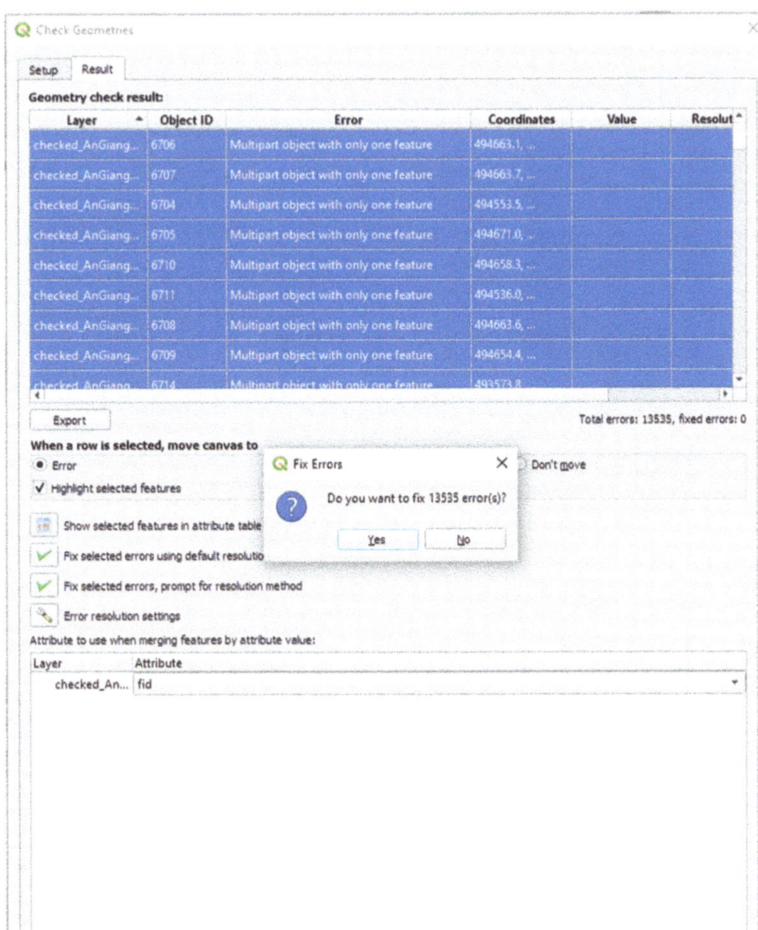

Source: Screenshot generated by the Asian Development Bank using QGIS 3.34.14. 2025.

(vii) After configuring, click on "Yes" to apply the settings.

While some errors can be automatically resolved, be cautious when configuring, as some resolutions may lead to other errors. The **Geometry Checker Plugin** in QGIS is particularly useful when there are multiple errors or invalid geometries that require automatic resolution, if the situation allows it. The optimal approach is to check individual errors, but circumstances do not always permit this, especially during surveys, when there may be multiple errors.

2. Using the Check Validity Tool

(i) Add the vector layer **AnGiang_exercise** by dragging and dropping it from the **Browser panel** into the **map area**. Alternatively, you can add the layer by clicking the Layer tab in the Menu **Toolbar**, then selecting **Add Layer** and **Add Vector Layer**. To check the polygons, right-click on the layer and select **Zoom to Layer**.

(ii) To identify invalid geometries, access the validity tool by selecting **Vector | Geometry Tools | Check Validity** (Figure 7.20). Alternatively, you can navigate to the **Processing menu** at the top of the QGIS interface. Open the **Processing Toolbox** and search for **Check Validity**.

Figure 7.20: Opening the Check Validity Tool

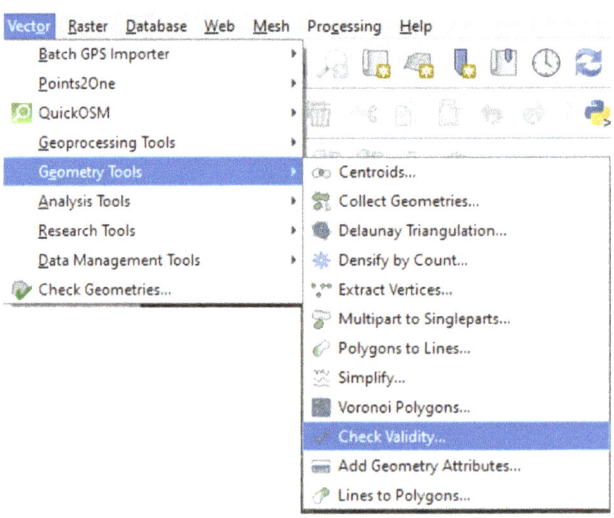

Source: Screenshot generated by the Asian Development Bank using QGIS 3.34.14. 2025.

(iii) A dialog box will appear with several options for configuring the check. By default, the current layer—in this example, the **AnGiang_exercise**—is already selected. You can also specify the geometry validity criteria, such as allowing or ignoring self-intersections. Click **Run** to execute the validity check (Figure 7.21).

Figure 7.21: Configuring the Check Validity Tool

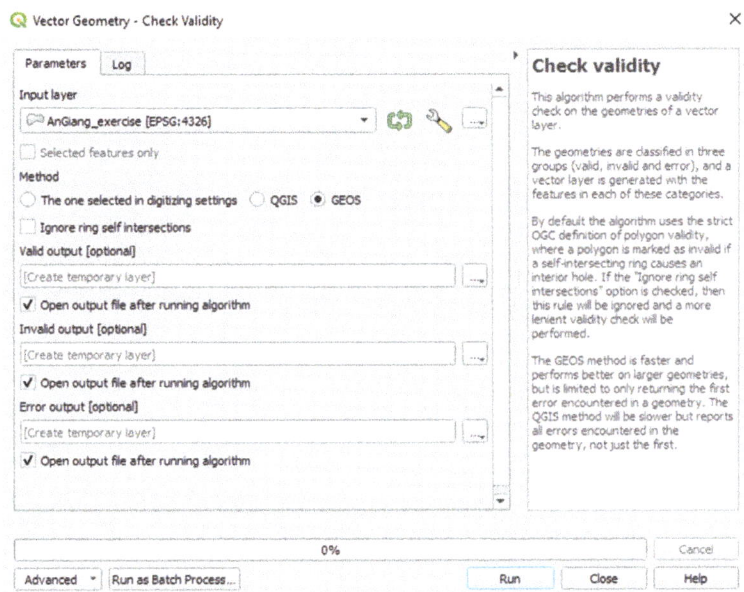

Source: Screenshot generated by the Asian Development Bank using QGIS 3.34.14. 2025.

(iv) After the tool has finished running, a new layer will be created showing the invalid geometries. You can inspect these geometries to present any problems (Figure 7.22). Depending on the errors found, you can manually edit the geometries or use other tools and plugins to automatically resolve certain types of errors.

Figure 7.22: Invalid Outputs Displayed as New Layers

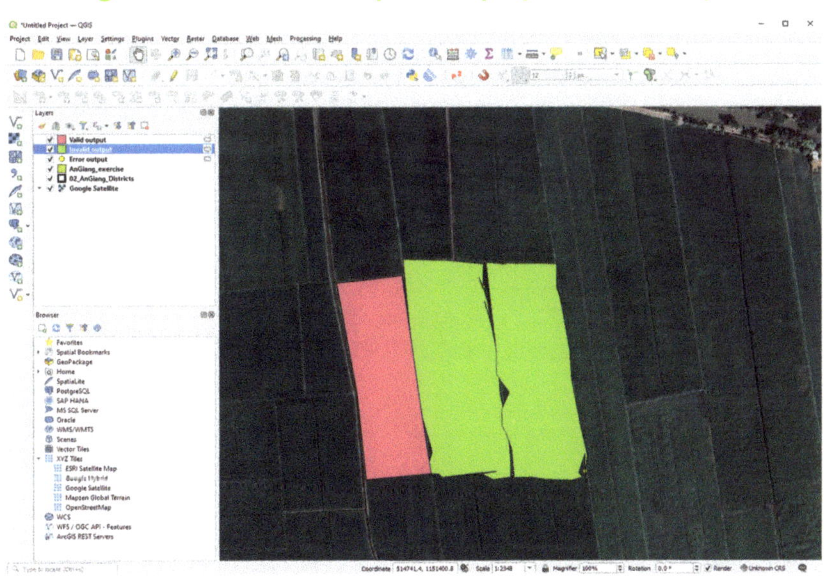

Source: Screenshot generated by the Asian Development Bank using QGIS 3.34.14. 2025.

To check for specific errors, open the **attribute table** and look for the **invalid output layer** (Figure 7.23).

Figure 7.23: Invalid Output Layer

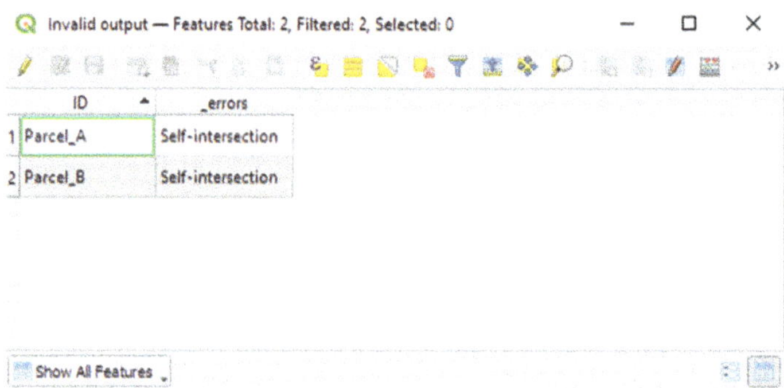

Source: Screenshot generated by the Asian Development Bank using QGIS 3.34.14. 2025.

Note that the **Check Validity tool** in QGIS does not check for duplicate nodes. If you need to check for duplicate nodes, you can use the **Geometry Checker Plugin**, which offers a more comprehensive set of geometry validity checks, including self-intersections and other common geometry issues.

3. Using the Fix Geometries Tool

(i) Navigate to the **Processing menu**, at the top of the QGIS interface. In the **Processing Toolbox**, search for **Fix Geometries** (Figure 7.24).

Figure 7.24: Opening the Fix Geometries Tool Through the Processing Toolbox

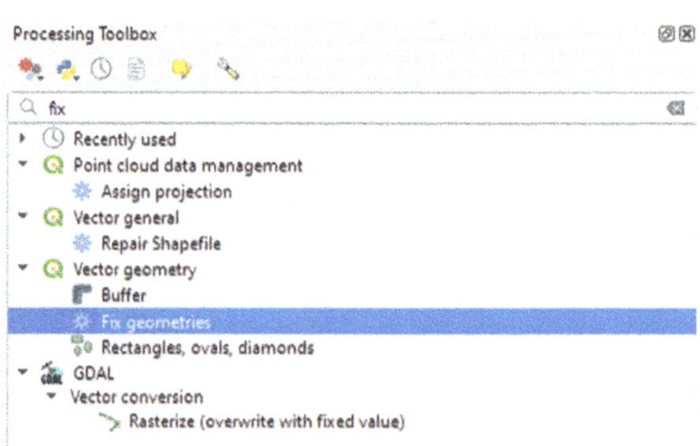

Source: Screenshot generated by the Asian Development Bank using QGIS 3.34.14. 2025.

(ii) Select the **AnGiang_exercise layer** from the **input layer drop-down menu**. You can leave the default settings as they are or adjust them if needed. Click **Run** to execute the **Fix Geometries tool** (Figure 7.25). The tool will analyze the selected layer and correct any invalid geometries.

Figure 7.25: Configuring the Fix Geometries Tool

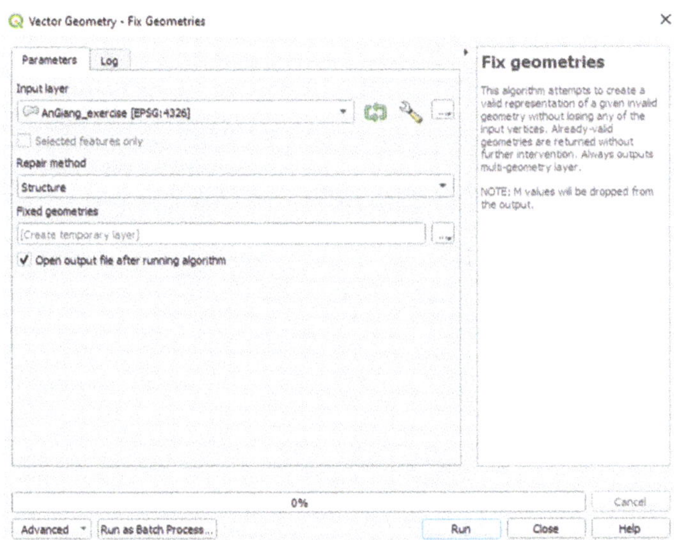

Source: Screenshot generated by the Asian Development Bank using QGIS 3.34.14. 2025.

(iii) After the tool has finished running, a new layer will be created with the fixed geometries, as illustrated in Figure 7.26. You can inspect these geometries to ensure that all the errors have been corrected.

Figure 7.26: New Layer with Fixed Geometries

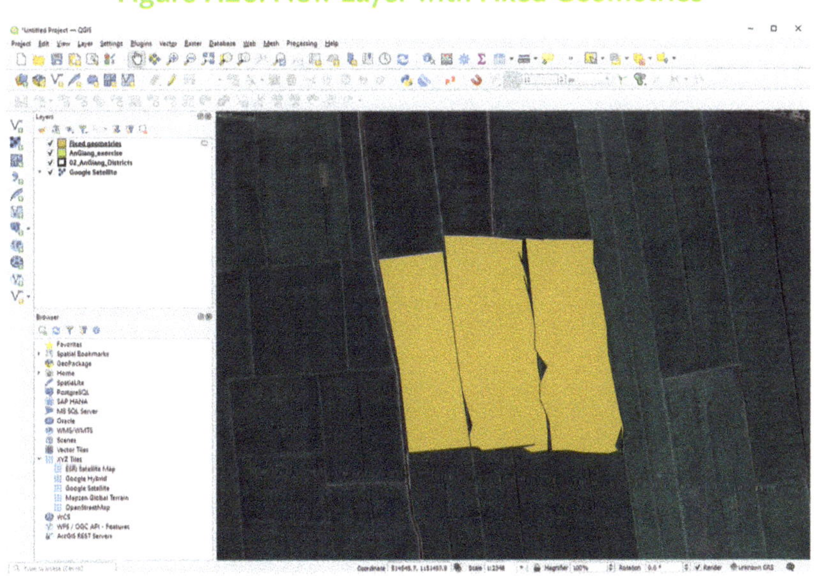

Source: Screenshot generated by the Asian Development Bank using QGIS 3.34.14. 2025.

The **Fix Geometries** tool in QGIS addresses common geometry errors such as self-intersections and duplicate nodes. While it effectively corrects many invalid geometries, it does not fix all types of errors, as some are better handled by the **Geometry Checker Plugin**. The Fix Geometries tool reconstructs valid geometries by either combining rings into lines (linework method) or maintaining the original structure while making rings valid (structure method). Despite the tool's capabilities, however, some complex errors may still require manual intervention or additional tools for resolution.

4. Using the Topology Checker Plugin

(i) Open the **Topology Checker plugin**. Note that this plugin is not installed by default in some QGIS versions (to install, click on **Plugins | Manage and Install Plugins | Topology Checker | Install**). Open the tool by clicking on **Vector**, then **Topology Checker** (Figure 7.27).

Figure 7.27: Opening the Topology Checker Plugin

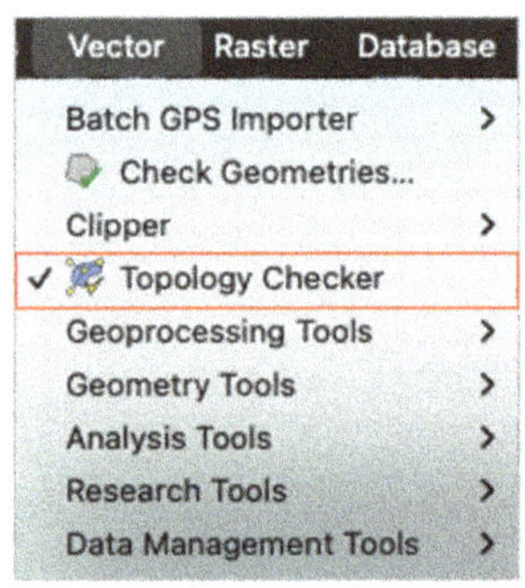

Source: Screenshot generated by the Asian Development Bank using QGIS 3.34.14. 2025.

(ii) In the **Topology Checker panel** (Figure 7.29) click the **Configure** icon, . In the dialog box that appears, click the **Add rule** icon, . Select the layer you want to check from the Layer drop-down menus. Choose the topology rules you want to apply (e.g., "must not have duplicates," "must not overlap," "must not have gaps," etc.). Click **OK** to add the rules (Figure 7.28).

Figure 7.28: Setting Topology Rules

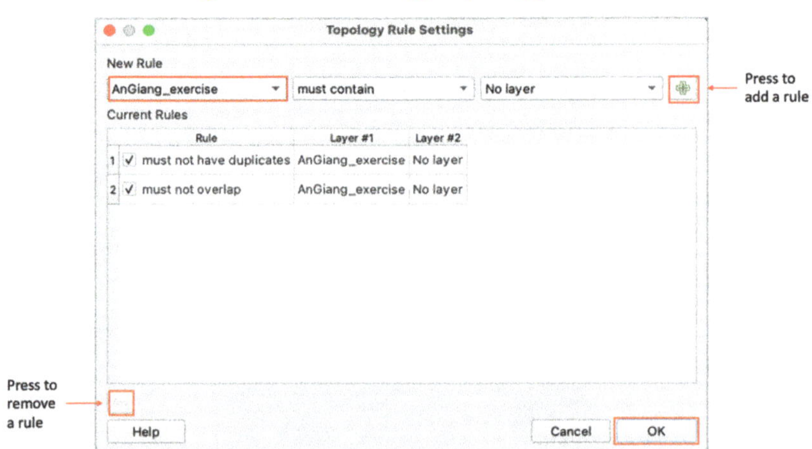

Source: Screenshot generated by the Asian Development Bank using QGIS 3.34.14. 2025.

(iii) After setting up your rules, click on the **Validate All** icon, ✓, in the Topology Checker panel (Figure 7.29). The tool will analyze your polygons and list any topological errors it finds.

Figure 7.29: Topology Checker Panel

Source: Screenshot generated by the Asian Development Bank using QGIS 3.34.14. 2025.

(iv) The errors will be displayed in the **Topology Checker window** (Figure 7.30). Click on an error to zoom in on its location on the map. Use the editing tools in QGIS to correct the errors manually.

Figure 7.30: Displaying Errors with the Topology Checker Plugin

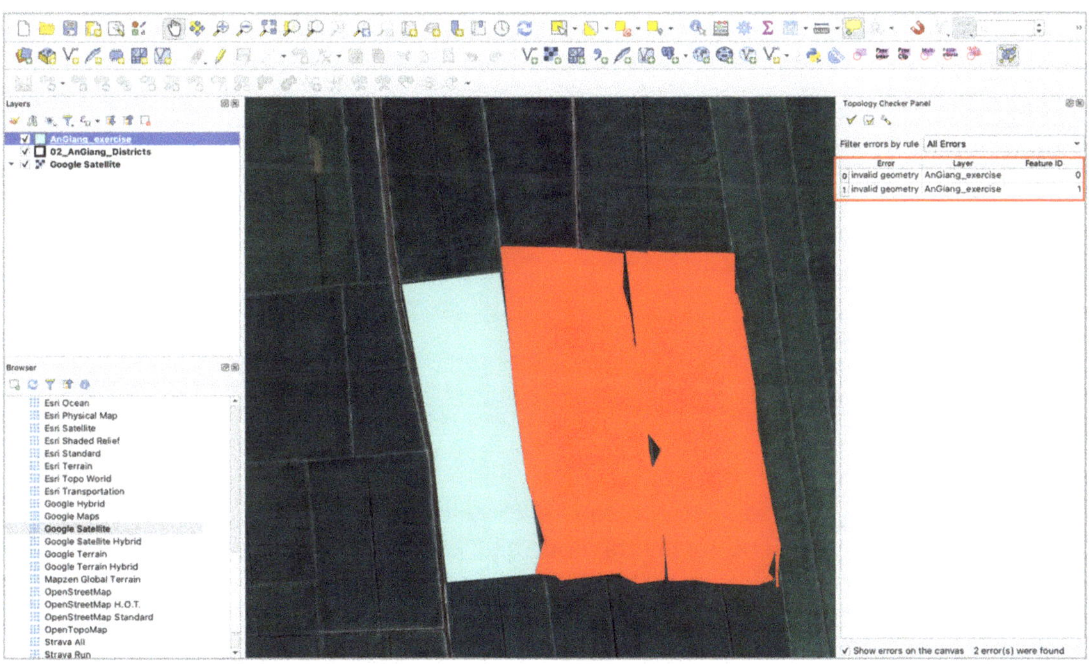

Source: Screenshot generated by the Asian Development Bank using QGIS 3.34.14. 2025.

7.2.2 Digitizing Features

Digitizing features is a critical step in maintaining the accuracy and integrity of spatial data, especially in agricultural mapping. While the previous section discussed tools and plugins for identifying and automatically fixing gaps, overlaps, and invalid geometries, there are instances where automatic fixes are not feasible. This section focuses on manual techniques for correcting these errors using QGIS's

and snapping toolbars. These tools are essential for precise data representation and accurate field boundaries, which are crucial for effective spatial analysis in agricultural applications.

In the following exercise, the invalid geometries identified in the previous section will be corrected using digitizing and snapping tools.

(i) Select the **AnGiang_exercise** from the **Layers panel** and enable the **Toggle editing** mode by clicking the icon in the **digitizing toolbar**. The **Snapping toolbar** will also be used for this exercise. To activate it, go to **Menu | View | Toolbars | Snapping Toolbar** (Figure 7.31).

Figure 7.31: Digitizing and Snapping Toolbars

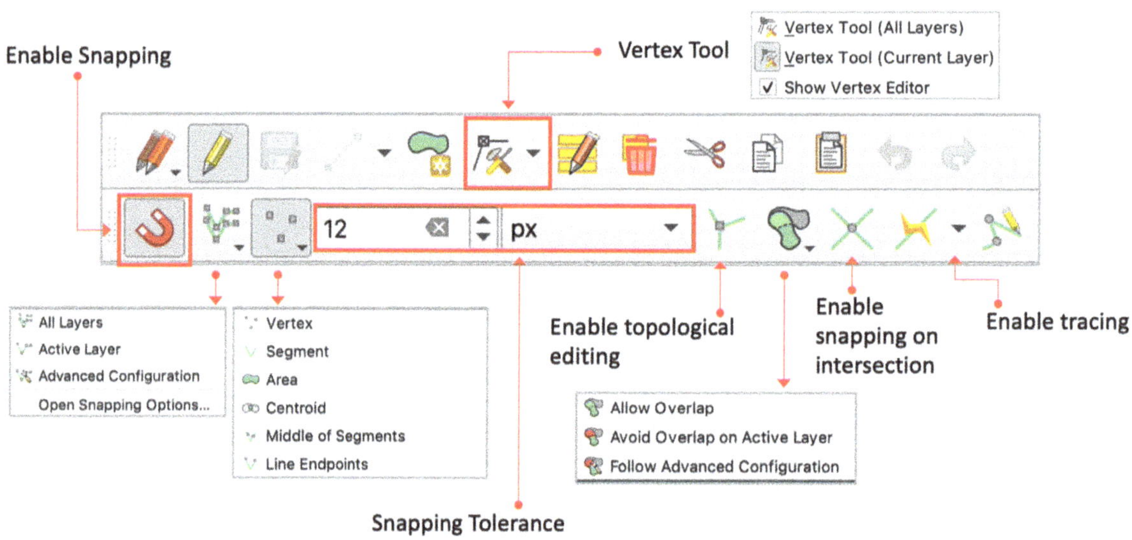

Source: Screenshot generated by the Asian Development Bank using QGIS 3.34.14. 2025.

The **Vertex tool**, , allows you to edit vector features. It offers two modes: All layers and Current layer. The **All layers** mode allows the movement and editing of vertices across all active layers simultaneously, while the **Current layer** mode restricts editing to the vertices of the currently active layer.

The **Snapping tool**, , facilitates the editing of vector geometries by automatically detecting and snapping to vertex locations. Selecting **All layers** enables snapping to vertices and segments across all visible layers, while **Current layer** restricts snapping to the active layer.

This tool offers several snapping modes (Figure 7.32), including the:

- **vertex**, which snaps to the endpoints of lines or polygon vertices;
- **segment**, which snaps along the entire length of a line segment;
- **middle of segment**, which snaps to the midpoint of a line segment;
- **centroid**, which snaps to the geometric center of a feature;
- **area**, which snaps to the centroid of polygon features; and
- **line endpoints**, which snap to the start or end points of line features.

When adding or moving a vertex, QGIS searches for the closest vertex or segment to connect to, based on the snapping tolerance. This tolerance can be adjusted in pixels or degrees. **Snapping tolerance** sets the distance within which snapping will occur, ensuring precise alignment during digitization.

The **snapping toolbar** includes several additional tools: **Enable topological editing** ensures that shared boundaries between features are maintained during editing. **Enable Tracing** allows users to create new features by tracing existing lines or shapes. **Advanced Snapping Options** provides additional settings for snapping, such as enabling snapping to intersections, configuring snapping priorities for different layers, and allowing overlaps.

Figure 7.32: Different Snapping Modes

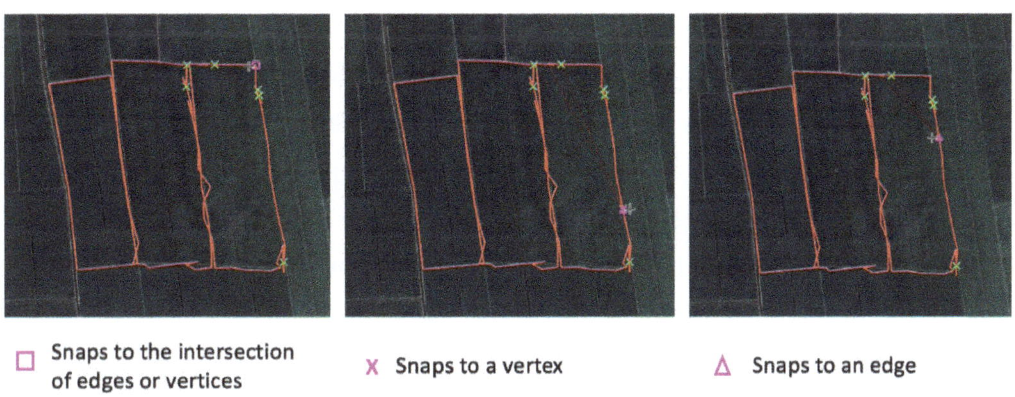

Source: Screenshot generated by the Asian Development Bank using QGIS 3.34.14. 2025.

(ii) Activate the **Vertex tool** from the **Digitizing toolbar** to select, move, add, or delete vertices. Zoom into the area of the parcel where the self-intersection occurs. Place the cursor over a vertex to enable editing (Figure 7.33), then click and drag it to the desired location.

Figure 7.33: Editing Polygons

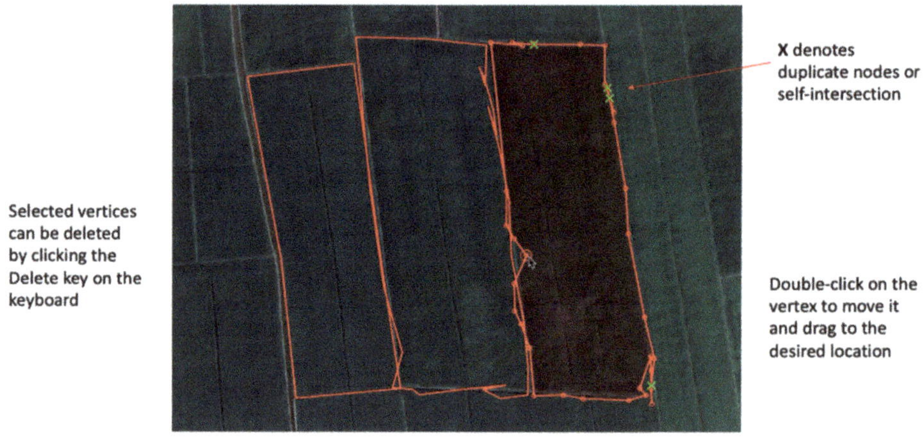

Source: Screenshot generated by the Asian Development Bank using 3.34.14. 2025.

Correct the self-intersection by clicking and dragging the problematic vertices to new positions. If necessary, add new vertices to reshape the parcel correctly by right-clicking to add a vertex. Remove any extra vertices that cause the intersection by selecting the vertex and pressing the **Delete** key.

(iii) For this exercise, move the vertices of "Parcel B" (rightmost parcel) along the boundaries of the rice field and remove the self-intersections and duplicate nodes of the polygon. Do the same for "Parcel A" (center). Ensure that the boundaries of both Parcel A and Parcel B do not overlap. When digitizing the adjacent sides of both parcels, turn off the **Enable snapping** mode and start moving the vertices (Figure 7.34).

Figure 7.34: Disabled Snapping Mode

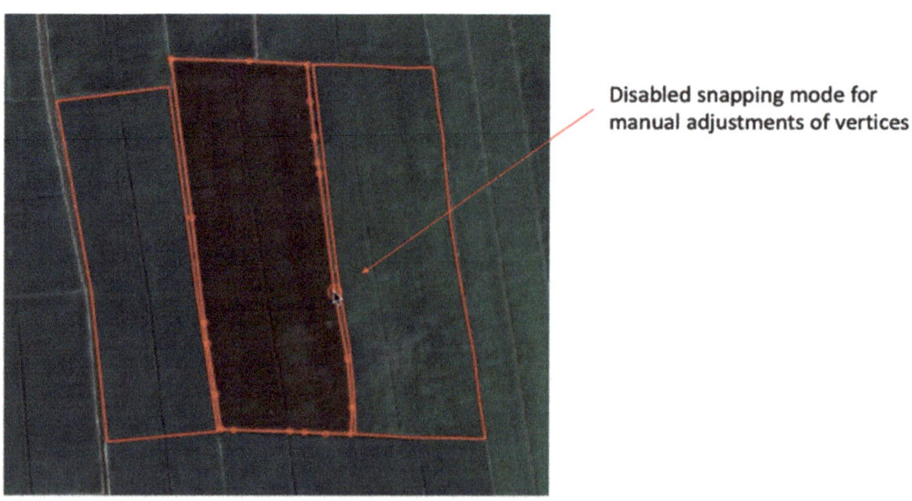

Source: Screenshot generated by the Asian Development Bank using QGIS 3.34.14. 2025.

Enabling the snapping mode helps create precise boundaries, connect features, and avoid gaps and overlaps between polygons. However, it can be less useful in areas with complex shapes, when exact manual adjustments are needed, or when editing existing features, as it may cause unintended changes or reduce precision.

(iv) Check the digitized polygons (Figure 7.35) and save the edited layer. To save the digitized polygon as a new layer, export it as a new shapefile. Right-click on the layer in the **Layers panel**, select **Export**, then **Save Feature as**. A dialog box will appear for further settings. For more information on saving a new layer, you may refer to Module 2, section 2.3.4, "Exporting Map Layers as New Layers."

Figure 7.35: Digitized Polygons

Source: Screenshot generated by the Asian Development Bank using QGIS 3.34.14. 2025.

7.3 Editing the Attribute Table

Module 4 covered the concepts of the attribute table and its management. The attribute table is essential for GIS vector analysis because it contains detailed information about each feature in a layer. This information extends beyond the spatial representation of features such as points, lines, and polygons to include attributes that describe various properties of these features.

For instance, with regard to agricultural parcel digitization, the attribute table might include details such as crop type, soil quality, and ownership information. By accessing and analyzing the attribute table, users can conduct more comprehensive analyses, such as querying specific attributes, symbolizing features based on their properties, and generating reports. This makes the attribute table a vital component for effective data management and analysis.

In the following exercise, we will use the dataset **AnGiang_points**, extracted from OpenStreetMap, which contains points of interest representing amenities, including places of worship, schools, hospitals, and other establishments in An Giang Province. Download these files from the shared folder.[a]

[a] For access to shapefiles and other open-access datasets referenced in the handbook, go to the GitHub repository: https://github.com/ADB-Data-Division/QGIS-Handbook-2025

1. Adding New Fields

(i) Add the vector layer **AnGiang_points** by dragging and dropping it from the **Browser panel** into the **map area**. Alternatively, you can add the layer by clicking the **Layer tab** in the **Menu Toolbar**, then selecting **Add Layer** and **Add Vector Layer**. Do the same for the vector layer **AnGiang_Districts** (Figure 7.36).

Figure 7.36: Points of Interest in An Giang Province

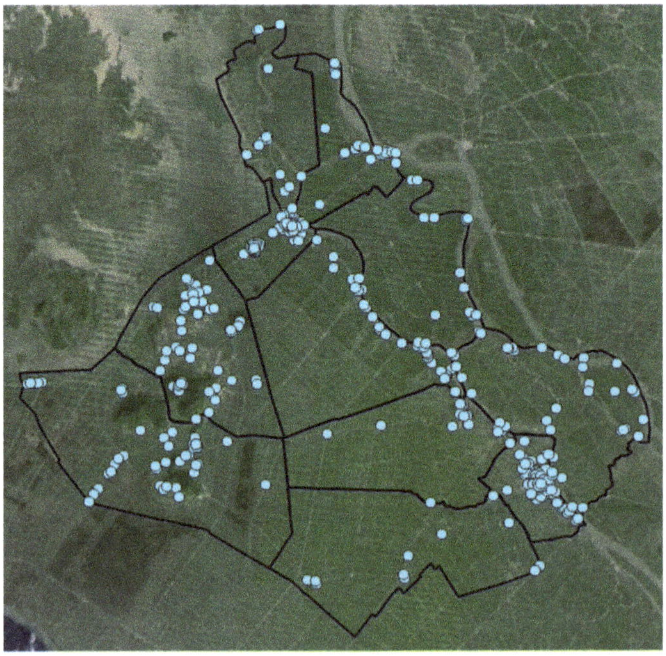

Source: Screenshot generated by the Asian Development Bank using QGIS 3.34.14. 2025.

(ii) Click the **Select Features** icon, , in the **Selection toolbar**, and pan over any district in An Giang Province. Open the **attribute table** and check the district name and district code. In the example shown in Figure 7.37, the selected district is Thoai Son, with District ID VN80521.

Figure 7.37: Selected District and Points of Interest

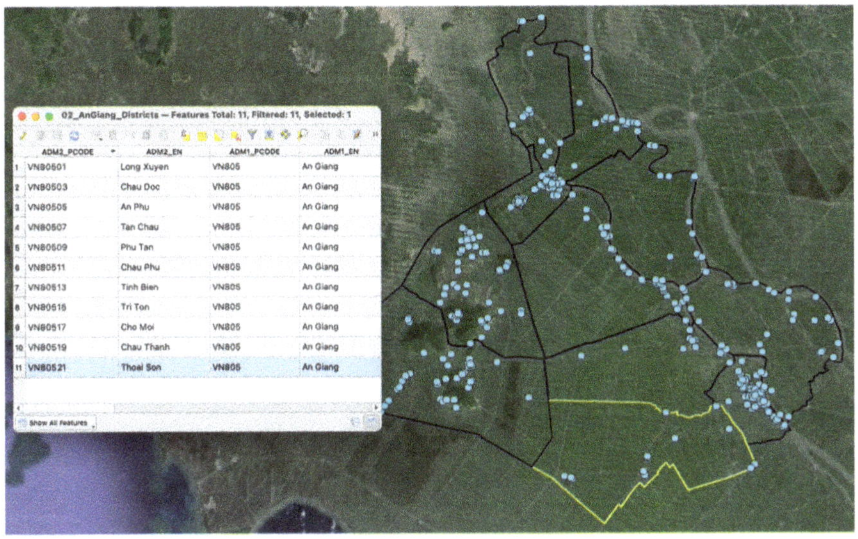

Source: Screenshot generated by the Asian Development Bank using QGIS 3.34.14. 2025.

(iii) Right-click on the **AnGiang_points** layer and open the **attribute table**, or simply click the icon in the **Attributes toolbar**. In the attribute table, enable the editing functionalities by clicking the **Toggle editing** icon, . Note that other icons are now enabled, including: **Add feature,** ; **Paste features from clipboard,** ; **New field,** ; and **Delete field,** (Figure 7.38).

Figure 7.38: Attribute Table with Toggle Editing Enabled

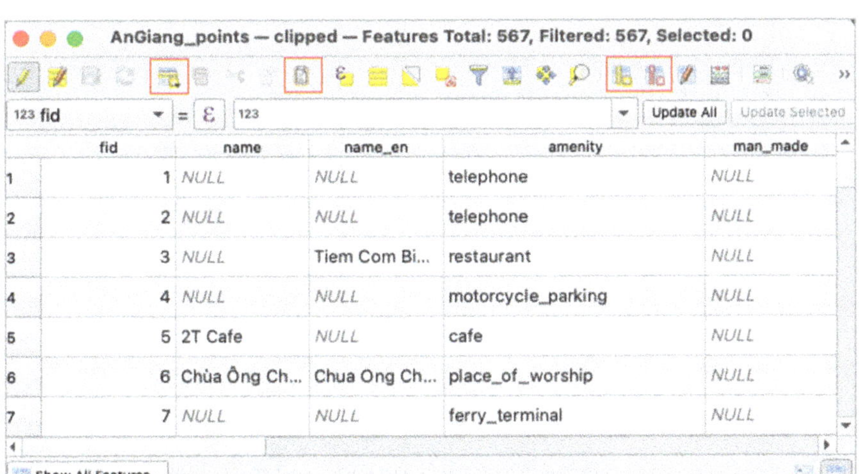

Source: Screenshot generated by the Asian Development Bank using QGIS 3.34.14. 2025.

(iv) Click the **New field** icon. In the dialog box that appears, enter the name of the field and choose the data type. In this example, add the District ID of each point of interest within the district. Select **Text (string)** as the data type (Figure 7.39), since the district ID (ADM2_PCODE) has a "VN" prefix.

Figure 7.39: Adding a New Field to the Attribute Table

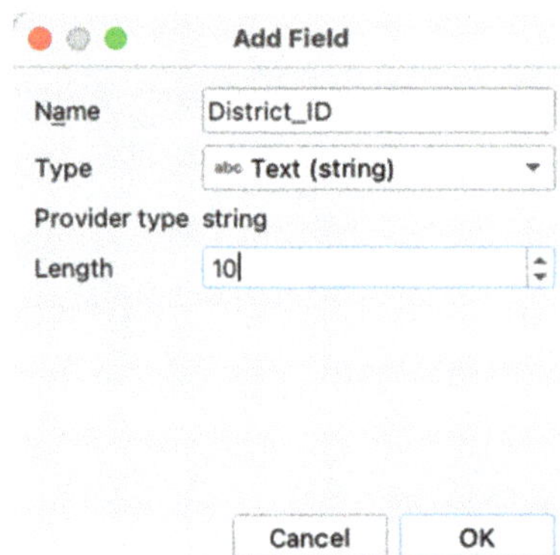

Source: Screenshot generated by the Asian Development Bank using QGIS 3.34.14. 2025.

Click **OK**. Check the newly added field in the attribute table, which now has attributes set to NULL.

(v) On the map, select all points within Thoai Son District. To do this, use the **Select Features** tool. Pan over the map to select multiple features at once, or click selected individual points while holding down the Shift key. The selected features will be highlighted in the attribute table (Figure 7.40).

Figure 7.40: Selected Features on the Map and in the Attribute Table

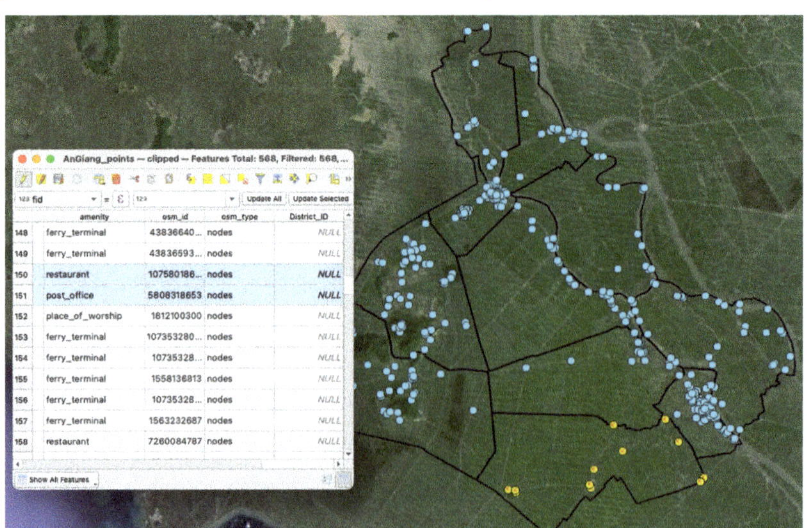

Source: Screenshot generated by the Asian Development Bank using QGIS 3.34.14. 2025.

(vi) To change the attribute values of the selected feature(s), double-click on the attribute-table cell to edit and type in the new value. In the example shown in Figure 7.41, type VN80521 for the District_ID. Repeat this for all the highlighted features in the same district.

Figure 7.41: Editing Selected Features in the Attribute Table

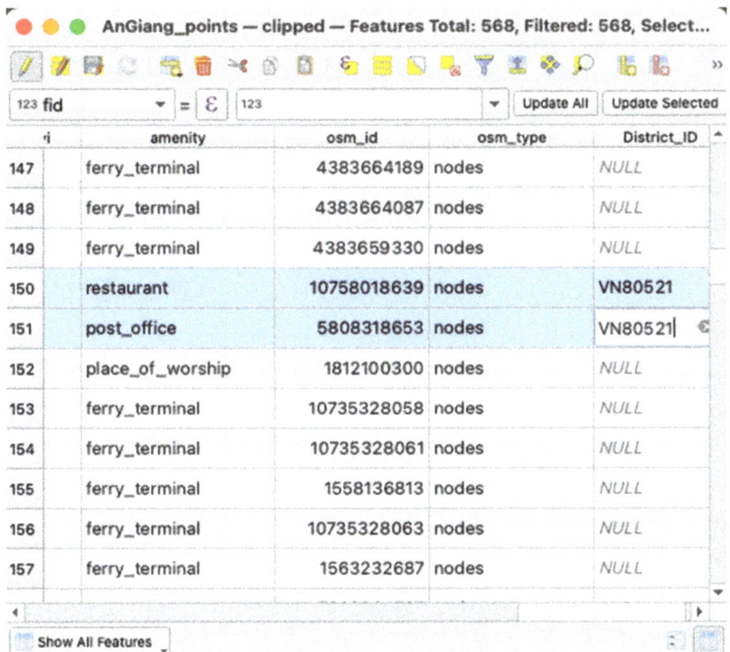

Source: Screenshot generated by the Asian Development Bank using QGIS 3.34.14. 2025.

Please note that while this is a straightforward method for editing attributes, it is only practical when working with a manageable number of features. For larger datasets, more efficient tools are available, such as **Vector Intersection** (Module 3, section 3.3.2) and **Join Attributes by Location** (Module 8, section 8.2.3).

(i) The **Field Calculator** can also be used to update multiple fields at once using expressions. For example, you could add another field for the district name. To do this, open the **Field Calculator** by clicking on 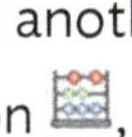, and check the box for **Create a new field**. Name the new field “District_name.” Select **Text (string)** as the data type, and indicate an appropriate output field length (e.g., 20). In the expression field, use a CASE statement to assign specific text values based on the District_ID. Refer to the CASE statement provided in Figure 7.42, then click **OK** to create the new field.

Figure 7.42: Editing Attributes Using the Field Calculator

Source: Screenshot generated by the Asian Development Bank using QGIS 3.34.14. 2025.

The expression will create a new **District_name field** that contains text descriptions based on the values in the **District_ID field**. The resulting attribute table is shown in Figure 7.43.

Figure 7.43: Attribute Table with Added Fields

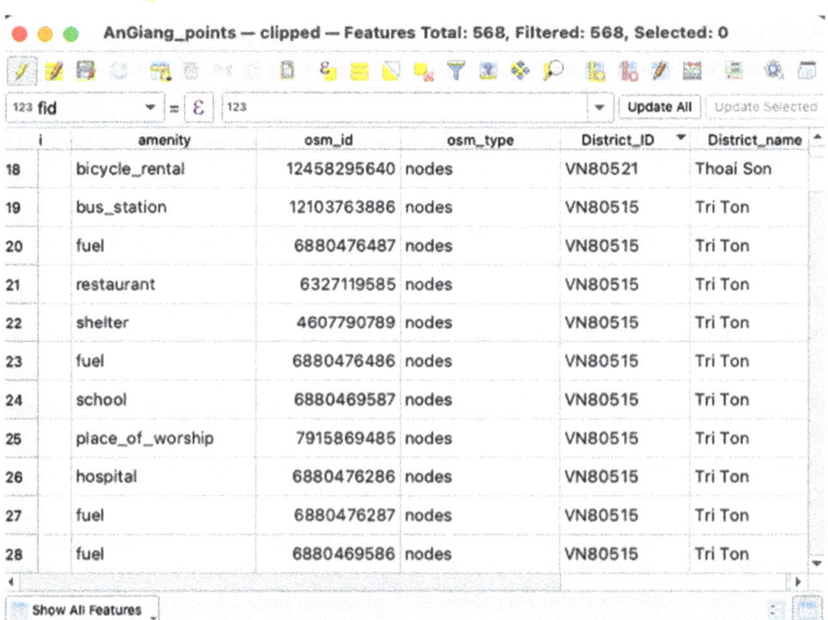

	amenity	osm_id	osm_type	District_ID	District_name
18	bicycle_rental	12458295640	nodes	VN80521	Thoai Son
19	bus_station	12103763886	nodes	VN80515	Tri Ton
20	fuel	6880476487	nodes	VN80515	Tri Ton
21	restaurant	6327119585	nodes	VN80515	Tri Ton
22	shelter	4607790789	nodes	VN80515	Tri Ton
23	fuel	6880476486	nodes	VN80515	Tri Ton
24	school	6880469587	nodes	VN80515	Tri Ton
25	place_of_worship	7915869485	nodes	VN80515	Tri Ton
26	hospital	6880476286	nodes	VN80515	Tri Ton
27	fuel	6880476287	nodes	VN80515	Tri Ton
28	fuel	6880469586	nodes	VN80515	Tri Ton

Source: Screenshot generated by the Asian Development Bank using QGIS 3.34.14. 2025.

QGIS offers a detailed reference of functions available in the Field Calculator, organized by category and accompanied by examples to guide utilization. You can explore the full list in the official documentation at:
https://docs.qgis.org/latest/en/docs/user_manual/expressions/functions_list.html.

2. Adding a New Feature

Adding a new feature follows the same procedure as that outlined in Module 5, section 5.1, which covers the use of digitizing tools to draw points, lines, and polygons. To review, the steps to create and add a new feature in a layer are as follows.

(i) Ensure that the point layer **AnGiang_points** is active by clicking on it in the **Layers panel**. Enable the **Toggle editing** mode. Click the **Add Point Feature** icon, found in the **Digitizing toolbar**. Click on the map canvas to place the point.

(ii) A dialog box will appear for entering the attribute values for the new feature (Figure 7.44). Click **OK**.

Figure 7.44: Adding a Feature on the Map Canvas

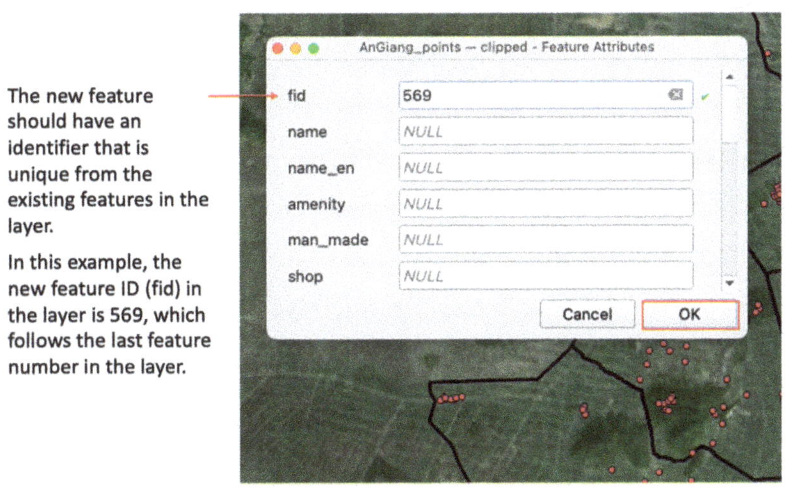

Source: Screenshot generated by the Asian Development Bank using QGIS 3.34.14. 2025.

(iii) Check the new feature on the map and its attributes in the **attribute table**.

3. Deleting Fields and Features

(i) Ensure that the point layer **AnGiang_points** is active by clicking on it in the **Layers panel**. Enable the **Toggle editing** mode. In the attribute table, locate the feature to be deleted. Click on the row number on the left-hand side to select the feature (Figure 7.45).

Figure 7.45: Selecting Features for Deletion in the Attribute Table

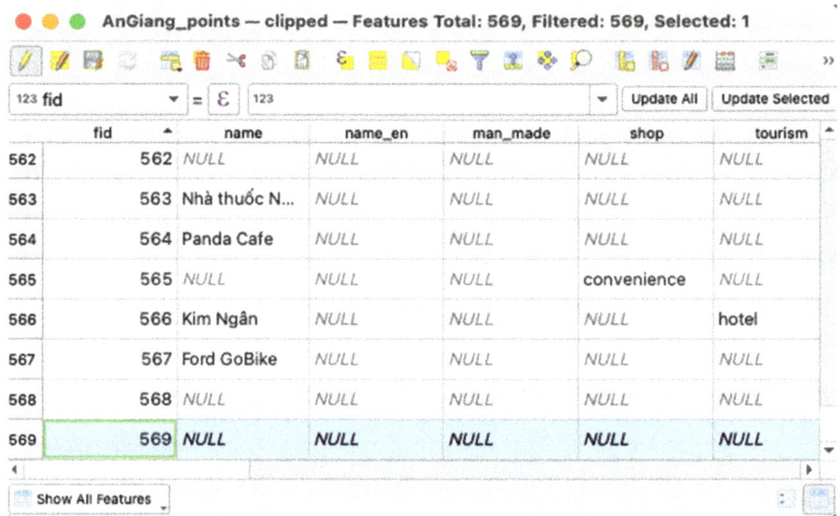

Source: Screenshot generated by the Asian Development Bank using QGIS 3.34.14. 2025.

Multiple features can also be selected using the **Filter** tool, , or the **Select by Expression** tool, . Refer to Module 4 for a step-by-step guide on using these tools.

(ii) To delete the selected features, click the **Delete selected features** button or the icon in the toolbar. Confirm the action before deleting any fields (Figure 7.46).

Figure 7.46: Deleting Selected Features

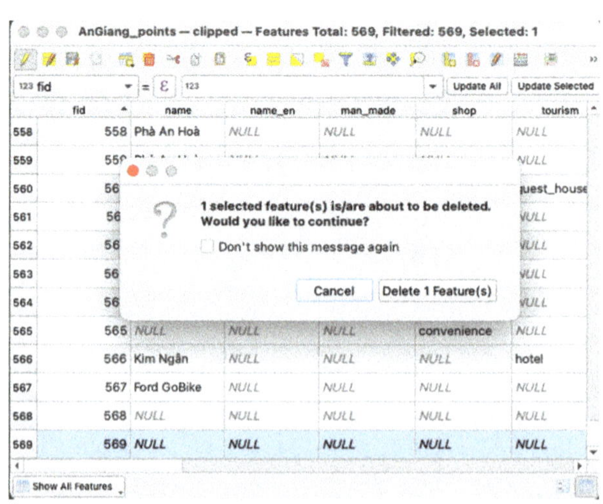

Source: Screenshot generated by the Asian Development Bank using QGIS 3.34.14. 2025.

(iii) To delete a field, click the **Delete Field** tool or button. A small pop-up window with a list of all fields will appear. Select the fields to delete, then click **OK** as shown in Figure 7.47.

Figure 7.47: Deleting Fields from the Attribute Table

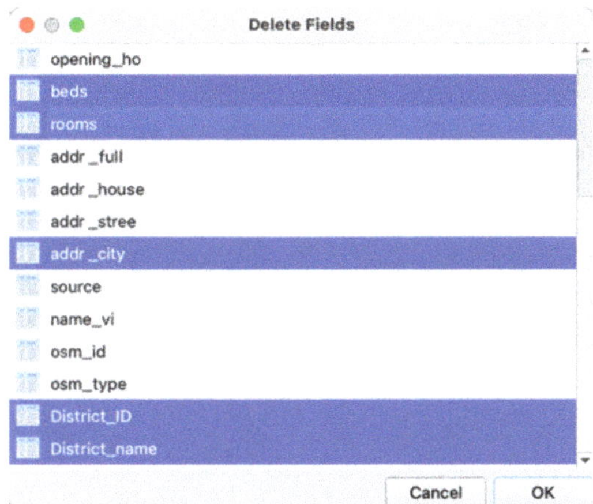

Source: Screenshot generated by the Asian Development Bank using QGIS 3.34.14. 2025.

(iv) To organize the **attribute table**, click on **Organize columns** or on the icon to open the dialog box where fields can be reordered, hidden, or renamed. Make the necessary adjustments (Figure 7.48).

Figure 7.48: Organizing Fields in the Attribute Table

Hide or unhide fields: Check or uncheck the box to hide or unhide a field or column in the table.

Rename fields: To rename a field, double-click on the field name and type in the new name. Note that in some versions of QGIS, this method may not be available. In such cases, go to Layer Properties, navigate to the "Fields" tab, and change the field name there.

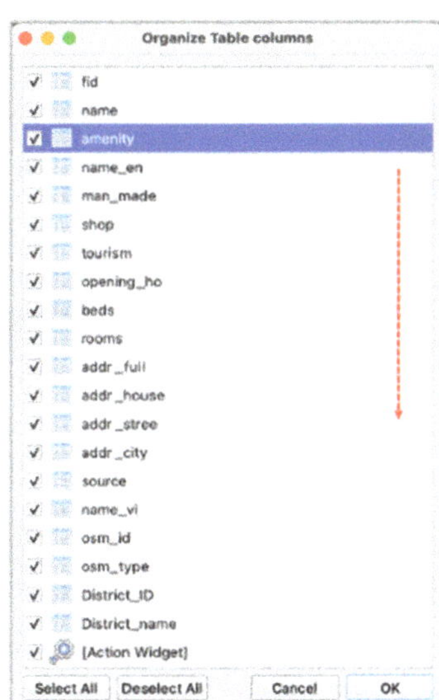

Reorder fields: Drag and drop the field to the desired position in the table.

Source: Screenshot generated by the Asian Development Bank using QGIS 3.34.14. 2025.

(v) To stop editing, click on the **Toggle editing** icon. Save the changes before exiting the editing mode.

7.4 Calculating Areas and the Number of Points Within Them

Accurately measuring the total area of enumeration zones and tallying the number of points they contain are fundamental steps in GIS mapping. These measures shed light on spatial distribution and density, both of which are key for designing efficient surveys and allocating resources effectively. While area calculation defines the size of each zone, point counts reveal patterns of population density and distribution.

Module 5, section 5.3 discusses area calculation using the **Field Calculator**. To review, the steps for calculating an area in the **attribute table** are laid out below.

Since there is no available open-source enumeration area map and no household data, this section will use the district maps (AnGiang_District) and points of interest (AnGiang_points) from OpenStreetMap data as examples to demonstrate these procedures in QGIS.

1. Calculating Area

(i) Reproject the **AnGiang_Districts** layer to the coordinate reference system (CRS). To do this, go to the **Main Menu | Vector | Data Management Tools | Reproject Layer** (Figure 7.49).

Figure 7.49: Reprojecting a Vector Layer

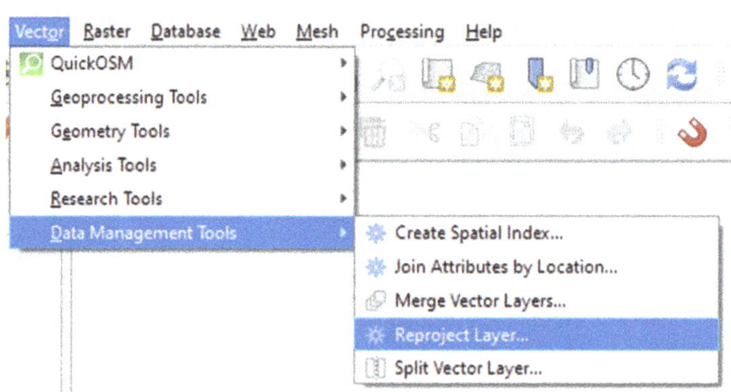

Source: Screenshot generated by the Asian Development Bank using QGIS 3.34.14. 2025.

(ii) Set the target CRS for Viet Nam to **EPSG:32648 – WGS 84 / UTM 48N**. If the target CRS does not appear in the drop-down menu, click the **Select CRS widget,** , to open the CRS selector. In the CRS selector, use the **Filter tab** to search for the CRS. Research the appropriate CRS for the locality; for An Giang Province, it is **EPSG:32648**. Note that it will be highlighted on the map. Select this CRS. The reprojected layer can be saved as a temporary file or as a file in a designated folder. Be sure to select **Open output file after running algorithm**, then click **Run** to reproject the layer. In the example shown in Figure 7.50, the data is saved as a temporary file, and will appear as **Reprojected**.

Figure 7.50: Reprojecting the AnGiang_Districts Layer to the Coordinate Reference System

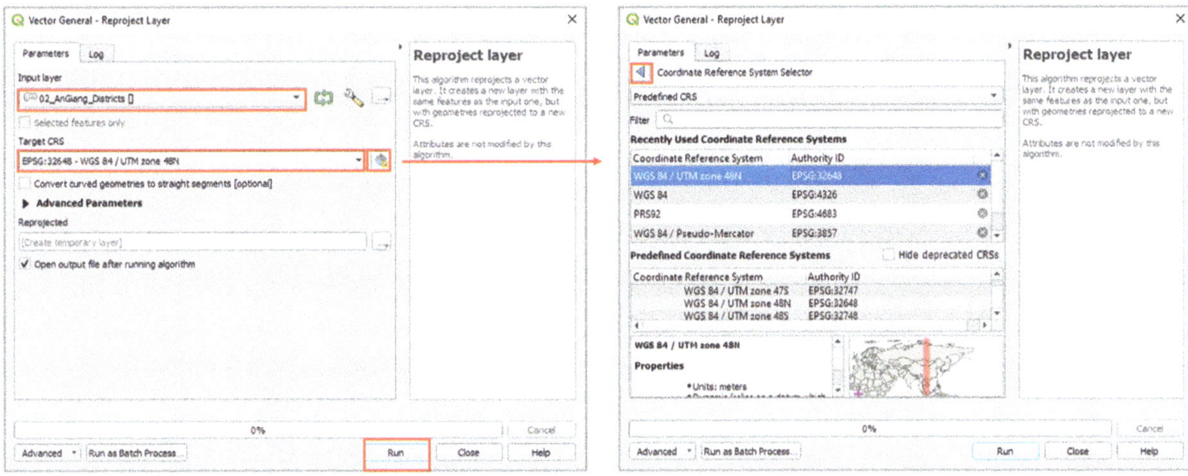

Source: Screenshot generated by the Asian Development Bank using QGIS 3.34.14. 2025.

(iii) Open the **attribute table** for the new **reprojected** layer by right-clicking and selecting **Open Attribute Table**. Click the button to open the **Field Calculator**. Click **Create a new field**. For the **Output field name**, type “Area.” For the **Output field type**, select “Decimal Number (real).” For **Output field length**, select 10, and set **Precision** to 3. In the **Expression box**, type the expression **$area** and double-click on the $area function. The default unit of measure is in square meters. To convert, multiply the function by a conversion factor to get the area in your desired units (e.g., hectares, acres). In the example in Figure 7.51, the unit of measure is converted from square meters to hectares.

Figure 7.51: Area Calculation Using the Field Calculator

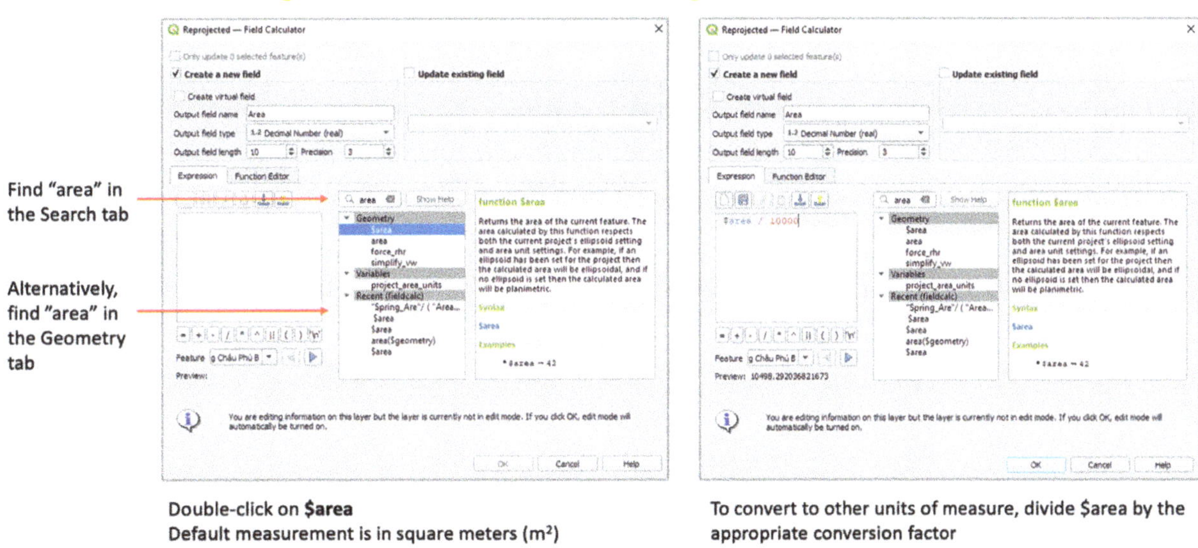

Source: Screenshot generated by the Asian Development Bank using QGIS 3.34.14. 2025.

(iv) Click **OK** to run the calculation. Close the dialog box after the calculation is complete. A new column with the area values for each polygon will appear in the attribute table, as shown in Figure 7.52.

Figure 7.52: Attribute Table with District Area Estimates

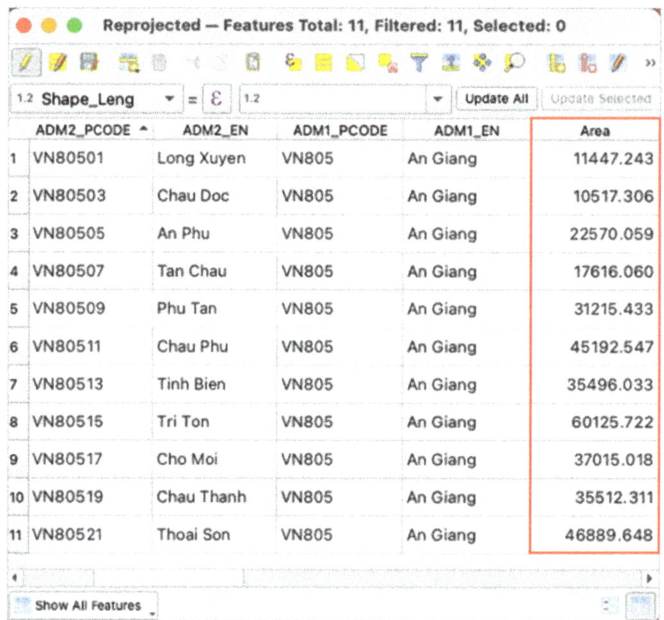

Reprojected — Features Total: 11, Filtered: 11, Selected: 0

	ADM2_PCODE	ADM2_EN	ADM1_PCODE	ADM1_EN	Area
1	VN80501	Long Xuyen	VN805	An Giang	11447.243
2	VN80503	Chau Doc	VN805	An Giang	10517.306
3	VN80505	An Phu	VN805	An Giang	22570.059
4	VN80507	Tan Chau	VN805	An Giang	17616.060
5	VN80509	Phu Tan	VN805	An Giang	31215.433
6	VN80511	Chau Phu	VN805	An Giang	45192.547
7	VN80513	Tinh Bien	VN805	An Giang	35496.033
8	VN80515	Tri Ton	VN805	An Giang	60125.722
9	VN80517	Cho Moi	VN805	An Giang	37015.018
10	VN80519	Chau Thanh	VN805	An Giang	35512.311
11	VN80521	Thoai Son	VN805	An Giang	46889.648

Source: Screenshot generated by the Asian Development Bank using QGIS 3.34.14. 2025.

(v) Click the **Toggle editing** button to stop editing. Ensure you have saved your changes before exiting the editing mode.

2. Counting the Number of Points Within an Enumeration Area

(i) **Reproject** the **AnGiang_points** layer. To do this, follow the process described in the prior subsection for enumeration area calculation. A new reprojected layer will appear in the **Layers panel**. In the example shown in Figure 7.53, the layer is saved as a temporary layer and renamed **Reprojected_points**, while the reprojected polygon layer (**AnGiang_Districts**) is renamed **Reprojected_district** to distinguish between the two.

Figure 7.53: Map View of Reprojected Layers

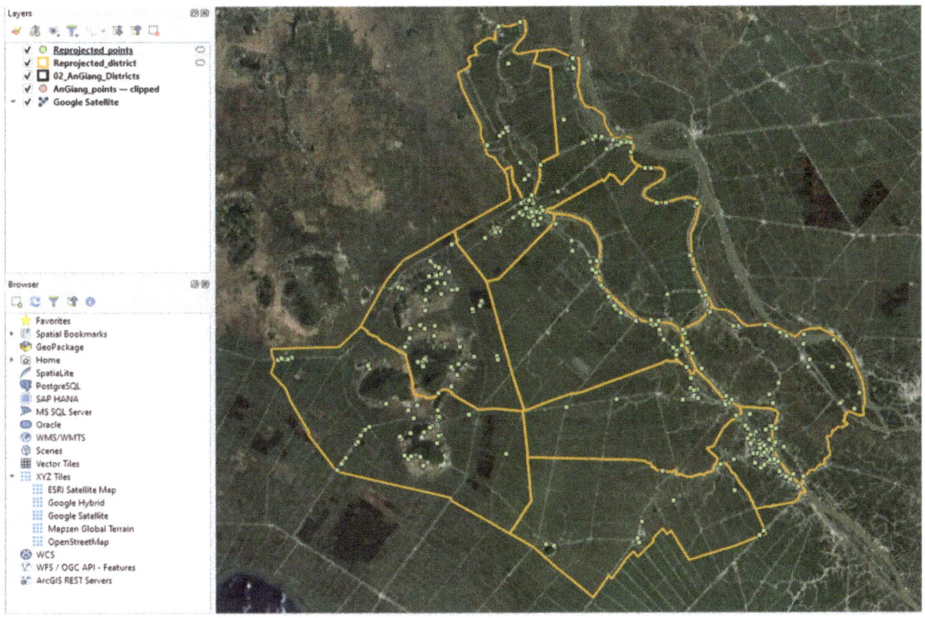

Source: Screenshot generated by the Asian Development Bank using QGIS 3.34.14. 2025.

(ii) To calculate the number of points within each polygon, go to **Vector**, click **Analysis Tools**, then **Count Points in Polygon**. This will open the Count Points in Polygon dialog box (Figure 7.54).

Figure 7.54: Analysis Tool for Counting Points in a Polygon

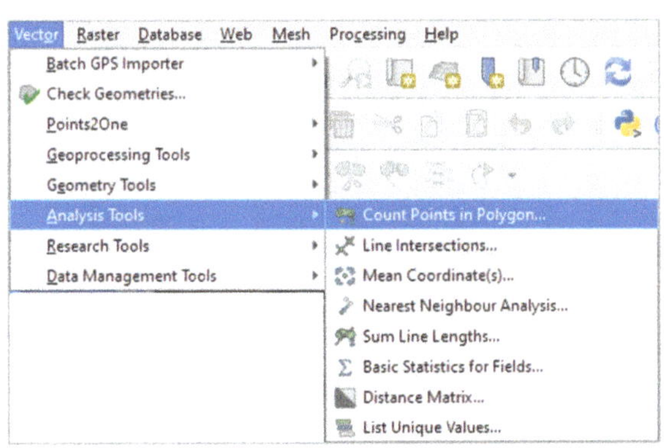

Source: Screenshot generated by the Asian Development Bank using QGIS 3.34.14. 2025.

(iii) Choose the **Reprojected_district layer** as the input polygon layer and the **Reprojected_points layer** as the input points layer. The resulting reprojected layer can be saved as a temporary file or as a file in a designated folder. In the example shown in Figure 7.55, it is saved as a temporary file.

To save it in a folder, navigate to the desired location by clicking [...] beside the **Count** field and selecting **Save to File**. Enter a name for the output layer. Click **Run** to execute the tool.

Figure 7.55: Defining Parameters in the Count-Points-in-Polygon Tool

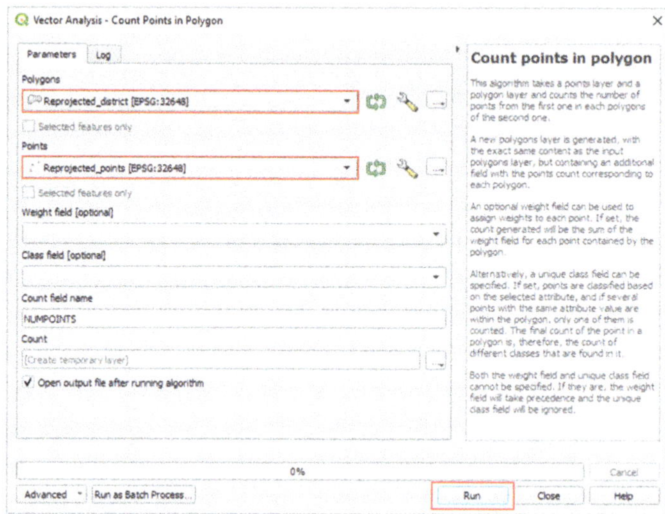

Source: Screenshot generated by the Asian Development Bank using QGIS 3.34.14. 2025.

(iv) A new **Count** layer should appear in the **Layers panel**. Open the **attribute table** to see a new column titled "**NUMPOINTS**," which indicates how many points of interest there are in each polygon (Figure 7.56).

Figure 7.56: Attribute Table with the Number of Points of Interest in Each District

Count — Features Total: 11, Filtered: 11, Selected: 0

	ADM2_PCODE	ADM2_EN	ADM1_PCODE	ADM1_EN	Area	NUMPOINTS
1	VN80501	Long Xuyen	VN805	An Giang	11447.243	137
2	VN80503	Chau Doc	VN805	An Giang	10517.306	136
3	VN80505	An Phu	VN805	An Giang	22570.059	19
4	VN80507	Tan Chau	VN805	An Giang	17616.060	32
5	VN80509	Phu Tan	VN805	An Giang	31215.433	20
6	VN80511	Chau Phu	VN805	An Giang	45192.547	35
7	VN80513	Tinh Bien	VN805	An Giang	35496.033	60
8	VN80515	Tri Ton	VN805	An Giang	60125.722	55
9	VN80517	Cho Moi	VN805	An Giang	37015.018	37
10	VN80519	Chau Thanh	VN805	An Giang	35512.311	20
11	VN80521	Thoai Son	VN805	An Giang	46889.648	16

Show All Features

Source: Screenshot generated by the Asian Development Bank using QGIS 3.34.14. 2025.

(v) The results can now be visualized and analyzed using QGIS tools discussed in Module 5, such as symbology, labeling, and statistics.

Module 8: Research and Sampling Tools in QGIS

This module focuses on the research and sampling tools in QGIS that are essential for agricultural statistics. These tools are fundamental for improving the accuracy and efficiency of data collection and analysis in agricultural surveys, particularly when implementing area sampling methods. Area sampling involves dividing land into distinct geographical units and selecting a sample of these areas for study.

The objective of this module is to guide the implementation of a multistage sampling design using QGIS. The process starts with the selection of primary area units, followed by the selection of secondary area units, such as households or points of interest, within each selected primary sampling unit. The final step involves exporting the list of selected units to Microsoft Excel for further processing and analysis. This module demonstrates how various sampling methods and tools can be utilized in QGIS to enhance the efficiency and accuracy of agricultural surveys.

Moreover, integrating open-source satellite data into the sampling process enhances the timeliness and objectivity of data collection. Open-source satellite imagery provides up-to-date and high-resolution data, which are invaluable for monitoring agricultural areas and making informed decisions. Utilizing these resources enables the collection of agricultural statistics with greater efficiency and precision, ensuring that the data reflect current land conditions and emerging trends.

In this module, the following topics will be covered:

8.1 Overview of Area Sampling Concepts
8.2 Sampling Methods and Tools in QGIS
- 8.2.1 Random Selection of Features
- 8.2.2 Random Points Within Boundaries
- 8.2.3 Creating and Sampling Grids and Lines

8.3 Multistage Sampling in QGIS
- 8.3.1 Reclassifying Raster Data (Grouping Land Cover Types)
- 8.3.2 Extracting the Extent of the Study Area
- 8.3.3 Raster Analysis for Stratified Sampling by Land Use

By the end of this module, you will be able to implement a multistage sampling design in QGIS using open-source land cover data from the Japan Aerospace Exploration Agency (JAXA) for stratification. The sampling design follows three main stages in our example. First, the An Giang Province is divided into grid cells, each 1-square-kilometer (km^2) in size. From these, a few grids are selected. Within each selected grid, four smaller sub-grids of 200 square meters (m^2) are chosen. Finally, two 50 m^2 segments are selected from each sub-grid. These segments serve as the final sampling units for ground validation and analysis.

This approach supports stratified sampling and is especially useful for area frame designs, which help ensure that agricultural data collection is both representative and efficient. Upon completing this module, you will also be able to build a map based on this area frame design to support the ground truthing of remote sensing data. This is essential for validating agricultural statistics, such as the example shown in Figure 8.1, which shows a sample grid map used for ground validation. The resulting file of comma-separated values (CSV) will include key attributes such as segment identifiers and centroid coordinates, making it suitable for field validation and further spatial analysis.

Figure 8.1: Multistage Sampling Grid for Agricultural Ground Truthing

Source: Screenshot generated by the Asian Development Bank using QGIS 3.34.14. 2025.

8.1 Overview of Area Sampling Concepts

Area sampling is a multistage sampling technique that uses maps or aerial boundaries as the sampling frame. It is often used when a list of individual elements in the population is not available or is out of date, or when the sampling cost is prohibitive. Area sampling can be applied to various populations, such as dwellings, households, buildings, crops, or other units that can be associated with a geographic location.

An area frame is a division of the target area into smaller, mutually exclusive units, called "segments" or "subareas." The boundaries of the segments should be clearly defined and easily recognized, such as roads, rivers, or landmarks. The segments should also be numbered and mapped, and their size and population characteristics should be recorded. The area frame can be constructed using a geographic information system (GIS), aerial photos, satellite images, or current maps.

Area sampling involves selecting a sample of segments from the area frame using a random or systematic approach. The objectives and resources of the study will determine the sample size and selection probability. Within each selected segment, either a complete enumeration or a further subsampling of units can be carried out. The subsampling can be done using simple random sampling, stratified sampling, cluster sampling, or other methods. The subsampling can also be repeated at multiple stages until the final units are reached.

Compared with other sampling techniques, area sampling has several advantages. The method can be applied to the whole population without any duplications or omissions. It can also provide objective estimates for a number of variables at once, including income level, population density, and land use. When targeted at segments with comparable qualities and proximity to one another, area sampling can lower sampling costs and errors. Moreover, the method can adapt to population changes over time by routinely updating the area frame.

However, there are also some limitations and challenges to area sampling. To build an area frame, accurate and up-to-date maps or aerial photos are needed. This is one of the first requirements in the construction of an area frame. In some places, there may also be challenges in defining and identifying segment boundaries. An area frame that contains overlapping segments, or one that does not completely cover the target area, may create issues in selection. For example, an area frame that contains overlapping segments may cause an area to have a higher probability of selection, and gaps in the area frame may lead to inadequate coverage. In some cases, large sample sizes may be necessary to provide the appropriate level of accuracy and reliability. Lastly, because population density and segment size vary, area sampling could introduce bias and variability.

8.2 Sampling Methods and Tools in QGIS

Sampling is a common technique used to gather information about a population or area by examining a representative subset. Various sampling methods can be employed, such as stratified, random, and systematic sampling. This section will concentrate on using QGIS tools to randomly select points and polygon features.

This module will rely on the district (ADM2) boundaries within An Giang Province, Viet Nam, because the enumeration-area boundaries are not available. In addition, a dataset named "AnGiang_points," extracted from OpenStreetMap, will be used to demonstrate key procedures in QGIS. The dataset contains points of interest, or point features, representing various amenities and public services, including places of worship, school, hospitals, markets, and other establishments across the province. It will be used to demonstrate how spatial data can be integrated into sampling workflows in QGIS. Download all the required files from the shared folder[a] to proceed with the exercise.

[a] For access to shapefiles and other open-access datasets referenced in the handbook, go to the GitHub repository: https://github.com/ADB-Data-Division/QGIS-Handbook-2025

8.2.1 Random Selection of Features

1. Selecting Points

Point sampling aims to randomly select a subset of points from a larger population. In the context of agricultural surveys, this often starts with a dataset containing points that represent all agricultural households within the target population. From this dataset, a sample of points is selected for further study.

(i) Add the vector layer **AnGiang_points** by dragging and dropping it from the **Browser panel** into the **map area**. Alternatively, you can add the layer by clicking the **Layer tab** in the **Menu Toolbar**, then selecting **Add Layer** and **Add Vector Layer**. Do the same for the vector layer **AnGiang_District** (Figure 8.2).

Figure 8.2: Points of Interest in An Giang Province, Viet Nam

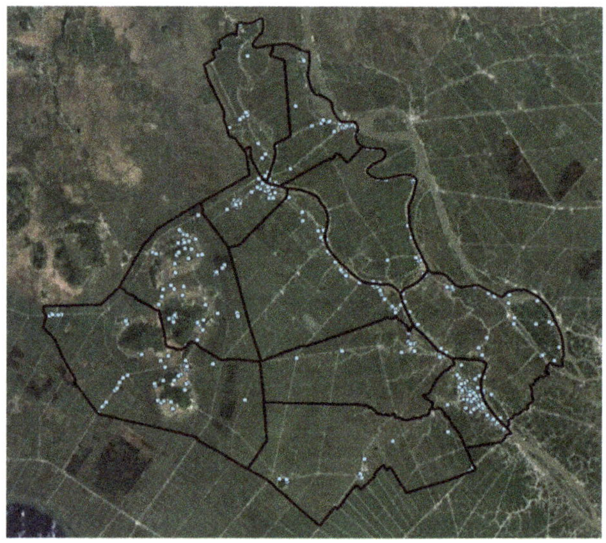

Source: Screenshot generated by the Asian Development Bank using QGIS 3.34.14. 2025.

(ii) To randomly select points in the **AnGiang_points** layer, go to **Vector | Research Tools | Random Selection** (Figure 8.3). Alternatively, go to **Processing** in the **main menu** and select **Toolbox**, then search for the **Random Selection** tool.

Figure 8.3: Opening the Random Selection Tool

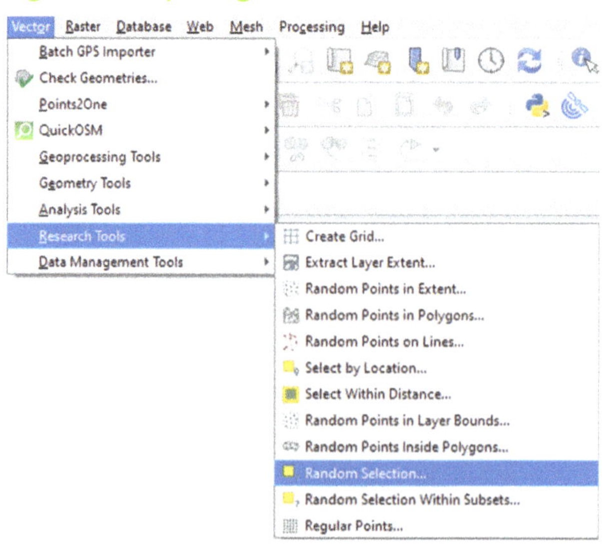

Source: Screenshot generated by the Asian Development Bank using QGIS 3.34.14. 2025.

In the dialog box, set the **input layer** to the **points layer**; in this example, that is the **AnGiang_points** layer. Specify the number of selected features or the percentage of selected features to randomly select.

The **percentage of selected features method** provides a variable subset size, depending on the size of the input layer, whereas the **number of selected features method** gives a constant subset size. In agricultural surveys, the number of selected features method is useful for obtaining a fixed sample size, such as selecting exactly 100 households from each district for consistent comparison. Conversely, the percentage of selected features method is advantageous when the sample size should be proportional to the total number of features, such as selecting 10% of fields in various provinces to ensure scalable and proportional representation. These methods provide flexibility and precision, allowing researchers to tailor their sampling strategies to the specific needs of their studies.

In the example shown in Figure 8.4, the **number of selected features method** was used, specifying 100 samples.

Figure 8.4: Defining the Parameters for Randomly Selecting Points

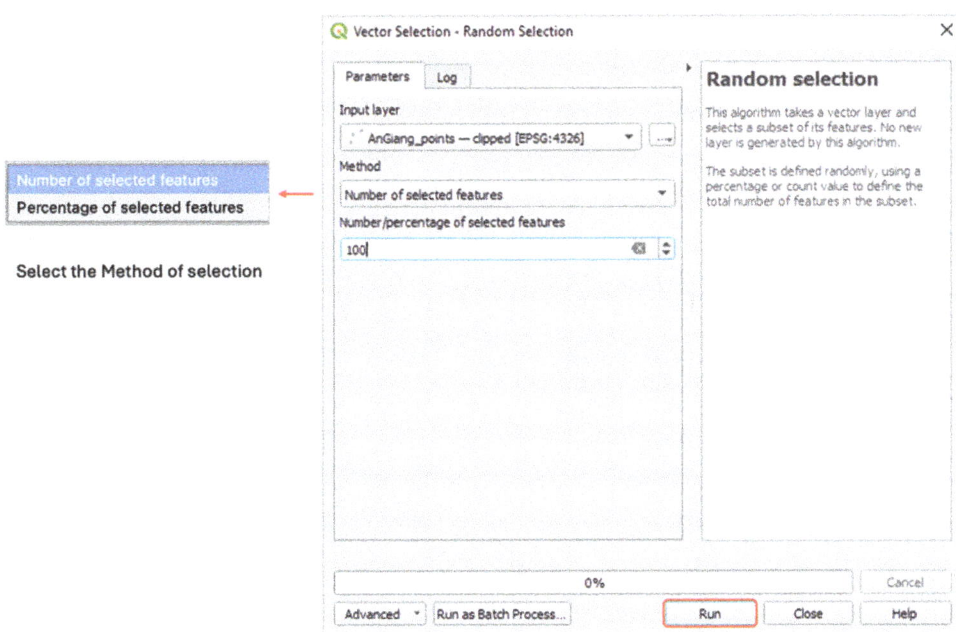

Source: Screenshot generated by the Asian Development Bank using QGIS 3.34.14. 2025.

Click **Run** to execute.

(iii) The randomly selected points will be highlighted in the **Map Canvas** (Figure 8.5).

Figure 8.5. Showing Selected Features in the Map Canvas

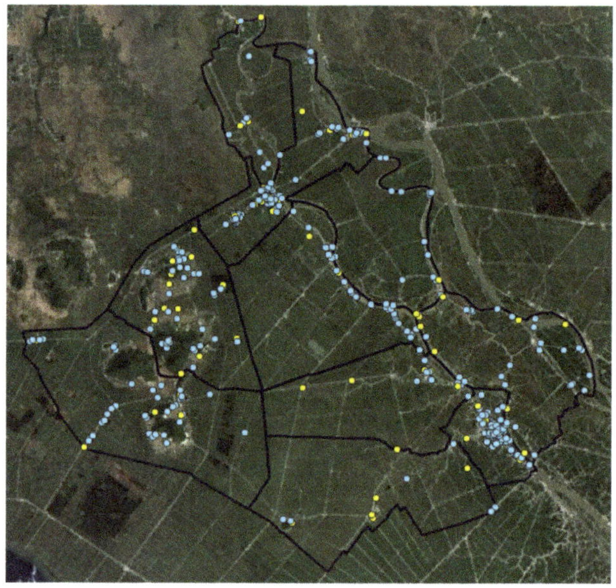

Source: Screenshot generated by the Asian Development Bank using QGIS 3.34.14. 2025.

(iv) Open the **attribute table** to check the attributes of the random points. Right-click on the layer, then on **Open Attribute Table**. Alternatively, click the icon in the menu. At the top of the attribute table, you will see the number of selected features (Figure 8.6).

Figure 8.6: Showing Selected Features in the Attribute Table

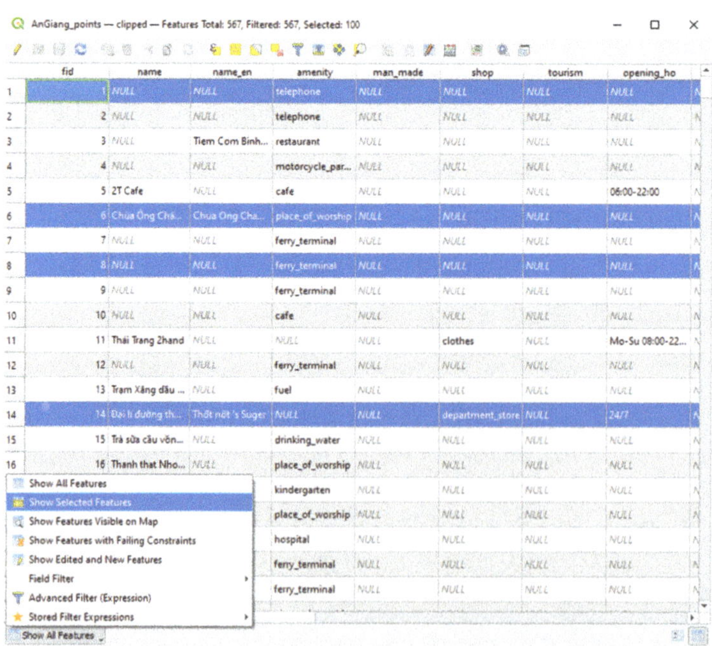

Source: Screenshot generated by the Asian Development Bank using QGIS 3.34.14. 2025.

In the **Show Features** menu at the bottom left of the **attribute table**, select **Show Selected Features** to view the randomly selected points (Figure 8.7).

Figure 8.7: Selected Features in the Attribute Table

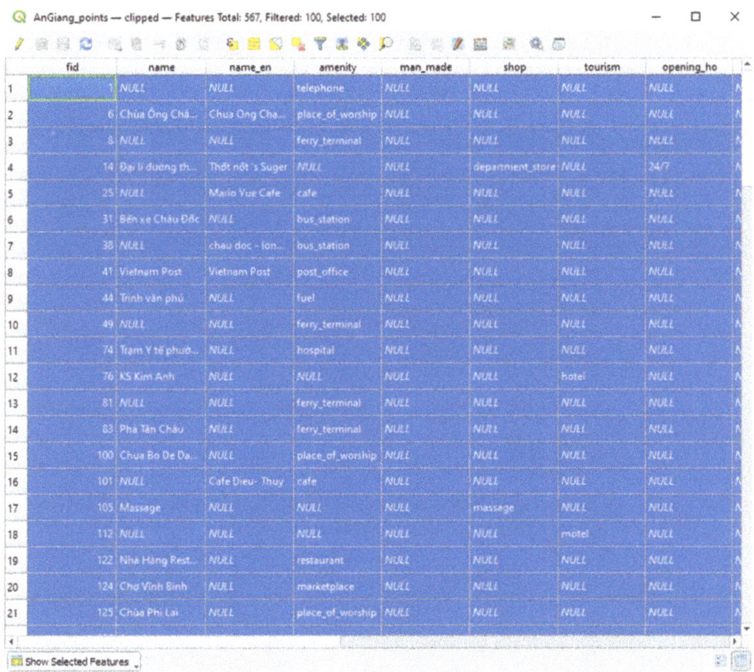

Source: Screenshot generated by the Asian Development Bank using QGIS 3.34.14. 2025.

(v) To export the selected points, right-click on the **AnGiang_points layer** in the **Layers panel**, choose **Export**, and then select **Save Selected Features As** (Figure 8.8).

Figure 8.8: Exporting Selected Points into a New File

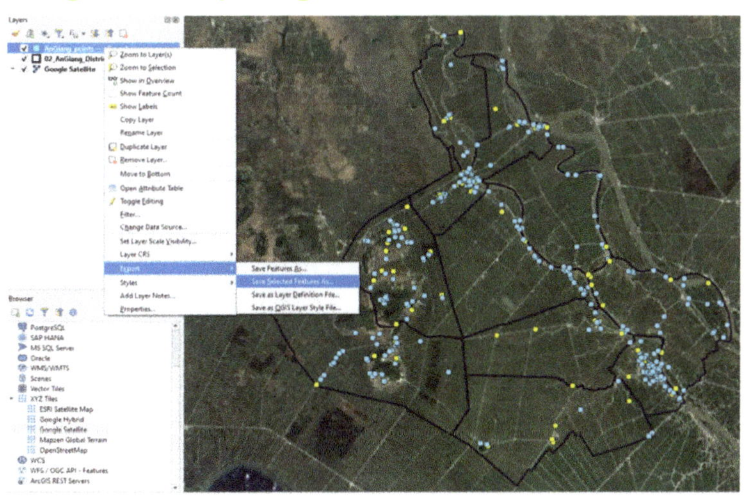

Source: Screenshot generated by the Asian Development Bank using QGIS 3.34.14. 2025.

(vi) Select the desired format and location in which to save the selected points (Figure 8.9). Refer to Module 2, section 2.3.4, for instructions on saving the vector layer.

Figure 8.9: Saving Selected Points as a New Vector Layer

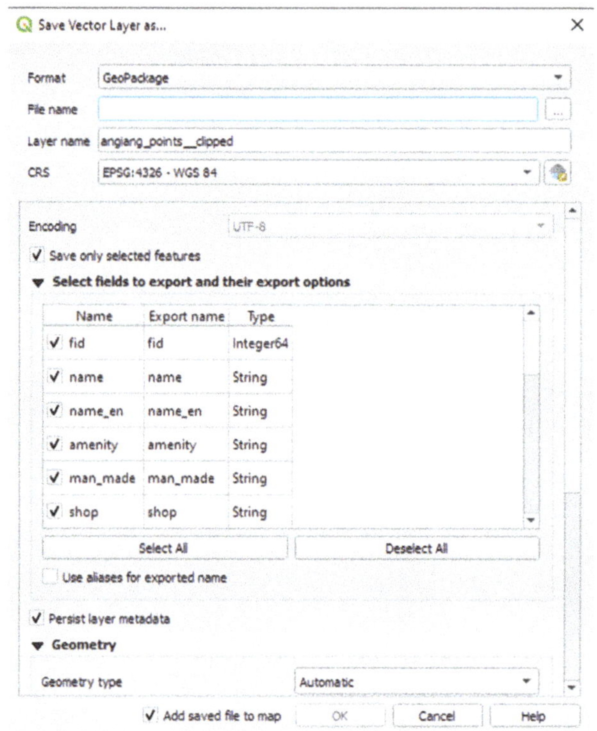

Source: Screenshot generated by the Asian Development Bank using QGIS 3.34.14. 2025.

3. Selecting Polygons

As in the case of points, described in the prior section, a subset of polygons or segments of a set of polygons can be selected randomly. This section will demonstrate how to use the same tool to randomly select a district within An Giang Province.

(i) Ensure that the **AnGiang_District** layer is activated by selecting it in the **Layers panel**. Follow the same procedure as before to open the tool: Go to **Vector | Research Tools | Random Selection**. Alternatively, go to **Processing** in the **main menu**, select **Toolbox**, and then search for the **Random Selection** tool (Figure 8.10).

Figure 8.10: Launching the Random Selection Tool from the Processing Toolbox

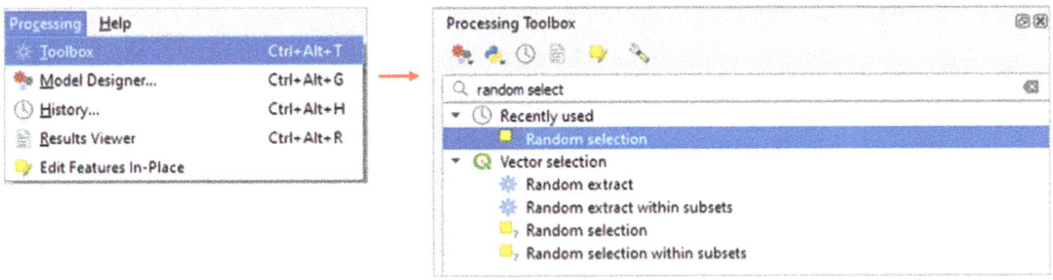

Source: Screenshot generated by the Asian Development Bank using QGIS 3.34.14. 2025.

(ii) To configure the tool, set the input layer to **AnGiang_District** and define the method of selection. For this exercise, choose **Percentage of selected features** and select **30** as the percentage (Figure 8.11).

Figure 8.11: Defining the Parameters for Randomly Selecting Polygons

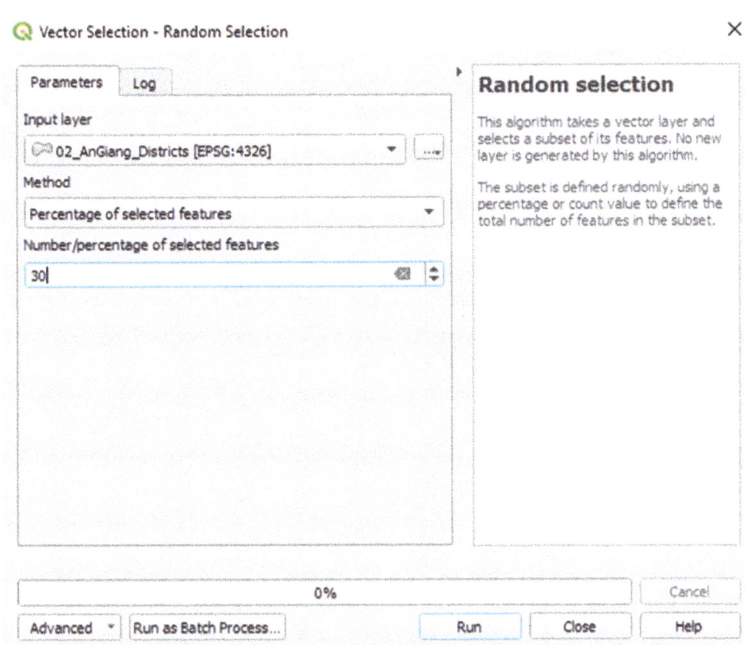

Source: Screenshot generated by the Asian Development Bank using QGIS 3.34.14. 2025.

Click **Run** to execute the random selection process.

(iii) Check the randomly selected polygons (Figure 8.12).

Figure 8.12: Randomly Selected Polygons

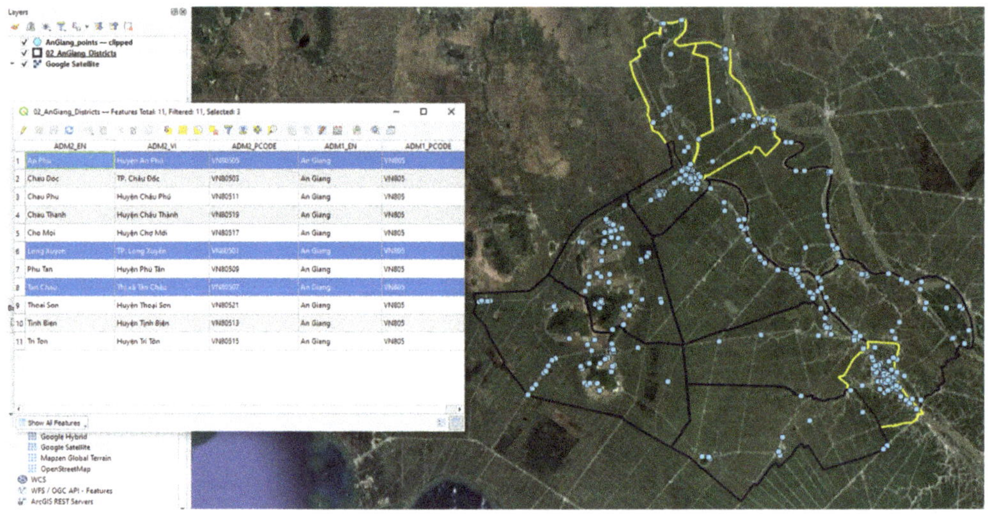

Source: Screenshot generated by the Asian Development Bank using QGIS 3.34.14. 2025.

Since no additional selection criteria have been defined, the districts have been selected at random. The **Random Selection** tool is used to extract a subset of features from the entire layer.

(iv) To **select a subset based on a specific attribute**, another tool can be used: **Random Selection Within Subsets**. This tool selects a subset of features from each category defined by an attribute field. For example, it can be used to randomly select districts from different provinces, with each province defined as a category.

To do this, go to **Vector | Research Tools | Random Selection Within Subsets** (Figure 8.13). Alternatively, go to **Processing** in the **main menu**, select **Toolbox**, and then search for the **Random Selection Within Subsets** tool.

Figure 8.13: Opening the Random Selection Within Subsets Tool

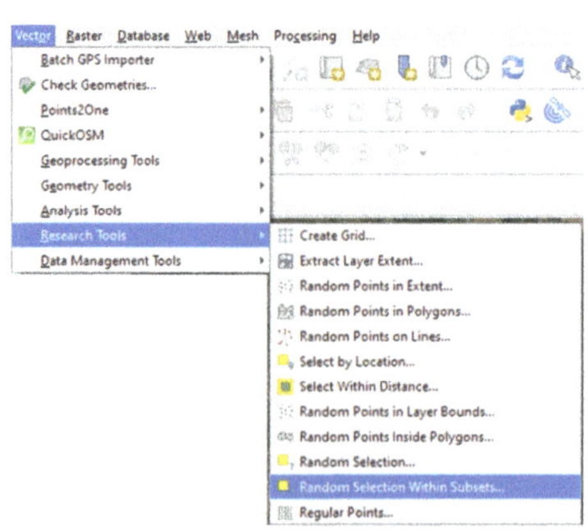

Source: Screenshot generated by the Asian Development Bank using QGIS 3.34.14. 2025.

(v) Select **AnGiang_District** as the **Input layer**. Ideally, the goal is to randomly select districts within each province. However, in this example dataset, only one province is represented, and multiple districts fall within it. To simulate the intended functionality—assuming a dataset with multiple provinces, each containing several districts—you must define the selection category using a province-level identifier.

In the attribute table, the fields **ADM1_EN** (province name) and **ADM1_PCODE** (province identification [ID]) can be used for this purpose, since both will identify a particular province. Choose **Percentage of selected features** as the method of selection. Then select **20** in the percentage field to randomly select 20% of the districts within each province. Click **Run** to execute the process (Figure 8.14).

Note: This procedure assumes the presence of multiple provinces in the dataset. When working with a single-province dataset, as in this example, the selection will be applied across all districts within that province, rather than stratified by province.

Figure 8.14: Defining the Parameters for Randomly Selecting a Subset of Polygons by Attribute

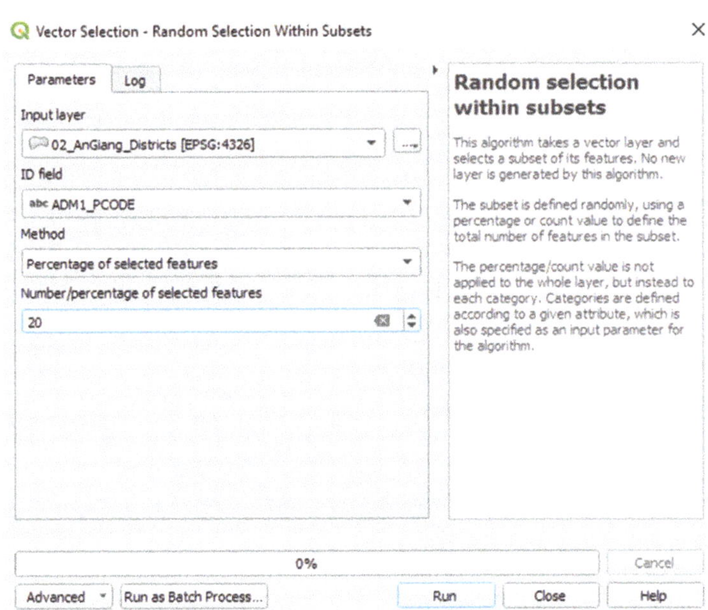

Source: Screenshot generated by the Asian Development Bank using QGIS 3.34.14. 2025.

(vi) Check the subset of polygons selected (Figure 8.15).

Figure 8.15: Subset of Polygons Randomly Selected by Attribute

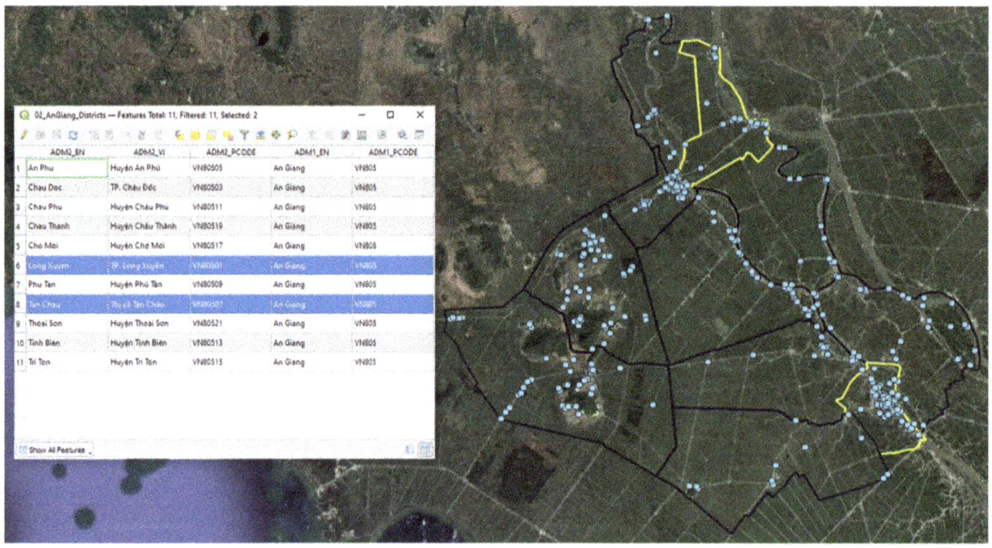

Source: Screenshot generated by the Asian Development Bank using QGIS 3.34.14. 2025.

The **Random Selection Within Subsets** tool can be particularly useful for agricultural surveys. For instance, when working with a dataset of farms categorized by district, this tool can be used to randomly select a sample of farms from each district. This ensures that the survey covers a representative sample from all districts, rather than concentrating on just a few of them.

8.2.2 Random Points Within Boundaries

In some cases, spatial analysis requires the creation of random points within defined boundaries. When there are no existing point locations, QGIS offers tools to generate random points based on specific criteria set by the user. These tools provide flexibility in restricting the locations of random points, using polygons, lines, or extents to delineate boundaries.

QGIS provides several options for creating random points, differing mainly in how you control the point count. Users can either specify a fixed number of points or set a density value to manage the distribution of points within the designated area. This feature is particularly beneficial for sampling, modeling, and conducting diverse spatial analyses in agricultural sampling and research.

1. Random Points in Polygons

This tool creates a new point layer with a given number of random points distributed within each polygon of the input layer.

(i) Go to **Vector | Research Tools | Random Points in Polygons** (Figure 8.16). Alternatively, go to **Processing** in the **main menu**, select **Toolbox**, and then search for the **Random Points in Polygons** tool.

Figure 8.16: Opening the Random Points in Polygons Tool

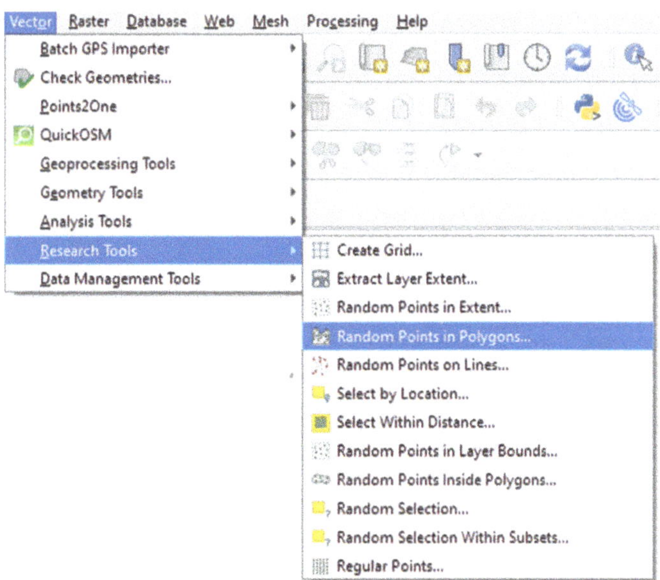

Source: Screenshot generated by the Asian Development Bank using QGIS 3.34.14. 2025.

Ensure that the **AnGiang_District layer** uses a projected coordinate system—for instance, Universal Transverse Mercator (UTM)—so that point spacing is accurately measured. For guidance on reprojecting layers to EPSG:32648 / UTM 48N, refer to Module 4, section 4.4, or to Module 7, section 7.4, which cover the necessary procedures for reprojection.

(ii) After you select **Random Points in Polygons**, a dialog box will appear. Under **Input polygon layer**, select **Reprojected [EPSG:32648]**. Next, under **Number of points for each feature**, enter the number of random points to generate per polygon—this will apply uniformly across all polygons. There is also the option of setting a minimum distance between points; since the layer uses a projected coordinate reference system (CRS), this ensures that the spacing is measured in meters or feet. For this example, set the total number of points to 5 and the minimum distance between points to 50 meters. Specify a file path for saving the output layer, or leave it as temporary if saving is not required. Click **Run** to start the algorithm (Figure 8.17).

Figure 8.17: Defining the Parameters for the Random Points in Polygons Tool

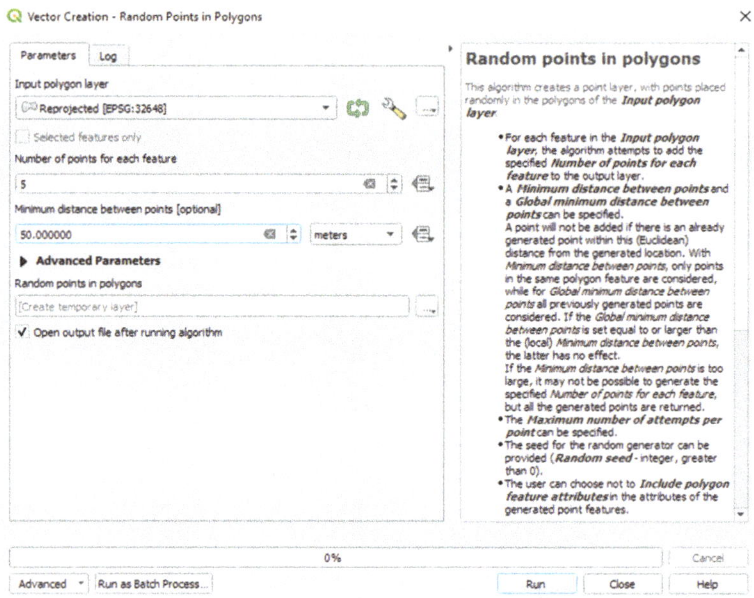

Source: Screenshot generated by the Asian Development Bank using QGIS 3.34.14. 2025.

A prompt will appear once the algorithm stops running. Close the window after the points are generated.

(iii) Check the new layer created with the random points generated (Figure 8.18).

Figure 8.18: Random Points in a Polygon

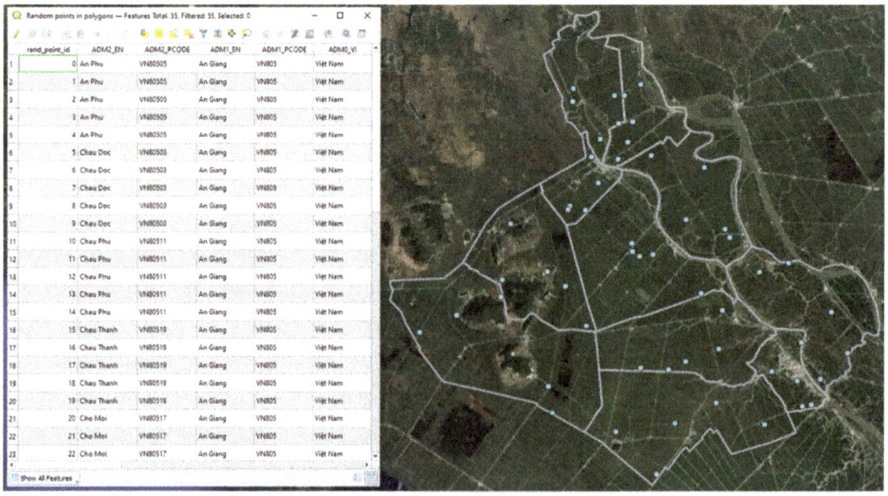

Source: Screenshot generated by the Asian Development Bank using QGIS 3.34.14. 2025.

Aside from the **Random Points in Polygons tool**, QGIS offers several other tools for generating random samples across spatial datasets, each tailored to different analytical needs and spatial contexts. Table 8.1 summarizes the key characteristics of other commonly used sampling tools in QGIS: **Random Points Inside Polygons, Random Points in Extents, Random Points in Layer Bounds, and Random Selection Within Subset**. It outlines each tool's main function, sampling strategy options, agricultural research use cases, and notable features, to identify the most suitable tool based on sampling goals, data structure, and analytical needs. While these tools differ in terms of function, sampling strategy, and application, they share a common procedural foundation.

The steps demonstrated in the preceding section for using the **Random Points in Polygons** tool can generally be applied when using the other tools.

Table 8.1: Overview of QGIS Sampling Tools

Sampling Tool	Main Function	Sampling Strategy Options	Use Case in Agricultural Research	Notable Features
Random Points in Polygons	Generates random points within each polygon	Fixed count only per polygon	Uniform sampling across fields or plots	• The setup is simple. • There are no density- or attribute-based variations.
Random Points Inside Polygons	Generates random points with flexible strategies	Count or density varying by polygon, depending on attributes	Stratified or intensity-based sampling (e.g., more samples in larger fields, or more variable fields)	• The user can set the number of points or density. • The number of points can vary by polygon, depending on the attribute-based settings. • This tool is ideal for area-proportional sampling, ensuring that larger polygons have more sample points; this helps achieve uniform coverage and representative sampling across polygons of varying sizes.
Random Points in Extents	Generates points within a user-defined rectangular extent	Fixed total count	Broad area sampling (e.g., regional surveys without defined field boundaries)	• This tool may include irrelevant areas unless masked. • It is good for exploratory and broad sampling.
Random Points in Layer Bounds	Generates random points within the spatial extent (outer limits) of all features in a layer	Fixed total count	Sampling across the spatial extent of a dataset	• This tool uses the overall spatial extent that encloses all features (not actual feature shapes) in the layer; this could be the administrative boundary or the outer limits of the dataset. • It generates points across the entire spatial extent of the layer, which can be useful for broad coverage, but may result in points falling outside the actual feature geometries, such as in gaps between fields or plots.

continued on next page

Table 8.1 continued

Sampling Tool	Main Function	Sampling Strategy Options	Use Case in Agricultural Research	Notable Features
Random Selection Within Subset	Selects a random subset of features from an already filtered or selected group within a layer	Fixed number of features Percentage of the subset	Useful for selecting sample locations from a filtered group (e.g., only communes in a certain district, or rice fields above a certain size)	• This tool works on selected features only, so prefiltering is essential. • It is great for stratified or conditional sampling. • It does not generate new geometries, it only selects existing ones.

Note: Tool features may vary in future releases.

Source: Asian Development Bank based on QGIS official documentation and observed functionalities (QGIS version 3.34.14).

8.2.3 Creating and Sampling Grids and Lines

1. Creating Grids

Grids are a basic spatial tool used in a geographic information system (GIS) to divide an area into equally sized cells or units. In QGIS, **Research Tools** allow users to create customizable grids that support systematic sampling, data collection, and spatial analysis. This approach is particularly useful in agricultural research, where spatial variability in soil, crop health, crop yields, and environmental conditions must be accurately assessed.

Grids help design well-structured sampling schemes by ensuring even coverage across a study area. Each grid can represent a sampling unit for collecting data on such characteristics as land use, soil moisture, nutrient levels, or crop yields.

In remote sensing, grids make it easier to match field data with satellite or drone imagery. By overlaying grids on raster datasets, researchers can extract pixel-level information, use Zonal Statistics, and correlate ground observations with spectral indices such as the Normalized Difference Vegetation Index (NDVI) or the Enhanced Vegetation Index (EVI).

To create a grid, take the following steps:

(i) Go to **Vector | Research Tools | Create Grid** (Figure 8.19). Alternatively, go to **Processing** in the **main menu**, select **Toolbox**, and then search for the **Create Grid tool**.

Figure 8.19: Opening the Create Grid Tool from the Vector Toolbar

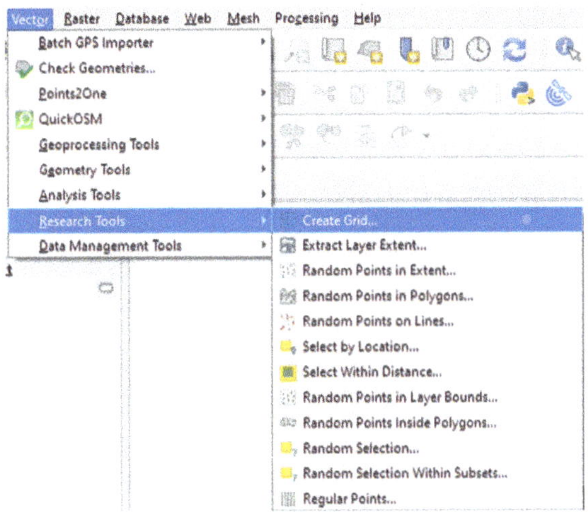

Source: Screenshot generated by the Asian Development Bank using QGIS 3.34.14. 2025.

(ii) Select the **Grid type**, which can be a rectangle (polygon), line, or point, depending on the sampling requirements. Then, define the **Grid extent** by selecting an existing layer or manually entering the bounding coordinates. Specify the **Horizontal and Vertical spacing** to set the dimensions of each grid cell based on the desired sampling resolution. For this example, choose "Rectangle (polygon)" as the grid type and set both the horizontal and vertical spacing to 100 meters. Ensure that the CRS matches your project's CRS, such as UTM Zone 48N for Viet Nam (Figure 8.20).

Note that larger grid extents may result in longer processing times, as the number of cells increases with the area. Grid generation time is also influenced by several factors, including the grid type, overlay size, and cell dimensions. Smaller cell sizes, larger overlays, and more complex grid geometries (i.e., hexagons) typically require more processing time. The tool must generate and store more features in the output layer, which can impact performance.

Figure 8.20: Defining the Parameters for the Create Grid Tool

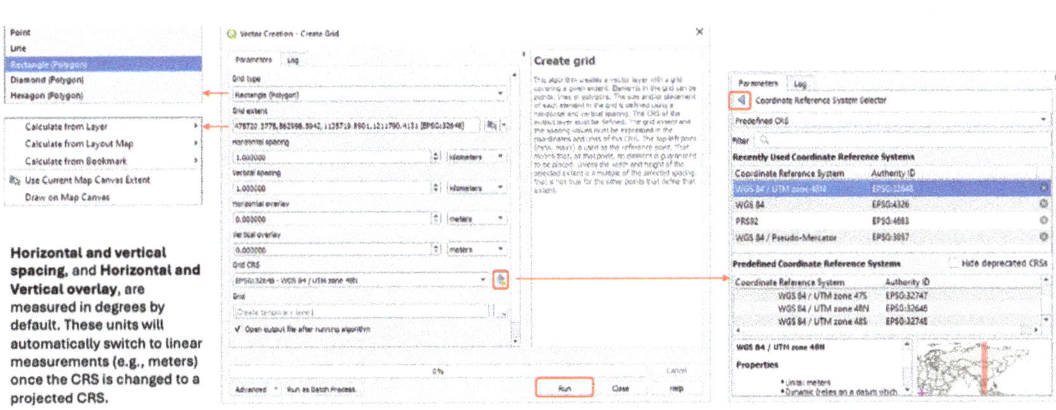

Source: Screenshot generated by the Asian Development Bank using QGIS 3.34.14. 2025.

Once all parameters are set, click **Run** to generate the grid, which will be added as a new layer in the project. A prompt will appear once the algorithm stops running. Close the window after the grids are generated.

(iii) Check the newly created grid layer. To improve the visibility of the basemap and other ground features, adjust the layer properties as needed. For instance, you could set the outline style to "Simple Line" with a red color, with the result shown in Figure 8.21.

Figure 8.21: Grid Overlay on Map of An Giang Province, Viet Nam

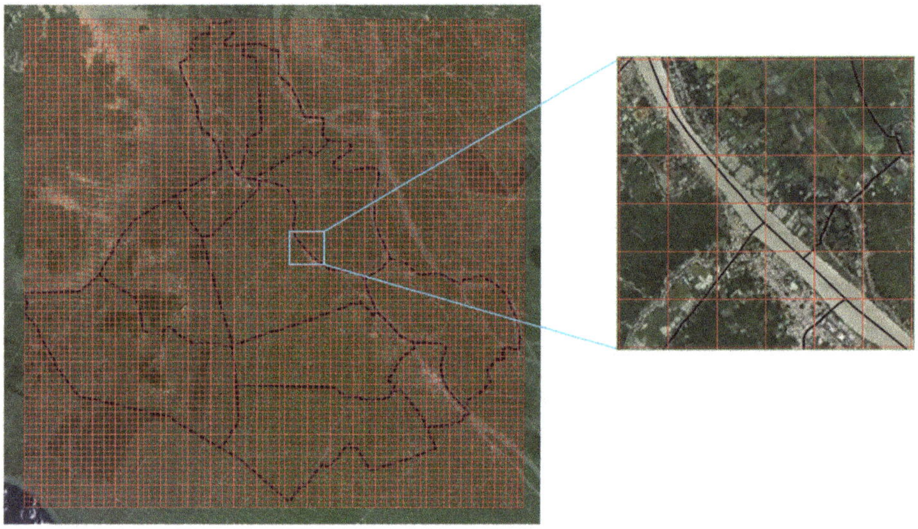

Source: Screenshot generated by the Asian Development Bank using QGIS 3.34.14. 2025.

(iv) Open the **attribute table** to check the attributes of the grids. In the example, the grids have the following attributes:

- **id:** a unique identifier for each grid cell. This is usually an auto-incremented number or a unique string assigned to distinguish each cell in the grid.
- **left:** the x-coordinate (longitude or easting) of the left boundary of the grid cell. This marks the western edge of the rectangle.
- **top:** the y-coordinate (latitude or northing) of the top boundary of the grid cell. This marks the northern edge of the rectangle.
- **right:** the x-coordinate of the right boundary of the grid cell. This marks the eastern edge of the rectangle.
- **bottom:** the y-coordinate of the bottom boundary of the grid cell. This marks the southern edge of the rectangle.

In QGIS, grid cells are numbered starting from the top-left corner of the grid extent. The numbering proceeds vertically down each column, then shifts to the next column to the right. This systematic arrangement is important for sampling designs, as it ensures a predictable and spatially coherent order of observations. Understanding this pattern allows for the consistent selection of sample units, whether for systematic sampling, stratification, or spatial analysis.

To check the grid numbering in your project, zoom into the top-left corner of the grid extent and use the **Select Features** tool to highlight the grid cells in the first column as demonstrated in Figure 8.22. Then, open the **attribute table** and observe the highlighted rows. If the table is sorted by grid ID, the selected features should appear in order, confirming the vertical numbering pattern.

Figure 8.22: Grid Numbering Sequence

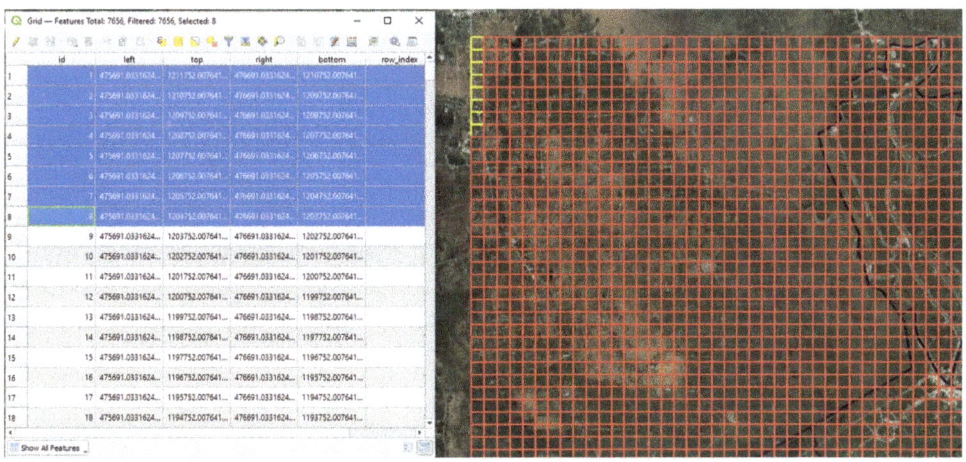

Source: Screenshot generated by the Asian Development Bank using QGIS 3.34.14. 2025.

2. Using the Join Attributes by Location Tool

Notice that the newly created **grid layer** does not contain the location attributes from the **input layer**. For this exercise, district names and IDs need to be added to the grid locations. This can be done by merging the attributes of the **grid layer** with those of the **reprojected layer**, which are both in the same CRS.

(i) Go to **Vector |Data Management Tools | Join Attributes by Location** (Figure 8.23).

Figure 8.23: Opening the Join Attributes by Location Tool

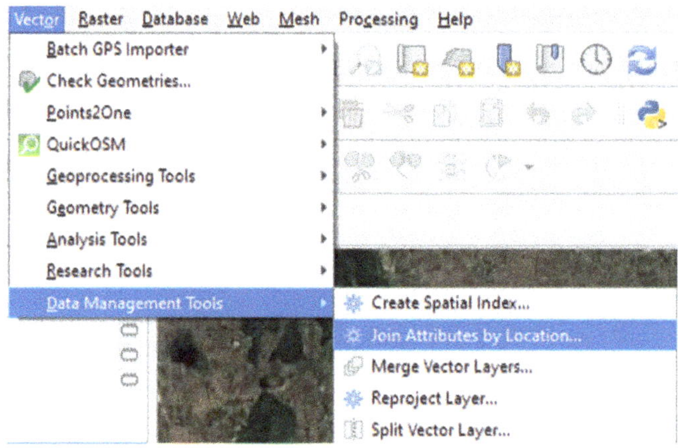

Source: Screenshot generated by the Asian Development Bank using QGIS 3.34.14. 2025.

(ii) Select the **layer** that will receive attributes under '**Join to features in**', which in this case is the **grid layer**. Under '**By comparing to**', choose the layer providing attributes, which in this case is the **reprojected layer**—this is where the additional attributes will be sourced. Define the **geometric predicate** to specify how the layers should relate spatially. For this example, as shown in Figure 8.24, select a predicate such as **intersect**, using the district boundaries. Under **Join type**, select **Take attributes of the feature with the largest overlap only (one-to-one)** to ensure that each grid cell is assigned a distinct district. In **Fields to add**, choose **attributes to join** (e.g., district name, district ID).

Figure 8.24: Defining the Parameters for the Join Attributes by Location Tool

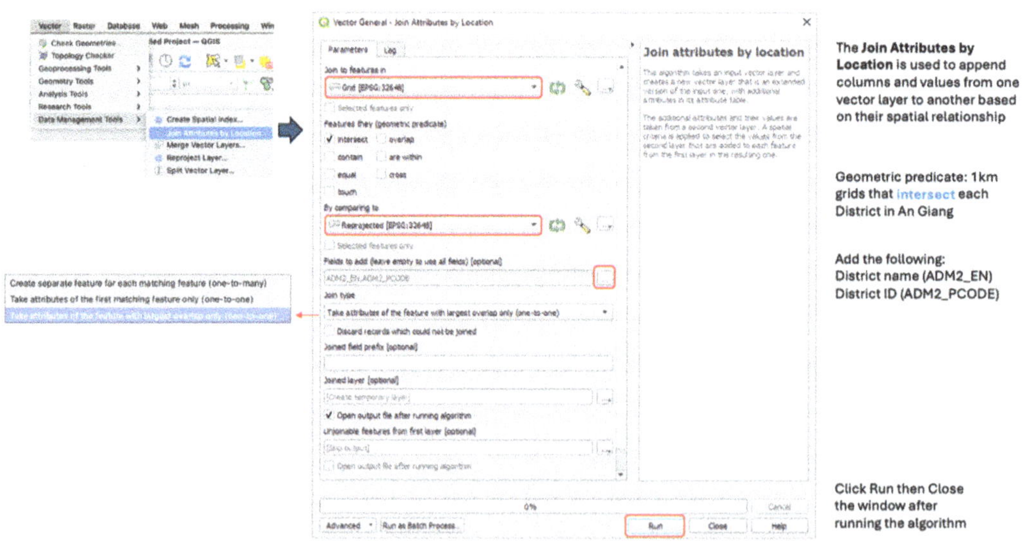

Source: Screenshot generated by the Asian Development Bank using QGIS 3.34.14. 2025.

Click **Run** to start the process.

The message “No spatial index exists for joined layer, performance will be severely degraded” means that the joined layer lacks a spatial index, a structure that speeds up spatial operations. Without it, processing may be slower, especially with large datasets. However, this does not affect the accuracy of the output.

Close the window after running the tool.

2. Vector Geoprocessing Tools

A new **joined layer** is created. Open the **attribute table**, which now includes the district name and ID, for each grid. The grid now covers the rectangular extent of all of An Giang Province. The next step is to clip the grid to the boundaries of the province.

(i) Go to **Vector | Geoprocessing Tools | Clip** (Figure 8.25).

Figure 8.25: Vector Clip Tool

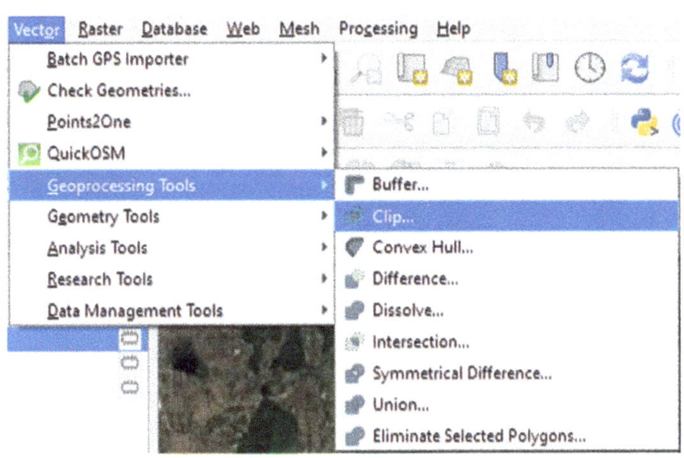

Source: Screenshot generated by the Asian Development Bank using QGIS 3.34.14. 2025.

(ii) Select the **Input layer** to be clipped, which in this case is the **joined layer**. Then select the **Overlay layer**, which is the polygon layer to be used for clipping. In this example, it is the province boundary, represented by the **reprojected layer** that shares the same CRS. Specify the output location or choose to keep the result as a temporary layer. Click **Run** to execute the tool (Figure 8.26).

Figure 8.26: Defining the Clip Tool Parameters

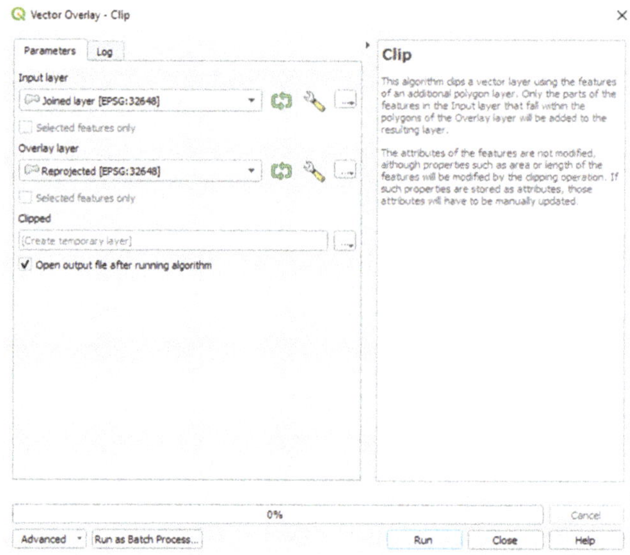

Source: Screenshot generated by the Asian Development Bank using QGIS 3.34.14. 2025.

The output will be a new layer containing only the portions of the input layer that fall within the overlay polygon.

(iii) Place the **clipped layer** above the **joined** and **grid layers** in the **Layers panel** to make the grids along the boundaries of the province visible.

Note: Optionally, the district shapefile may be placed above the clipped layer to display district boundaries. To avoid obscuring the clipped layer, the district layer's styling should be set to a simple line, as illustrated in Figure 8.27.

Figure 8.27: Clipped Grids over the Extent of An Giang Province, Viet Nam

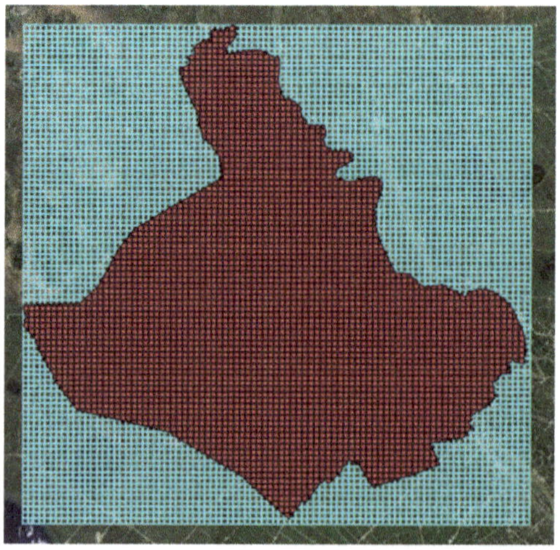

Source: Screenshot generated by the Asian Development Bank using QGIS 3.34.14. 2025.

(iv) Open the **attribute table**. The grids can now be selected based on attributes such as district name and ID or commune name and ID (Figure 8.28).

Figure 8.28: Grid Selection by Attributes

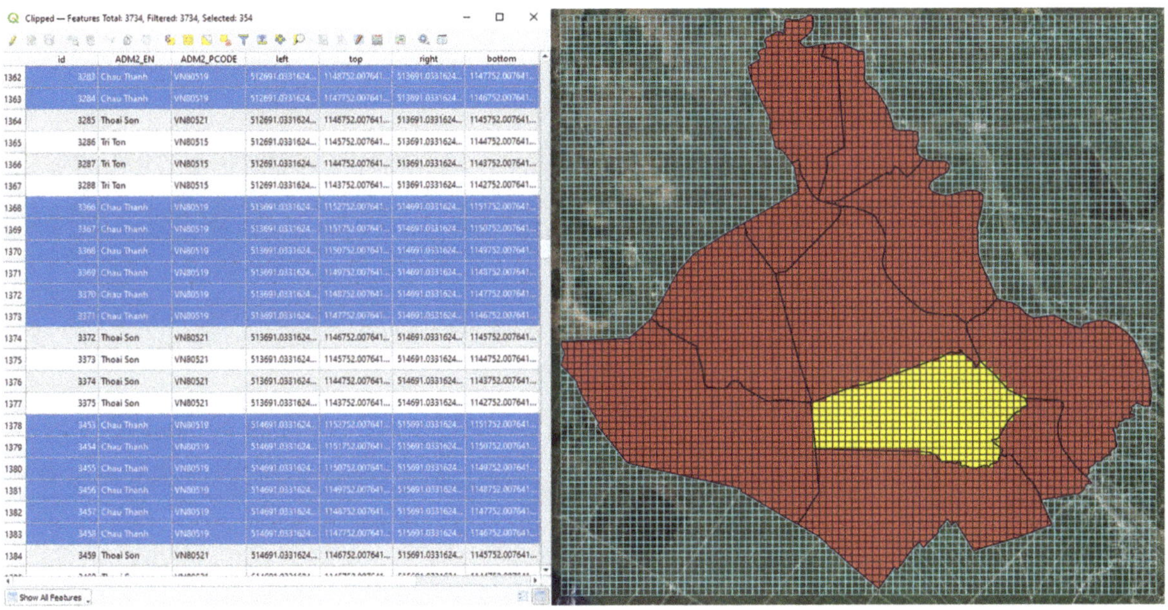

Source: Screenshot generated by the Asian Development Bank using QGIS 3.34.14. 2025.

In addition to the Clip tool, there are several other vector geoprocessing tools useful for spatial sampling. These include **Buffer**, **Intersect**, **Union**, **Difference**, and **Dissolve**, among others.

Table 2 presents a selection of commonly used tools, outlining their main functions, explaining how they manage attribute data, and providing examples of their use in agricultural sampling. These tools help define sampling areas, combine or separate features, and analyze spatial relationships between layers.

The steps for using these tools are generally similar to those demonstrated for the Clip tool, which involved selecting the input and overlay layers, setting parameters, and running the tool through the Processing Toolbox or Vector menu options.

Table 2: Common Geoprocessing Tools: Functions, Attribute Handling, and Sampling Examples

Tool Name	Definition	Attribute Handling	Example of Application in Agriculture
Clip	Extracts features from one layer that fall within the boundaries of another	Attributes from the input layer are retained, but those from the clipping layer are not.	Clip soil sample indicating a specific farm boundary
Buffer	Creates zones around features at a specified distance	Attributes from the input layer are copied to the output layer.	Identifies farms or sampling plots within a certain radius of irrigation sources
Intersect	Combines features from two layers where they overlap	Attributes from both layers are combined in the output layer.	Analyzes overlap between crop zones and soil types
Union	Merges two layers and retains all geometries and attributes	All attributes from both layers are preserved.	Combines land use and land cover maps
Difference	Removes overlapping areas of one layer from another	Attributes from the input layer are retained, but attributes from the subtracting layer are not.	Excludes protected areas from potential agricultural zones
Dissolve	Merges features based on a shared attribute value	Only the dissolve field(s) are retained, unless statistics (e.g., sum, mean) are applied.	Combines adjacent plots with the same crop type into one feature

Note: Tool features may vary in future releases.
Source: Asian Development Bank based on QGIS official documentation and observed functionalities (QGIS version 3.34.14).

3. Sampling Grids

There are multiple methods for sampling grids in QGIS. The process is nearly identical to that used for randomly selecting points, as demonstrated in the previous section. To randomly select sample grids, the **Random Selection Within Subsets** tool will be used.

(i) Go to **Vector | Research Tools | Random Selection Within Subsets** (Figure 8.29). Alternatively, go to **Processing** in the **main menu**, select **Toolbox**, and then search for **Random Selection Within Subsets**.

Figure 8.29: Launching the Random Selection Within Subsets Tool from the Processing Toolbox

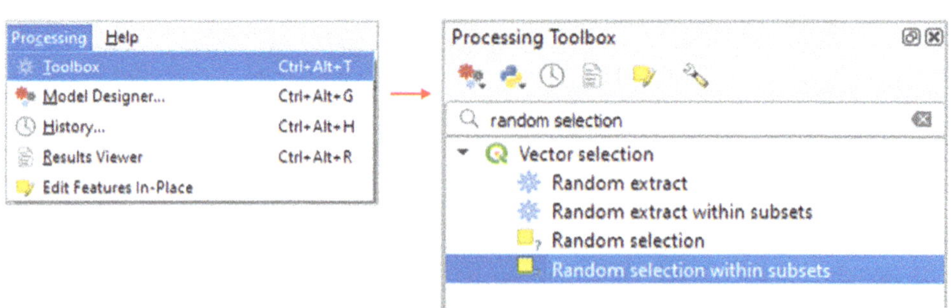

Source: Screenshot generated by the Asian Development Bank using QGIS 3.34.14. 2025.

(i) Select the **clipped layer** as the **Input layer**. Define the subset field as **ADM2_PCODE** under **ID field**, which represents the district ID. Then choose the selection **Method**. For this example, select **Percentage of selected features**. Enter **10** in the percentage field. This value will be applied individually to each category (i.e., each district), meaning that 10% of grids will be randomly selected from each district (Figure 8.30).

Figure 8.30: Defining the Parameters for the Random Selection Within Subsets Tool

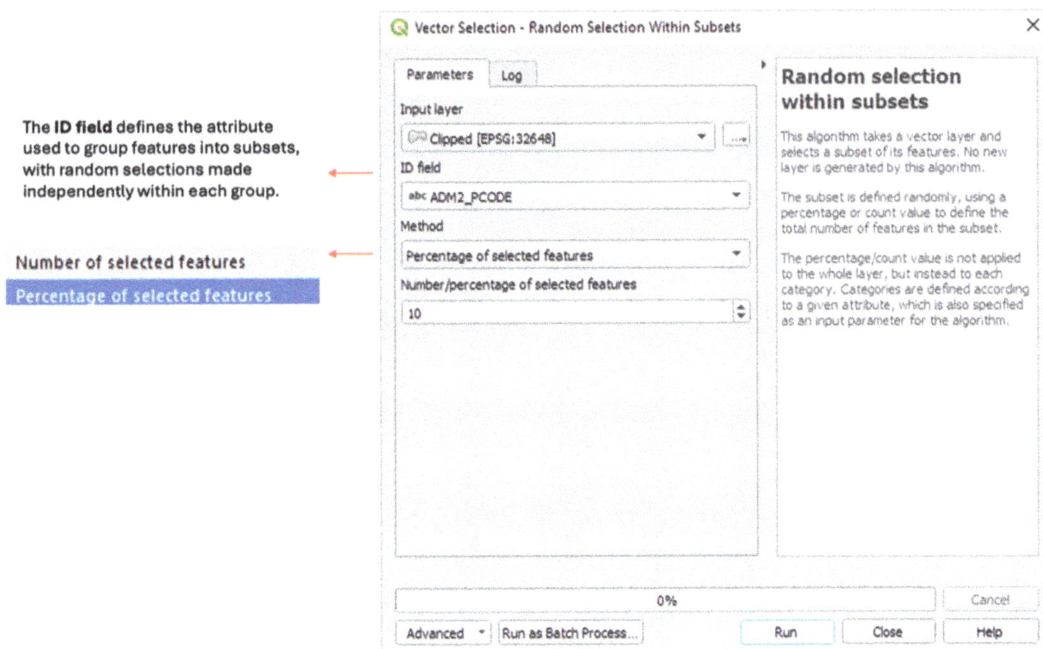

Source: Screenshot generated by the Asian Development Bank using QGIS 3.34.14. 2025.

Click **Run** to execute the process.

A prompt will appear once the algorithm stops running. Close the window once the random grids are selected.

(ii) Check the randomly selected grids (Figure 8.31). To save the selection, right-click on the layer in the **Layers panel** and choose **Save Selected Features As**. For detailed instructions on exporting selected features from a layer, refer to Module 3, section 3.2.

Figure 8.31: Randomly Selected Grids by District in An Giang Province, Viet Nam

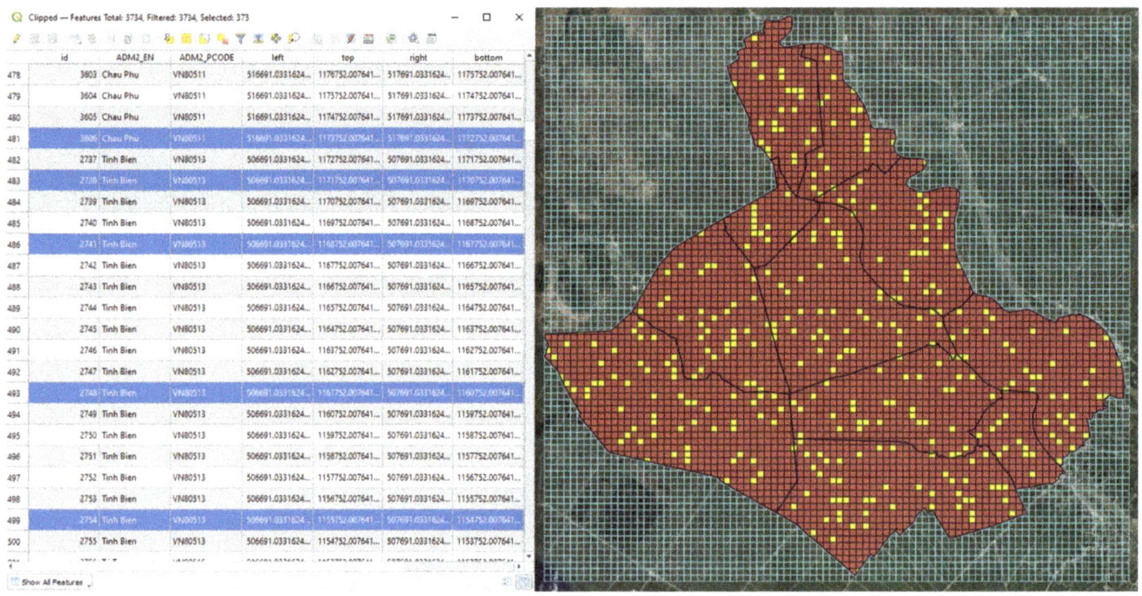

Source: Screenshot generated by the Asian Development Bank using QGIS 3.34.14. 2025.

4. Sampling Lines

Several methods are available for sampling line features. The procedure closely resembles the approaches described in earlier sections of this module for randomly selecting point and polygon features. The tools **Random Selection, Random Selection Within Subsets**, and **Random Extract Within Subsets** can be accessed through the **Processing Toolbox** or the **Vector** menu. Table 3 provides a summary of these tools, highlighting their functionality and potential applications.

Table 3: Random Selection Tools for Linear Features

Tool Name	Location in QGIS	Function	Output	Agriculture Use Case
Random Selection	Processing Toolbox or Vector selection	Randomly selects a specified number or percentage of features in the active layer	Selected features in a layer	Samples of irrigation canals or farm roads for inspection or maintenance
Random Selection Within Subsets	Processing Toolbox or Vector selection	Randomly selects features within defined attribute-based subsets	Selected features in a layer	Ensures balanced sampling across road types or farm zones
Random Extract	Processing Toolbox	Creates a new layer with a random selection of features	New layer with features	Generates a separate dataset of randomly selected farm roads for pilot studies

continued on next page

Table 8.3 continued

Tool Name	Location in QGIS	Function	Output	Agriculture Use Case
Random Extract Within Subsets	Processing Toolbox	Creates a new layer with randomly selected features from each subset	New layer with features	Extracts road segments grouped by farm zones for targeted infrastructure planning
Random Points on Lines (plugin)	Processing Toolbox, Vector selection, or plugin (installed via Plugin Manager)	Generates random points along line features	Points along lines	Simulates sampling points along irrigation channels for water quality testing or flow monitoring
PyQGIS Script (custom)	Python Console	Allows custom logic for random selection using Python scripting	Selected features in layer	Supports advanced modeling or simulation of agricultural transport or irrigation systems

Note: The main difference between **Random Selection Within Subsets** and **Random Extract Within Subsets** lies in the output: Random Selection Within Subsets selects features directly within the original layer, while Random Extract Within Subsets creates a new layer containing only the selected features; Tool features may vary in future releases.

Source: Asian Development Bank based on QGIS official documentation and observed functionalities (QGIS version 3.34.14).

Road networks and other linear features (**AnGiang_roads_lines**) extracted from OpenStreetMap data will be used to demonstrate the process in QGIS. Download the file from the shared folder.[a]

[a] For access to shapefiles and other open-access datasets referenced in the handbook, go to the GitHub repository: https://github.com/ADB-Data-Division/QGIS-Handbook-2025

(i) Go to **Processing | Toolbox**, and then search for the **Random Extract Within Subsets** tool (Figure 8.32).

Figure 8.32: Opening the Random Extract Within Subsets Tool

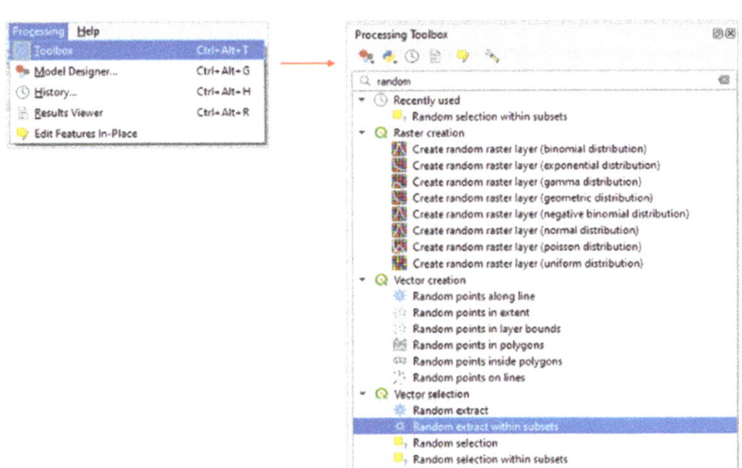

Source: Screenshot generated by the Asian Development Bank using QGIS 3.34.14. 2025.

(ii) Set the **Input layer** to **AnGiang_roads_lines** and define the subset under **ID field** as **highway**, which categorizes roads by type (e.g., primary, secondary, residential). For this example, under **Method** select **Number of selected features** and then enter **10** to randomly extract 10 features from each highway category. Specify a file path for saving the output layer, or leave it as temporary if saving is not required (Figure 8.33).

Figure 8.33: Defining the Parameters for the Random Extract Within Subsets Tool

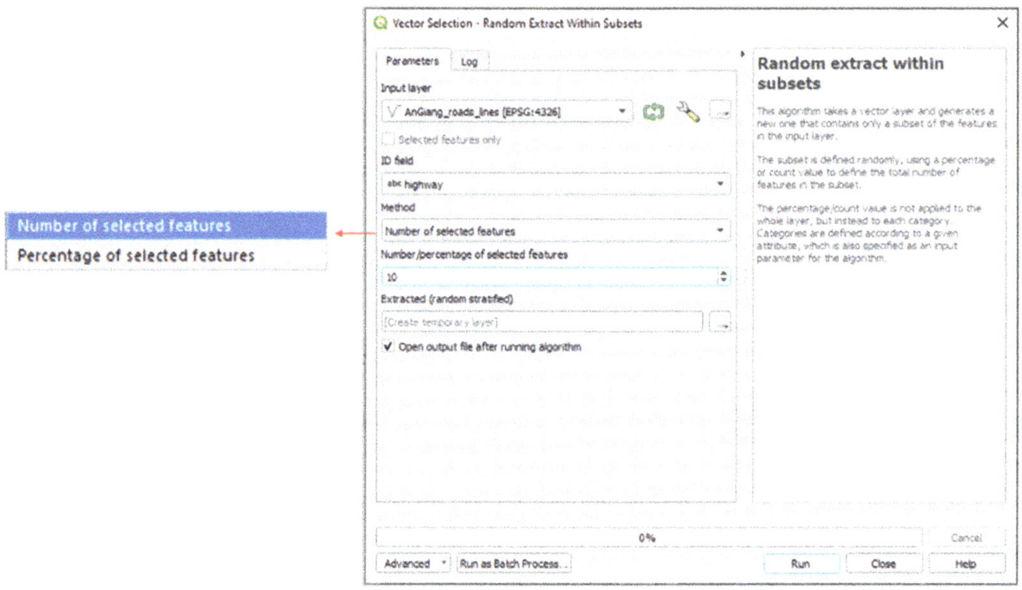

Source: Screenshot generated by the Asian Development Bank using QGIS 3.28. 2023.

Running the tool will generate a new, randomly extracted stratified layer containing the selected features, which can then be saved or exported for further analysis.

(iii) Check the **attribute table** of the extracted layer to confirm that each highway category contains the specified number of randomly selected features, noting that categories with fewer available features will have smaller samples (Figure 8.34).

Figure 8.34: Randomly Selected Highway Categories—An Giang Province, Viet Nam

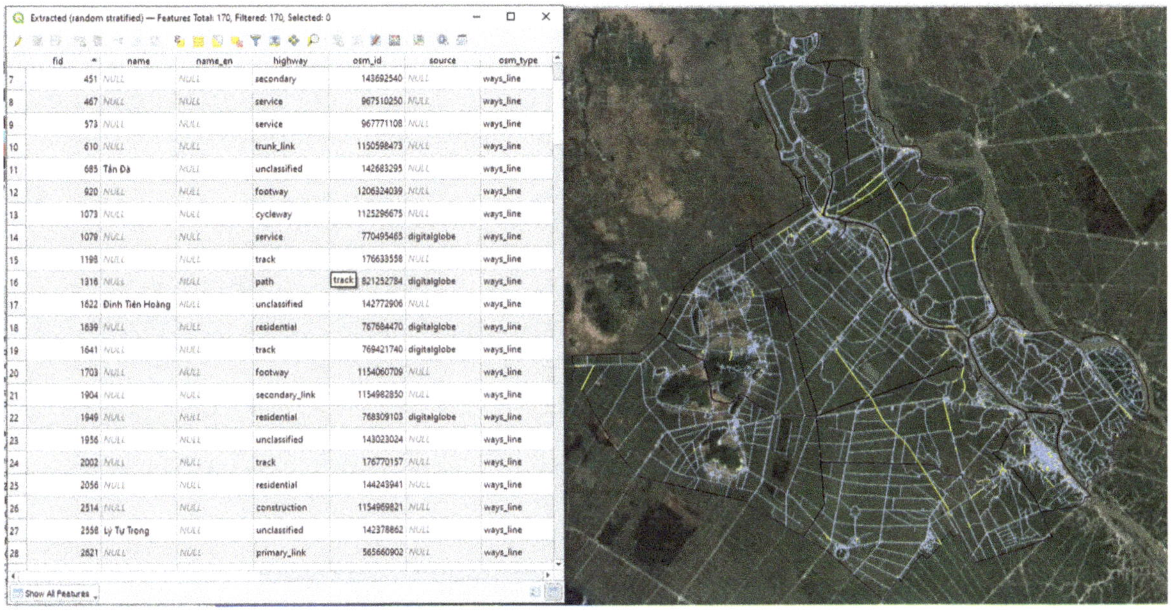

Source: Screenshot generated by the Asian Development Bank using QGIS 3.34.14. 2025.

8.3 Multistage Sampling in QGIS

The previous sections of this module introduced area sampling concepts and outlined various types of spatial units used in area sampling, including points, lines, polygons, and grids. Several sampling methods and tools available in QGIS were also demonstrated. This section presents the implementation of a multistage sampling design, applying concepts and tools covered earlier.

Multistage sampling within an area-sampling frame is a technique for selecting spatial units from a larger population using maps as the sampling frame. This approach is commonly applied in agricultural research, where spatial distribution and heterogeneity, such as cropping patterns and land use, are critical. It helps reduce sampling complexity and data-collection costs across large, heterogeneous areas while maintaining sample representativeness and accuracy.

In this section, an area frame will be constructed using administrative boundaries and land use data for An Giang Province. Multistage sampling involves stratifying the population into distinct subgroups, or strata, and selecting samples from each stratum, using a combination of random and nonrandom techniques. Stratification ensures that the sample adequately represents the population under study. The specific stratification and sampling strategy may vary depending on the research objectives and available resources. The first step in building the area frame is to define the strata. For this exercise, stratification will be based on land use.

The following exercise utilizes the 2020 land use and land cover (LULC) map of An Giang Province from the Japan Aerospace Exploration Agency (JAXA) to develop a multistage sampling design in QGIS. A comprehensive, step-by-step guide for downloading and preprocessing the raster data is available in **Module 3, Section 3.2.** You must follow the instructions in that module to obtain and prepare the dataset required for this exercise. If you have already completed Module 3, you may use the output file named "**AnGiang_Landuse.tiff**," which is a clipped and vectorized LULC map derived from JAXA's open-source dataset. This map includes 12 land cover classifications and serves as the basis for the sampling procedures in this module.

To support the implementation of a multistage sampling design, the dataset must be reclassified into four simplified categories: **Rice, Other Crops, Urban** (including settlements), and **Water** (covering all types of water features).

The district shapefile (**ADM2 - AnGiang_District**), used in earlier sections, is also required. It provides the administrative boundaries needed for stratification and spatial analysis.

Overview of the Sampling Design for Demonstration

This demonstration outlines a multistage sampling design intended to support LULC analysis, with a specific focus on mapping and estimating rice cultivation areas during the cropping season. While the design is based on sound statistical principles and operational considerations, it is presented here only to serve as an illustration for the handbook.

The sampling design consists of three stages:

- **Primary stage:** The entire province is segmented into a grid of 1 km^2 cells.
- **Secondary stage:** Within each selected 1 km^2 grid, four sub-grids measuring 200 m^2 are selected.
- **Tertiary stage:** From each 200 m^2 sub-grid, two 50 m^2 segments are selected. These represent the ultimate sampling units for ground validation and analysis.

To enhance representativeness across different land use types, the 1 km^2 grids were classified into distinct strata based on the dominant land use category. From these strata, a sample of 22 grids classified as **Rice** was randomly selected (2 each from 11 districts). In addition, 6 grids were selected for **Other Crops**, 2 for **Urban**, and 4 for **Water**—ensuring diversity and coverage across major land use categories.

The design thus includes a total of 34 selected 1 km^2 grids: comprising 22 Rice grids and 12 grids from other land use strata. This yields 136 sub-grids and 272 segments.

Note: For this section, perform the exercise based on the tools and techniques learned in the previous sections unless new tools are introduced. If new tools are to be applied, a step-by-step guide will be provided, but you should also continue using the methods already covered.

8.3.1 Reclassifying Raster Data (Grouping Land Cover Types)

In agricultural research, reclassifying a raster layer helps simplify complex land cover data into broader categories that are more relevant for analysis, making it easier to identify suitable zones for sampling. For example, multiple vegetation types can be grouped into a single "cropland" class to streamline site selection for soil or crop studies.

For the current exercise, the classification scheme will be applied to the JAXA LULC raster dataset containing **12 land cover classes**. These classes will be reclassified as follows:

- **Rice:** including all land areas identified as rice, originally assigned to class 3;
- **Other Crops:** including all other forms of vegetation, corresponding to classes 4–11;
- **Urban:** including all settlement areas, originally assigned to class 2; and
- **Water:** including all water-related features, originally assigned to classes 1 and 12.

(i) Add the **AnGiang_Landuse.tiff** raster layer by dragging it from the **Browser panel** into the **map area**. Alternatively, navigate to the **Layer** in the **Menu Toolbar**, then select **Add Layer | Add Raster Layer**. When added to the map, the layer is displayed in **singleband gray** by default (Figure 8.35).

Figure 8.35: Raster-Based Land Cover Map of An Giang Province, Viet Nam

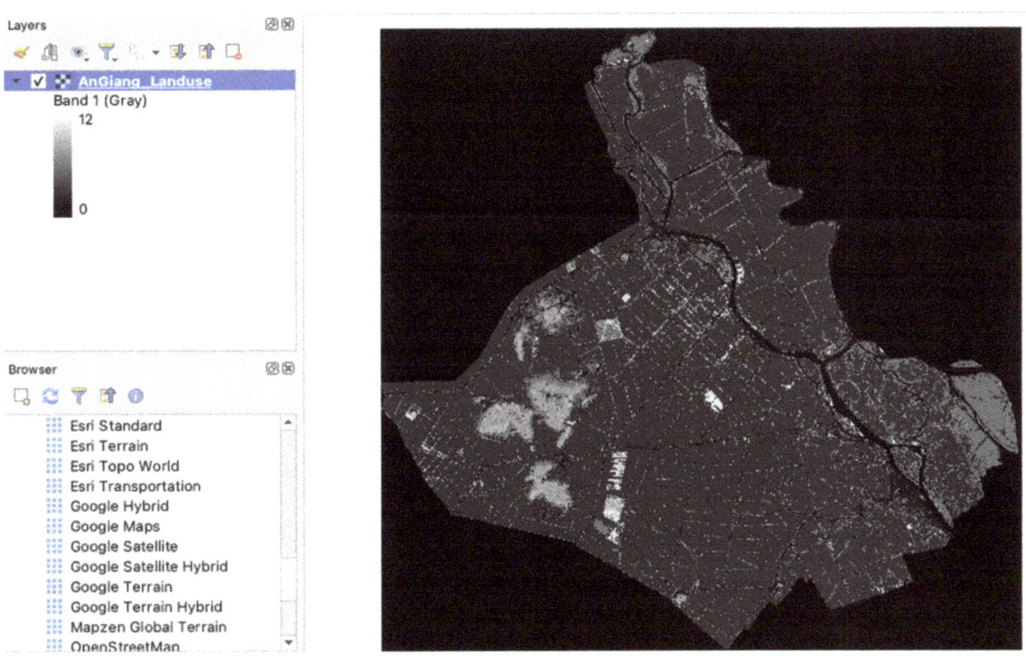

Source: Screenshot generated by the Asian Development Bank using QGIS 3.34.14. 2025.

(ii) To visualize the distribution of land use types across the province, apply a pseudocolor scheme (Figure 8.36). Right-click on the **AnGiang_Landuse.tiff** raster layer and select **Properties**. Go to the **Symbology** tab and change **Render type** from **Singleband gray** to **Paletted/Unique values**. Click **Classify**. Double-click on the color for the value 0, and set its opacity to 0%. Refer to Module 3, section 3.2.3 to learn more about rendering.

Figure 8.36: Land Use and Land Cover Classes in An Giang Province, Viet Nam

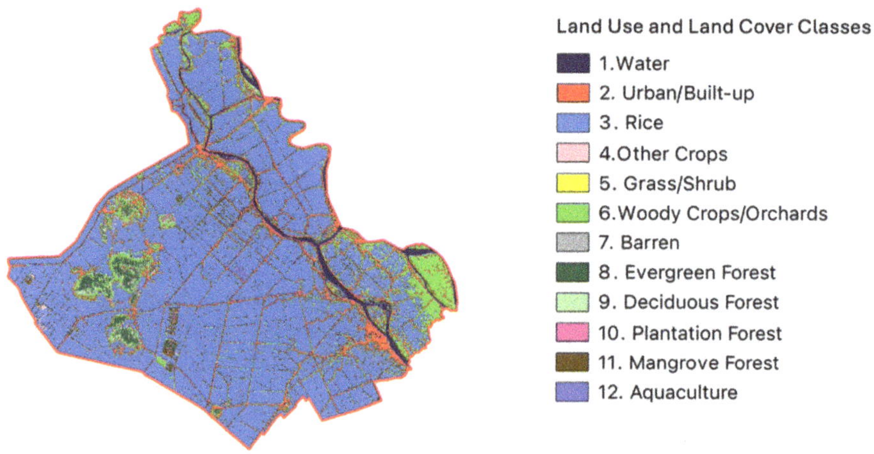

Source: Screenshot generated by the Asian Development Bank using QGIS 3.34.14. 2025.

Note that the color assigned to each land use type may vary between users, depending on individual settings or how the pseudocolor scheme is applied. These color assignments can also be manually adjusted to accommodate specific visualization requirements.

(iii) To implement the reclassification described above, navigate to **Processing | Toolbox | Reclassify by table** (Figure 8.37). Define the reclassification ranges (Figure 8.38).

Figure 8.37: Reclassification by Table Tool

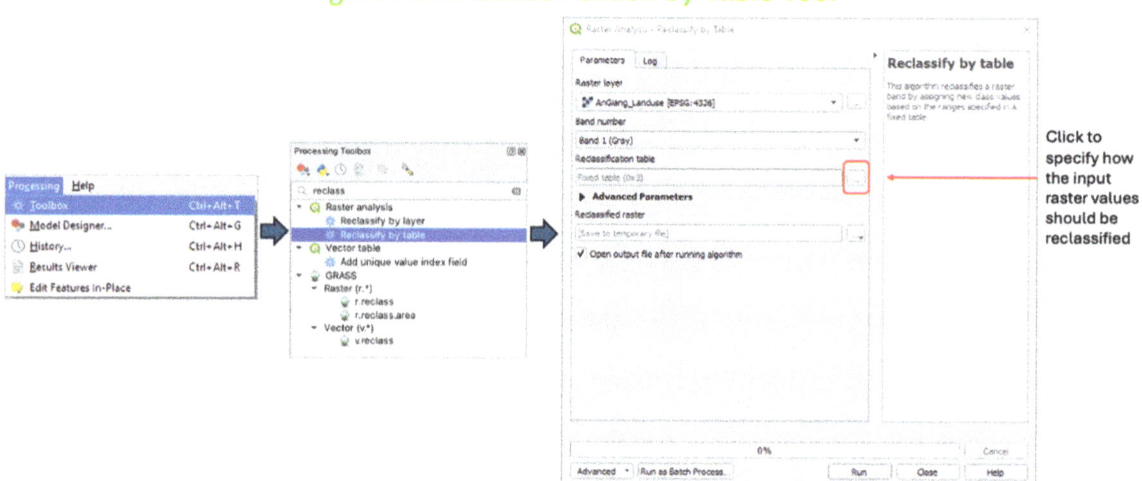

Source: Screenshot generated by the Asian Development Bank using QGIS 3.34.14. 2025.

Figure 8.38: Defining the Reclassification Ranges

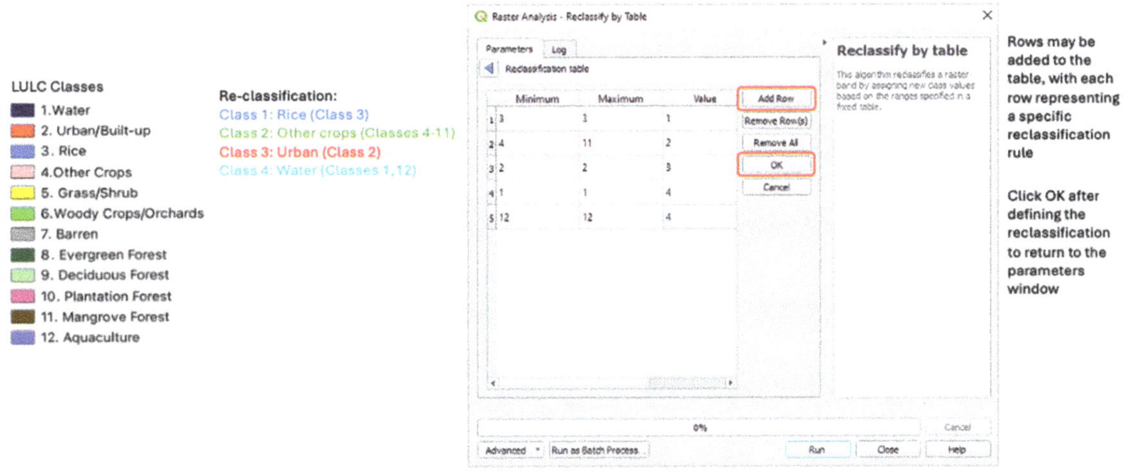

Source: Screenshot generated by the Asian Development Bank using QGIS 3.34.14. 2025.

(iv) Set the other reclassification parameters (Figure 8.39).

Figure 8.39: Setting Other Reclassification Parameters

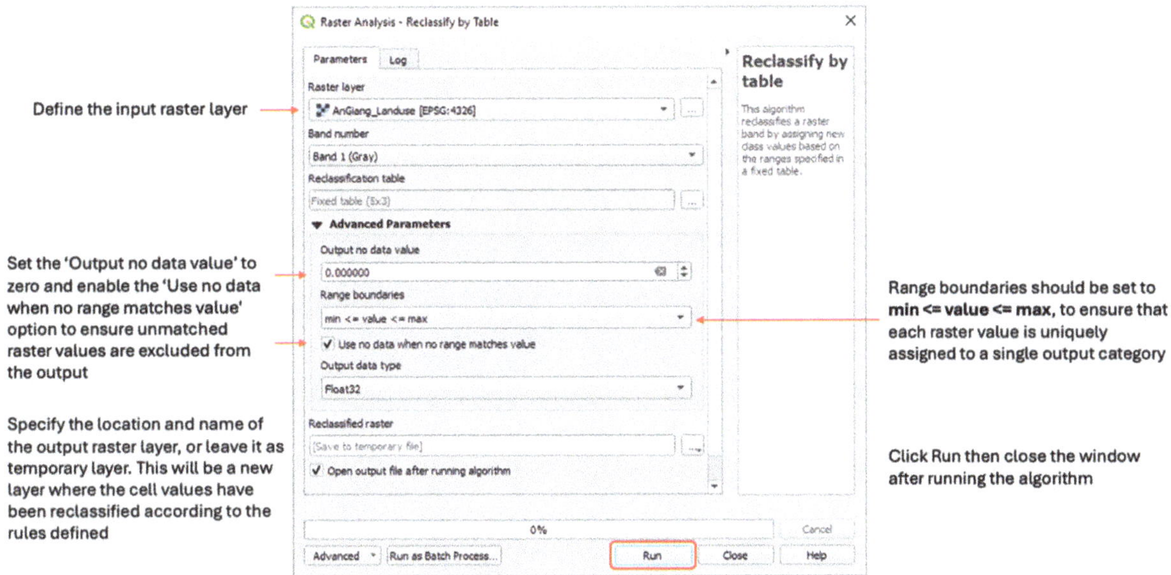

Source: Screenshot generated by the Asian Development Bank using QGIS 3.34.14. 2025.

(v) A new **reclassified raster layer** will be added to the **Layers panel**, now containing the four classes as illustrated in Figure 8.40.

Figure 8.40: Reclassified Raster Layer

Source: Screenshot generated by the Asian Development Bank using QGIS 3.34.14. 2025.

Reclassification can also be carried out using the **Field Calculator**; however, this requires converting the raster data into a vector layer beforehand. Refer to Module 3, section 3.3.5, for detailed instructions on this process.

(vi) Apply pseudocolor rendering to visualize the distribution of each land use class within the study area, following the color scheme below. The resulting layer should closely resemble the one shown in Figure 8.41. Rename this layer as **AnGiang_reclass**.

Figure 8.41: Applying Pseudocolor Symbology to a Reclassified Raster Layer

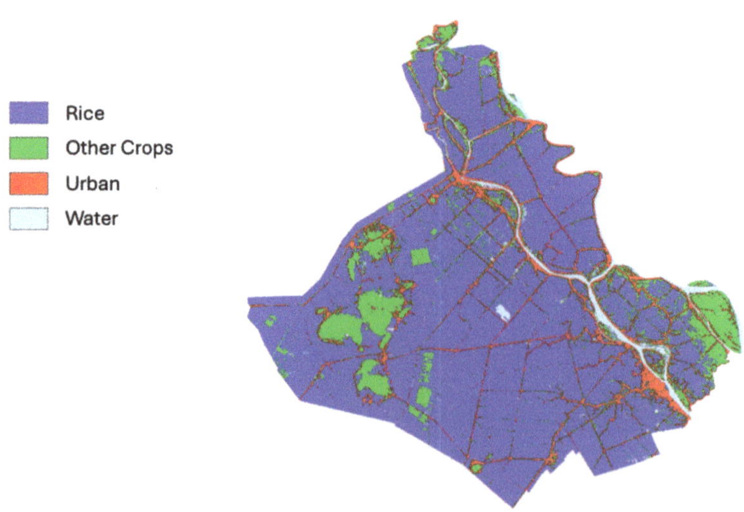

Source: Screenshot generated by the Asian Development Bank using QGIS 3.34.14. 2025.

8.3.2 Extracting the Extent of the Study Area

(i) Add the **AnGiang_District** vector layer by dragging it from the **Browser panel** into the **map area**. Alternatively, navigate to the **Layer tab** in the **Menu Toolbar**, then select **Add Layer | Add Vector Layer**.

(ii) Generate a grid with **1 km² cells** covering the extent of the **AnGiang_reclass** layer. Use the layer's extent as the basis for the grid coverage (Figure 8.42). Follow the step-by-step guide for creating a grid in section 8.2.3 of this module.

Figure 8.42: Grids over the Extent of An Giang Province, Viet Nam

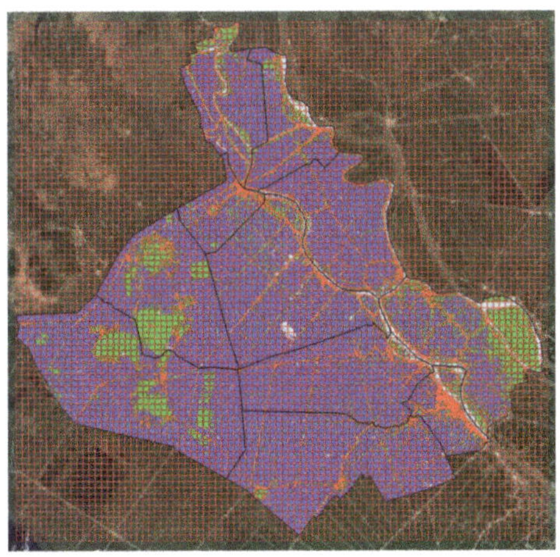

Note: Each grid cell represents 1-square-kilometer (km^2).
Source: Screenshot generated by the Asian Development Bank using QGIS 3.34.14. 2025.

(iii) **Clip** the grid layer using the boundaries of the **AnGiang_District** shapefile to ensure that the grid aligns with the administrative boundary limits (Figure 8.43).

Figure 8.43: Grids Clipped Based on the Boundaries of An Giang Province, Viet Nam

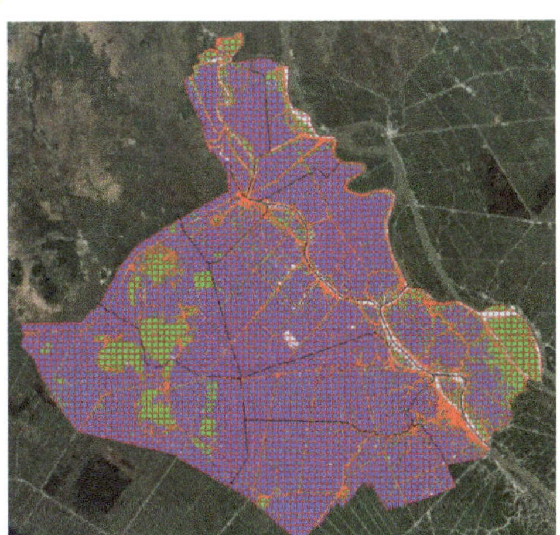

Note: Each grid cell represents 1-square-kilometer (km^2).
Source: Screenshot generated by the Asian Development Bank using QGIS 3.34.14. 2025.

8.3.3 Raster Analysis for Stratified Sampling by Land Use

In area frame sampling for agricultural surveys, pixel-level data from remote sensing provides a detailed and objective basis for delineating sampling units. Stratified sampling relies on this granularity to ensure proportional representation across diverse land use types such as cropland, urban areas, and bodies of water. QGIS tools such as **Zonal Statistics** and **Zonal Histogram** support this process by aggregating pixel-based information within defined zones, enabling the identification of intrazonal variability and guiding the allocation of sample points. This approach enhances the precision and relevance of survey design by aligning sampling frames with actual land cover and crop distribution patterns.

1. Zonal Statistics

Zonal Statistics calculates summary values such as the mean, minimum, and maximum for raster data within defined land use zones. This helps identify areas with greater variability or specific characteristics, guiding the allocation of sample points to ensure coverage across diverse conditions (Chatterjee 2018).

(i) To calculate **Zonal Statistics**, open the **Processing Toolbox** from the **main menu** and search for **Zonal Statistics**. Select the **Clipped grid** from the previous step as the **input layer** and the **Reclassified raster** layer as the **input raster**. Set the desired statistical parameters (e.g., mean, maximum, minimum), then click **Run** (Figure 8.44).

Figure 8.44: Calculating Statistics Within a Grid

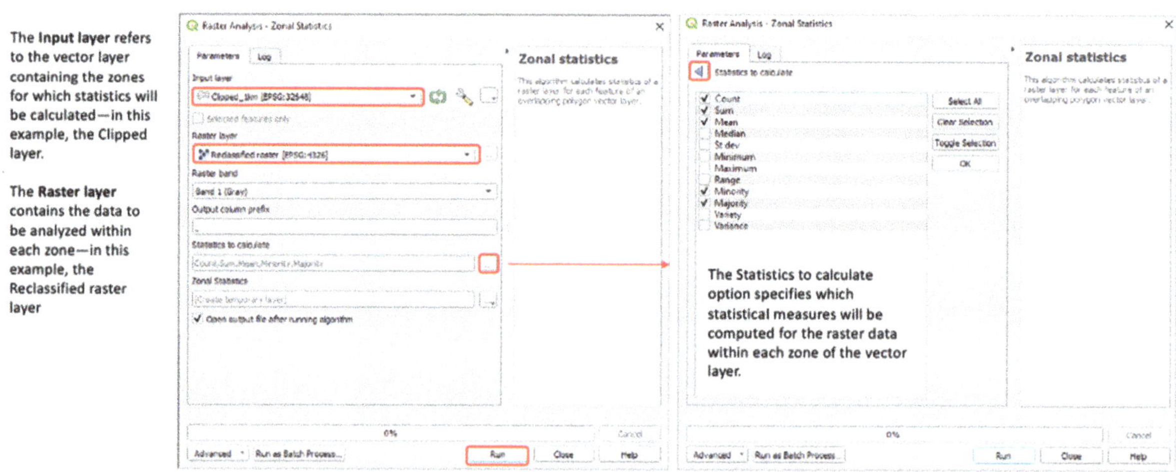

Note: Each grid cell represents 1-square-kilometer (km^2).
Source: Screenshot generated by the Asian Development Bank using QGIS 3.34.14. 2025.

(ii) Open the **attribute table** to examine the results of the **Zonal Statistics** analysis (Figure 8.45).

Figure 8.45: Attribute Table Displaying Land Use Statistics per Grid

Zonal Statistics — Features Total: 3734, Filtered: 3734, Selected: 0

	id	left	top	right	bottom	row_index	col_index	_count	_sum	_mean	_minority	_majority
1	2357	502691.0331624...	1204752.007641...	503691.0331624...	1203752.007641...	7	27	2302	2302	1	1	1
2	2358	502691.0331624...	1203752.007641...	503691.0331624...	1202752.007641...	8	27	4176	4176	1	1	1
3	2359	502691.0331624...	1202752.007641...	503691.0331624...	1201752.007641...	9	27	1478	1478	1	1	1
4	2442	503691.0331624...	1206752.007641...	504691.0331624...	1205752.007641...	5	28	415	846	2.038554216867...	3	2
5	2443	503691.0331624...	1205752.007641...	504691.0331624...	1204752.007641...	6	28	283	604	2.134275618374...	1	2
6	2444	503691.0331624...	1204752.007641...	504691.0331624...	1203752.007641...	7	28	8772	21750	2.479480164158...	4	3
7	2445	503691.0331624...	1203752.007641...	504691.0331624...	1202752.007641...	8	28	10302	17255	1.674917491749...	3	1
8	2446	503691.0331624...	1202752.007641...	504691.0331624...	1201752.007641...	9	28	10302	11029	1.070568821588...	4	1
9	2447	503691.0331624...	1201752.007641...	504691.0331624...	1200752.007641...	10	28	8129	8285	1.019190552343...	4	1
10	2448	503691.0331624...	1200752.007641...	504691.0331624...	1199752.007641...	11	28	3316	3517	1.060615199034...	4	1
11	2449	503691.0331624...	1199752.007641...	504691.0331624...	1198752.007641...	12	28	81	81	1	1	1
12	2527	504691.0331624...	1208752.007641...	505691.0331624...	1207752.007641...	3	29	279	789	2.827956989247...	4	3
13	2528	504691.0331624...	1207752.007641...	505691.0331624...	1206752.007641...	4	29	280	675	2.410714285714...	3	2
14	2529	504691.0331624...	1206752.007641...	505691.0331624...	1205752.007641...	5	29	8223	18849	2.29222911346224	1	2
15	2530	504691.0331624...	1205752.007641...	505691.0331624...	1204752.007641...	6	29	9106	19337	2.123544915440...	3	2
16	2531	504691.0331624...	1204752.007641...	505691.0331624...	1203752.007641...	7	29	10251	12754	1.244171300360...	4	1
17	2532	504691.0331624...	1203752.007641...	505691.0331624...	1202752.007641...	8	29	10302	16835	1.634148708988...	4	1
18	2533	504691.0331624...	1202752.007641...	505691.0331624...	1201752.007641...	9	29	10302	20924	2.031061929722...	4	2

Show All Features

Note: Each grid represents 1-square-kilometer (km^2).

Source: Screenshot generated by the Asian Development Bank using QGIS 3.34.14. 2025.

Zonal Statistics summarizes raster data within defined zones—in this case, each 1 km^2 grid cell. It calculates the number of 10 m^2 pixels per land use category (i.e., Rice, Other Crops, Urban, and Water) inside each grid cell. These counts can be converted into area measurements, providing a clear picture of land use distribution that is useful for spatial analysis.

2. Zonal Histograms

Zonal Histograms show the distribution of raster values across land use zones, showing how frequently certain values occur (Esri 2012). This is particularly useful in stratified sampling, as it reveals the internal distribution of variables such as the vegetation index across different land cover types. Understanding these distributions helps researchers ensure that the sampling captures the diversity within each land use category, leading to more representative data collection.

(i) To create a **Zonal Histogram**, go to the **Processing Toolbox** and search for **Zonal Histogram**. Configure the parameters by selecting **Reclassified raster** under **Raster layer**. Then, choose the **vector layer** that defines the zones; in this example, that means selecting the **Zonal Statistics** layer from the previous step to append additional fields. Click **Run** to execute the tool (Figure 8.46).

Figure 8.46: Setting Parameters for the Zonal Histogram Tool

The **raster layer** contains the land use values to be analyzed, typically represented as categories like croplands, urban, or water.

The **vector layer** defines the zones, such as a 1 km grid, within which the tool counts how many raster pixels fall into each category.

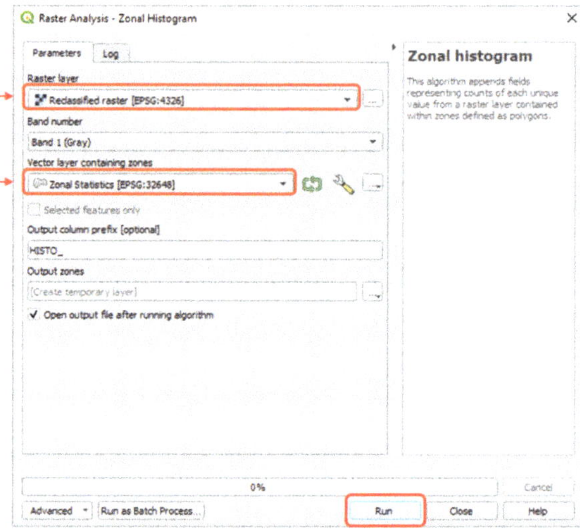

Source: Screenshot generated by the Asian Development Bank using QGIS 3.34.14. 2025.

(ii) Open the **attribute table** of the **output vector layer** (Figure 8.47).

Figure 8.47: Output Zones

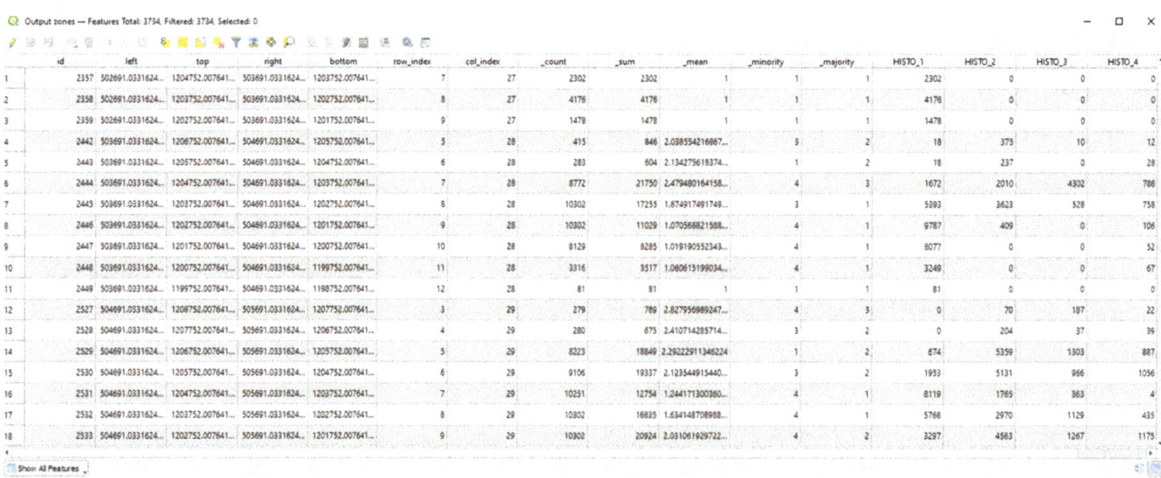

	id	left	top	right	bottom	row_index	col_index	_count	_sum	_mean	_minority	_majority	HISTO_1	HISTO_2	HISTO_3	HISTO_4
1	2357	502691.0331624...	1204752.007641...	503691.0331624...	1203752.007641...	7	27	2302	2302	1	1	1	2302	0	0	0
2	2358	502691.0331624...	1203752.007641...	503691.0331624...	1202752.007641...	8	27	4176	4176	1	1	1	4176	0	0	0
3	2359	502691.0331624...	1202752.007641...	503691.0331624...	1201752.007641...	9	27	1478	1478	1	1	1	1478	0	0	0
4	2442	503691.0331624...	1206752.007641...	504691.0331624...	1205752.007641...	5	28	415	846	2.038554216867...	3	2	18	375	10	12
5	2443	503691.0331624...	1205752.007641...	504691.0331624...	1204752.007641...	6	28	283	604	2.134275618374...	1	2	18	237	0	28
6	2444	503691.0331624...	1204752.007641...	504691.0331624...	1203752.007641...	7	28	8772	21750	2.479480164158...	4	3	1672	2010	4302	788
7	2445	503691.0331624...	1203752.007641...	504691.0331624...	1202752.007641...	8	28	10302	17255	1.674917491749...	3	1	5393	3623	528	758
8	2446	503691.0331624...	1202752.007641...	504691.0331624...	1201752.007641...	9	28	10302	11029	1.070568821588...	4	1	9787	409	0	106
9	2447	503691.0331624...	1201752.007641...	504691.0331624...	1200752.007641...	10	28	8129	8285	1.019190552343...	4	1	8077	0	0	52
10	2448	503691.0331624...	1200752.007641...	504691.0331624...	1199752.007641...	11	28	3316	3517	1.060615199034...	4	1	3249	0	0	67
11	2449	503691.0331624...	1199752.007641...	504691.0331624...	1198752.007641...	12	28	81	81	1	1	1	81	0	0	0
12	2527	504691.0331624...	1208752.007641...	505691.0331624...	1207752.007641...	3	29	279	789	2.827956989247...	4	3	0	70	187	22
13	2528	504691.0331624...	1207752.007641...	505691.0331624...	1206752.007641...	4	29	280	675	2.410714285714...	3	2	0	204	37	39
14	2529	504691.0331624...	1206752.007641...	505691.0331624...	1205752.007641...	5	29	8223	18849	2.29222911346224	1	2	674	5359	1303	887
15	2530	504691.0331624...	1205752.007641...	505691.0331624...	1204752.007641...	6	29	9106	19337	2.123544915440...	3	2	1953	5131	966	1056
16	2531	504691.0331624...	1204752.007641...	505691.0331624...	1203752.007641...	7	29	10251	12754	1.244171300360...	4	1	8119	1765	363	4
17	2532	504691.0331624...	1203752.007641...	505691.0331624...	1202752.007641...	8	29	10302	16835	1.634148708988...	4	1	5768	2970	1129	435
18	2533	504691.0331624...	1202752.007641...	505691.0331624...	1201752.007641...	9	29	10302	20924	2.031061929722...	4	2	3297	4563	1267	1175

Source: Screenshot generated by the Asian Development Bank using QGIS 3.34.14. 2025.

New fields will appear, showing the count of each raster value within each zone. Every row represents a 1 km^2 grid cell, and each column indicates the frequency of a specific land use category—Rice, Other Crops, Urban, or Water—based on the distribution of 10 m^2 pixels.

(iii) Use the **Join Attributes by Location** tool to append district names and district IDs from the **AnGiang_District** layer to each individual grid cell in the **Output Zones** layer. Ensure that both layers are in the same coordinate reference system (CRS) before performing the join to avoid spatial mismatches.

Set the **geometric predicate** to **intersect**, and configure the join to take attributes with the largest area of overlap using a one-to-one relationship. This ensures that each grid cell receives a unique land use classification based on the dominant overlapping feature. This process may take some time to complete (Figure 8.48).

Figure 8.48: Configuring Join Parameters

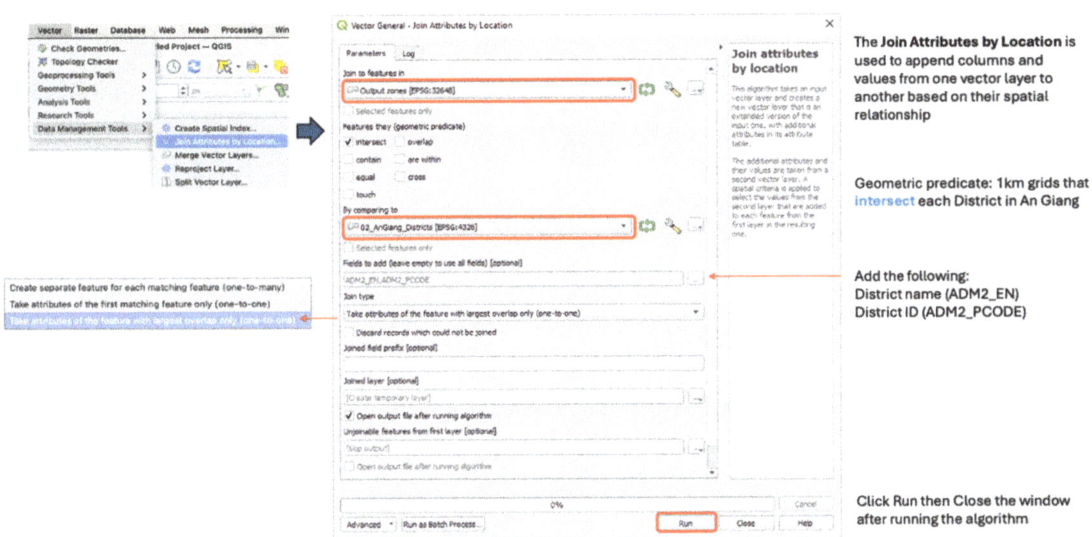

Source: Screenshot generated by the Asian Development Bank using QGIS 3.34.14. 2025.

The final output is a grid layer in which each cell contains the district name, district ID, and the corresponding land use classification (Figure 8.49).

Figure 8.49: Attributes of a Grid Layer

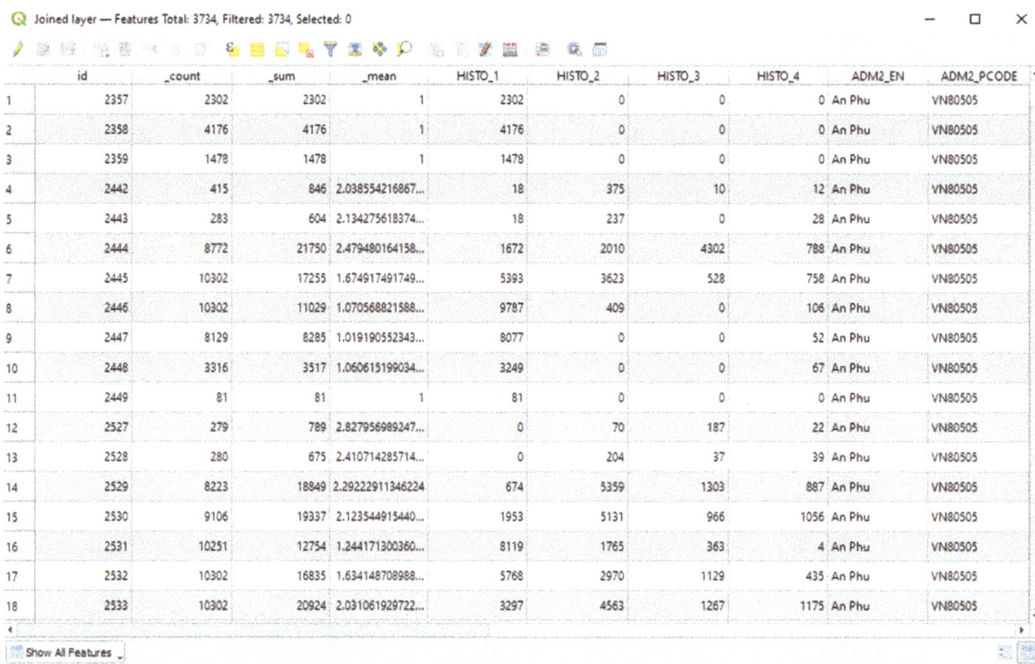

	id	_count	_sum	_mean	HISTO_1	HISTO_2	HISTO_3	HISTO_4	ADM2_EN	ADM2_PCODE
1	2357	2302	2302	1	2302	0	0	0	An Phu	VN80505
2	2358	4176	4176	1	4176	0	0	0	An Phu	VN80505
3	2359	1478	1478	1	1478	0	0	0	An Phu	VN80505
4	2442	415	846	2.038554216867...	18	375	10	12	An Phu	VN80505
5	2443	283	604	2.134275618374...	18	237	0	28	An Phu	VN80505
6	2444	8772	21750	2.479480164158...	1672	2010	4302	788	An Phu	VN80505
7	2445	10302	17255	1.674917491749...	5393	3623	528	758	An Phu	VN80505
8	2446	10302	11029	1.070568821588...	9787	409	0	106	An Phu	VN80505
9	2447	8129	8285	1.019190552343...	8077	0	0	52	An Phu	VN80505
10	2448	3316	3517	1.060615199034...	3249	0	0	67	An Phu	VN80505
11	2449	81	81	1	81	0	0	0	An Phu	VN80505
12	2527	279	789	2.827956989247...	0	70	187	22	An Phu	VN80505
13	2528	280	675	2.410714285714...	0	204	37	39	An Phu	VN80505
14	2529	8223	18849	2.29222911346224	674	5359	1303	887	An Phu	VN80505
15	2530	9106	19337	2.123544915440...	1953	5131	966	1056	An Phu	VN80505
16	2531	10251	12754	1.244171300360...	8119	1765	363	4	An Phu	VN80505
17	2532	10302	16835	1.634148708988...	5768	2970	1129	435	An Phu	VN80505
18	2533	10302	20924	2.031061929722...	3297	4563	1267	1175	An Phu	VN80505

Note: The grid frame is 1-square-kilometer (km^2) in size.
Source: Screenshot generated by the Asian Development Bank using QGIS 3.34.14. 2025.

3. Creating a New Grid ID Field

Assigning a unique grid ID to each zone is recommended to facilitate sampling and data management. To ensure that the grid ID reflects the spatial arrangement of the grids, the **Order by expression** tool from the **Processing Toolbar** can be used. This tool enables the sorting of grid features based on spatial location, often using expressions that prioritize the coordinates of grid centroids.

After ordering the grids, the **Field Calculator** is used on the resulting **ordered layer** to generate a new attribute field named "**Grid_ID**." This field contains sequential values that correspond to the spatial layout of the grids, supporting consistent and structured data handling.

This method ensures that the assigned IDs align with the intended spatial sequence, which is particularly useful for systematic sampling and spatial analysis.

(i) To create a **Grid_ID** field, open the **Processing Toolbox** and search for **Order by expression**. In the tool dialog box, under **Input layer**, select your grid layer (which, in this example, is the joined layer that resulted from the previous step). Under **Expression**, select the **id field** from the drop-down menu.

The **id** field is automatically assigned by QGIS when a vector layer is created or imported. It typically reflects the order in which features were created or stored. In grid-generation tools such as Create Grid, the order follows a systematic spatial pattern—often from top to bottom, left to right. Since the id field reflects the desired spatial order, no custom expression is needed.

Check **Sort ascending**. Choose a file location and name for saving the **output layer**, or opt to save it as a temporary layer. Then click **Run**. (Figure 8.50).

Figure 8.50: Configuring the Order by Expression Tool

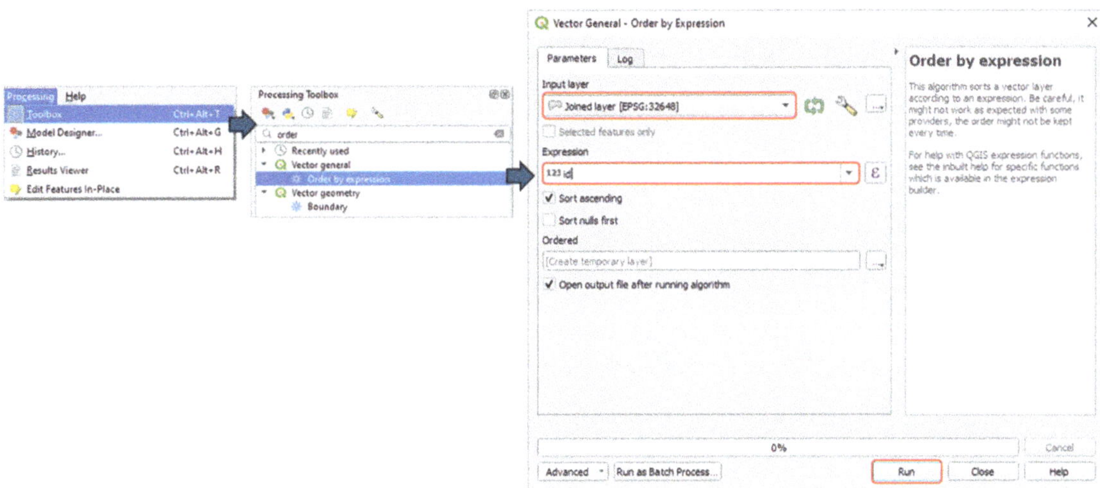

Source: Screenshot generated by the Asian Development Bank using QGIS 3.34.14. 2025.

This will create a new layer in which the features are sorted by the id field, thereby preserving the intended spatial layout.

(ii) Next, open the **attribute table** of the newly created **ordered layer** and click the **Field Calculator** icon, . In the dialog box, check **Create a new field**, name the output field **Grid_ID**, and set the output field type to **Whole Number (integer)**. In the **Expression box**, enter **@row_number**, then click **OK**. This will assign a unique sequential number to each feature based on the current order, generating a **Grid_ID** that aligns with the spatial structure of the grid.

Figure 8.51: Creating a Sequential Grid ID Using the Field Calculator

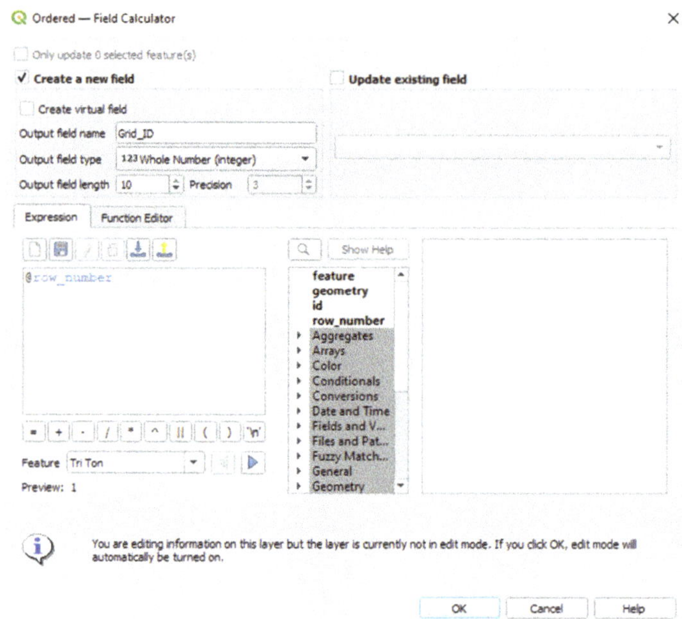

Source: Screenshot generated by the Asian Development Bank using QGIS 3.34.14. 2025.

(iii) Open the **attribute table** of the output lay0er to confirm that each feature has a unique sequential ID (Figure 8.52).

Figure 8.52: Output Layer with Sequential Grid Identifiers

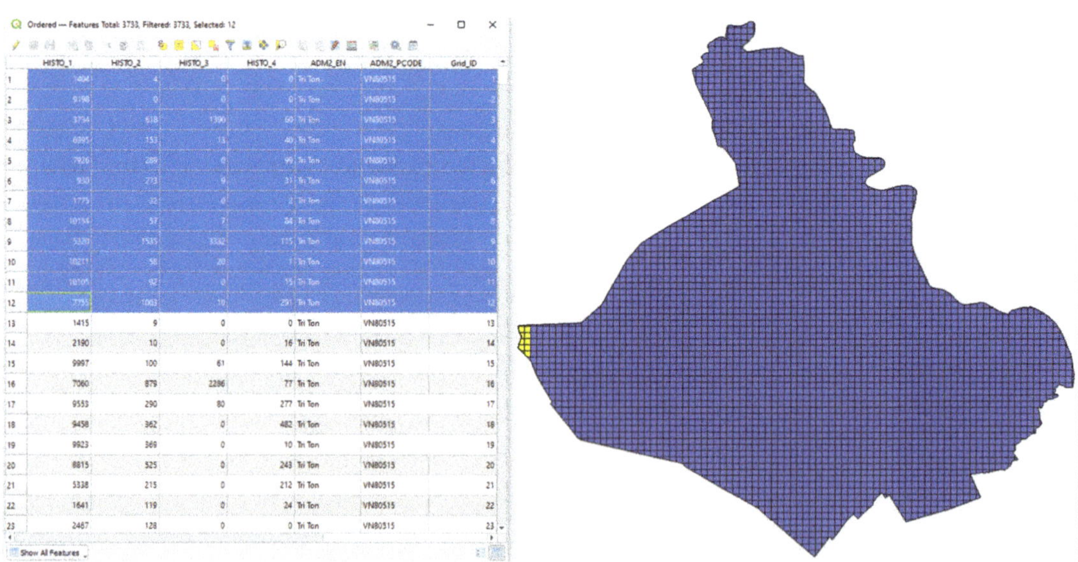

Source: Screenshot generated by the Asian Development Bank using QGIS 3.34.14. 2025.

4. Merging Attributes in the 1-Square-Kilometer Grid Frame

With land use statistics now summarized across defined spatial units, the outputs from the zonal-raster-analysis tools provide a basis for designing a statistically sound sampling framework. These results can support the development of an area frame survey that ensures representation across all land use categories, an essential component for remote sensing applications requiring ground truth validation.

Based on the available dataset, it is now possible to identify grid cells predominantly under rice cultivation by analyzing the proportion of area planted. Given that the classification is pixel-based, this information can be used to inform sampling strategies, typically through manual analysis, such as exporting attribute data for statistical review and sampling design.

The following exercise introduces a simplified method for labeling land use by identifying the dominant land use type within each grid cell. This approach is intended solely for demonstration purposes, to support the succeeding sections of the module. It is not intended to serve as a fixed methodology.

Users may opt to develop their own sampling design based on project needs and available data. Tools such as **Zonal Statistics** and **Zonal Histograms** can support this process. For reference, a sample sampling design is available in the ADB publication *Mapping Rice in Viet Nam: Estimating Rice Area Using Remote Sensing and Area Frame Survey Methods in An Giang Province* (Burgard et al. 2025).

To systematically assign land use categories at the grid level, a pixel-counting convention is applied. Within each 1 km^2 grid cell, the number of 10 m^2 pixels corresponding to each type of land cover—Rice, Other Crops, Urban, and Water—is calculated. The dominant land use category is then determined by identifying which class has the highest pixel count. This approach thus reflects the proportional representation of land cover types and enables consistent classification across spatial units.

This classification is implemented using the **Field Calculator**. A conditional expression is used to evaluate pixel counts and assign a land use label to each grid cell. The expression assumes that if a single land use type exceeds 50% of the total pixels within a grid, that class is designated as the dominant type of land use. However, this threshold may result in some grid cells remaining unclassified when no single category surpasses the 50% threshold. In such cases, manual review and adjustment are recommended to ensure accurate classification.

(i) To identify the dominant land use of a grid, start by opening the **attribute table** of the **ordered layer** from the previous step and launch the **Field Calculator**. Create a new output field, **LANDUSE**, with **Text (string)** as the output field type (Figure 8.53). In the **Expression** box, enter the **CASE** expression indicated in Figure 8.53 to classify each 1 km^2 grid cell based on the dominant land use type.

Figure 8.53: Assigning Grid-Level Land Use Classifications Using Conditional Expressions

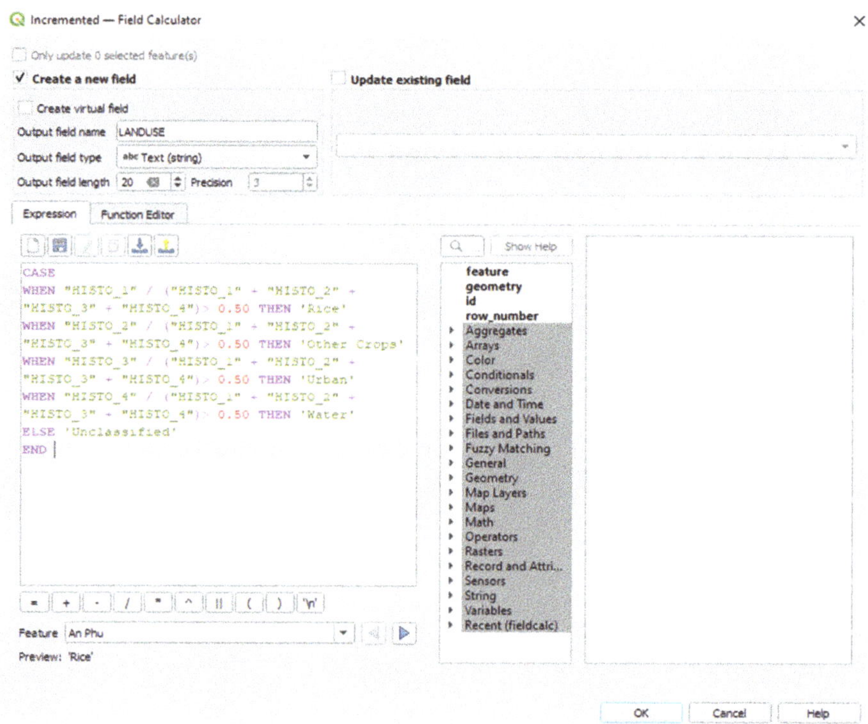

Source: Screenshot generated by the Asian Development Bank using QGIS 3.34.14. 2025.

(ii) The resulting attribute table (with the output field, **LANDUSE**) is shown in Figure 8.54, which presents the land use classifications based on the proportion of each land cover type within a 1 km^2 grid cell. Note the presence of grid cells labeled "**Unclassified**." These occur when no single land use category exceeds the 50% threshold.

For example, **Grid 6** is marked as unclassified due to a mixed distribution of land cover types. In such cases, it is recommended that the user manually inspect the pixel proportions and assign the land use class based on the highest proportion. In this example, the dominant category is **Urban**, and the grid should therefore be reclassified accordingly.

Figure 8.54: Land Use Classifications Based on Pixel Proportions

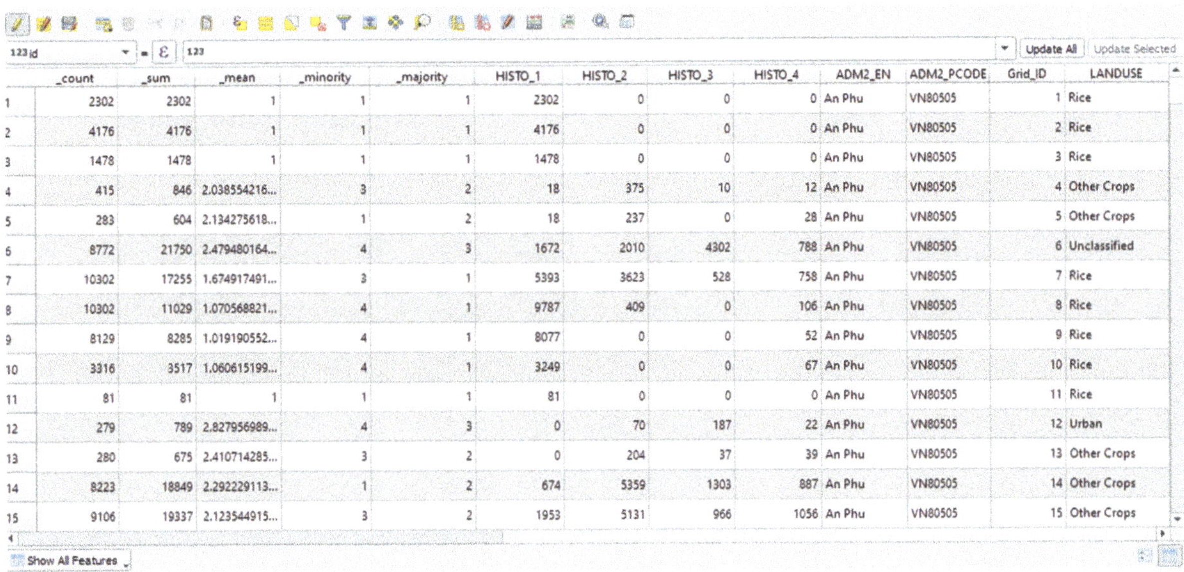

	_count	_sum	_mean	_minority	_majority	HISTO_1	HISTO_2	HISTO_3	HISTO_4	ADM2_EN	ADM2_PCODE	Grid_ID	LANDUSE
1	2302	2302	1	1	1	2302	0	0	0	An Phu	VN80505	1	Rice
2	4176	4176	1	1	1	4176	0	0	0	An Phu	VN80505	2	Rice
3	1478	1478	1	1	1	1478	0	0	0	An Phu	VN80505	3	Rice
4	415	846	2.038554216...	3	2	18	375	10	12	An Phu	VN80505	4	Other Crops
5	283	604	2.134275618...	1	2	18	237	0	28	An Phu	VN80505	5	Other Crops
6	8772	21750	2.479480164...	4	3	1672	2010	4302	788	An Phu	VN80505	6	Unclassified
7	10302	17255	1.674917491...	3	1	5393	3623	528	758	An Phu	VN80505	7	Rice
8	10302	11029	1.070568821...	4	1	9787	409	0	106	An Phu	VN80505	8	Rice
9	8129	8285	1.019190552...	4	1	8077	0	0	52	An Phu	VN80505	9	Rice
10	3316	3517	1.060615199...	4	1	3249	0	0	67	An Phu	VN80505	10	Rice
11	81	81	1	1	1	81	0	0	0	An Phu	VN80505	11	Rice
12	279	789	2.827956989...	4	3	0	70	187	22	An Phu	VN80505	12	Urban
13	280	675	2.410714285...	3	2	0	204	37	39	An Phu	VN80505	13	Other Crops
14	8223	18849	2.292229113...	1	2	674	5359	1303	887	An Phu	VN80505	14	Other Crops
15	9106	19337	2.123544915...	3	2	1953	5131	966	1056	An Phu	VN80505	15	Other Crops

Source: Screenshot generated by the Asian Development Bank using QGIS 3.34.14. 2025.

(iii) Repeat this process for all unclassified grids. Once the review and reclassifications are complete, the layer may be renamed as "**1km Grid Frame**" to facilitate identification and future reference for sample selection.

5. Sample Selection—First Stage

(i) Open the **attribute table** of the **1km Grid Frame layer**. Use the **Select by Expression tool**, , to select all grid cells classified as **Rice** under the **LANDUSE** field. Apply the **Random Extract Within Subsets** tool from the **Processing Toolbar** to randomly select two rice grids from each of the 11 districts within the Rice-only layer.

Ensure that the **Selected features only** option is checked. Under **ID field**, use either the district ID (**ADM2_PCODE**) or district name (**ADM2_EN**) as the grouping field. Set the number of selected features to 2 to randomly draw two Rice grids from each of the 11 districts. Then click **Run** (Figure 8.55).

Figure 8.55: Sample Selection Using the Random Extract Within Subsets Tool

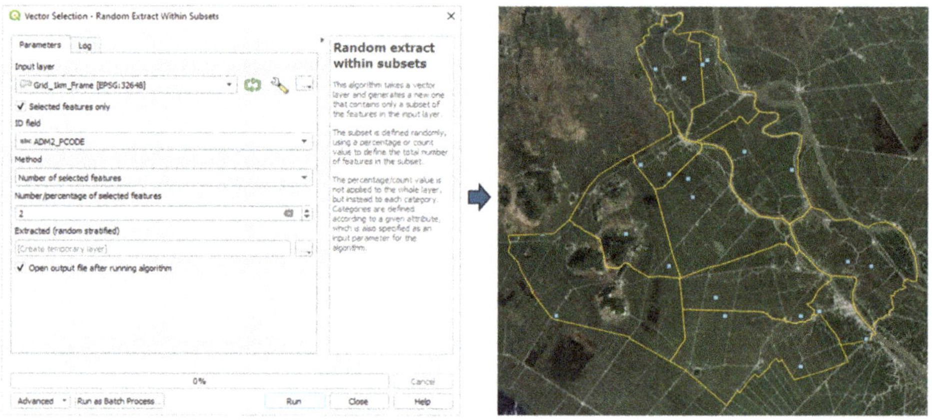

Source: Screenshot generated by the Asian Development Bank using QGIS 3.34.14. 2025.

This process ensures spatial coverage across administrative units. The result is a total of **22 Rice grids**, two for each of the 11 Rice-only districts in An Giang Province.

Since the numbers of grids and samples vary by land use category (Figure 8.56), repeat this process for the remaining classes—**Other Crops**, **Urban**, and **Water**—using the **Random Extract** tool, instead. These selections must be made across the entire province, not by district.

*Note: An alternative, though more manual, approach is to create individual layers for each land use class and perform the selections separately using the **Random Selection Within Subsets** tool for rice grids and the **Random Selection** tool for the other classes.*

Figure 8.56: Grid Layers by Land Classification

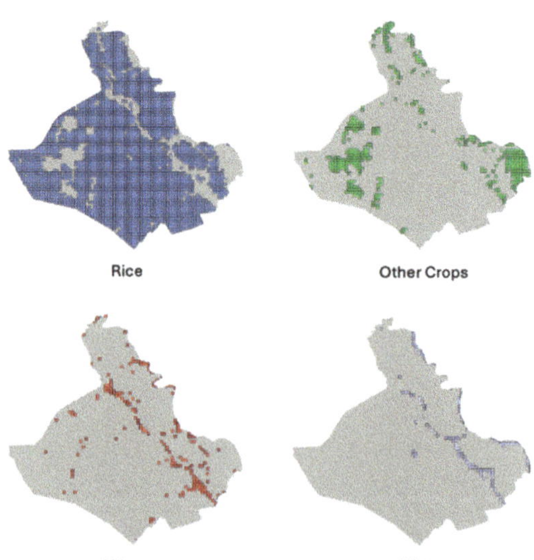

Note: Each grid cell represents 1-square-kilometer (km^2).
Source: Screenshot generated by the Asian Development Bank using QGIS 3.34.14. 2025.

(ii) After selecting the samples from all the land use classes, combine the individual layers into a single layer while preserving all the attribute information; this is done by using the **Merge Vector Layers** tool. To access the tool, go to **Processing | Toolbox**, and then search for **Merge Vector Layers**. Alternatively, you can navigate to **Vector | Data Management Tools | Merge Vector Layers**. Select the layers to be merged, which include the Rice layer (22 grids), Other Crops layer (6 grids), Urban layer (2 grids), and Water layer (4 grids).

Once the tool is executed, the result is a unified layer containing **34 features**, each retaining its original attribute data, as illustrated in the **attribute table** shown in Figure 8.57.

Figure 8.57: Merged Layer Created from All the Land-Classification Layers

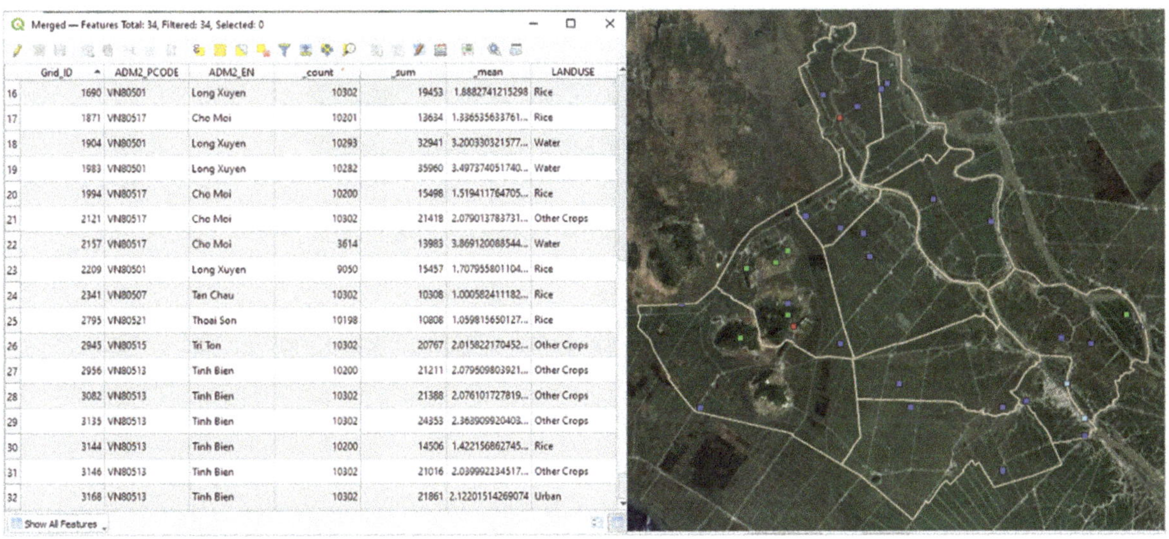

	Grid_ID	ADM2_PCODE	ADM2_EN	_count	_sum	_mean	LANDUSE
16	1690	VN80501	Long Xuyen	10302	19453	1.8882741215298	Rice
17	1871	VN80517	Cho Moi	10201	13634	1.336535633761...	Rice
18	1904	VN80501	Long Xuyen	10293	32941	3.200330321577...	Water
19	1983	VN80501	Long Xuyen	10282	35960	3.497374051740...	Water
20	1994	VN80517	Cho Moi	10200	15498	1.519411764705...	Rice
21	2121	VN80517	Cho Moi	10302	21418	2.079013783731...	Other Crops
22	2157	VN80517	Cho Moi	3614	13983	3.869120088544...	Water
23	2209	VN80501	Long Xuyen	9050	15457	1.707955801104...	Rice
24	2341	VN80507	Tan Chau	10302	10308	1.000582411182...	Rice
25	2795	VN80521	Thoai Son	10198	10808	1.059815650127...	Rice
26	2945	VN80515	Tri Ton	10302	20767	2.015822170452...	Other Crops
27	2956	VN80513	Tinh Bien	10200	21211	2.079509803921...	Other Crops
28	3082	VN80513	Tinh Bien	10302	21388	2.076101727819...	Other Crops
29	3135	VN80513	Tinh Bien	10302	24353	2.363909920403...	Other Crops
30	3144	VN80513	Tinh Bien	10200	14506	1.422156862745...	Rice
31	3146	VN80513	Tinh Bien	10302	21016	2.039992234517...	Other Crops
32	3168	VN80513	Tinh Bien	10302	21861	2.12201514269074	Urban

Source: Screenshot generated by the Asian Development Bank using QGIS 3.34.14. 2025.

6. Sample Selection—Second Stage

(i) Lay a **200 m² grid** over each of the selected 1 km² grid cells (Figure 8.58). Use the extent of the **AnGiang_District** layer, ensuring that it aligns with the extent of the grid cells. **Clip** the 200 m² grid layers to the merged 34-feature grid layer to ensure that only the areas within the selected sample grids are retained. These 200 m² grids will be referred to as the "**Sub-Grids**."

Figure 8.58: Sub-Grids Clipped into 1-Square-Kilometer Sample Grid Cells

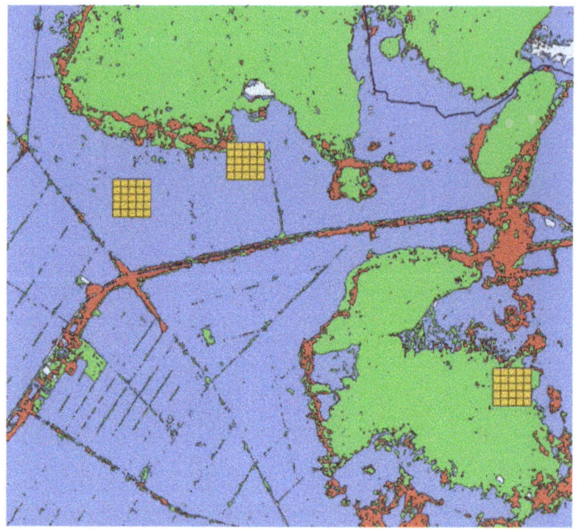

Note: Each sub-grid represents 200 square meters (m^2).
Source: Screenshot generated by the Asian Development Bank using QGIS 3.34.14. 2025.

(ii) The **resulting clipped layer** does not contain attributes, such as land use, district ID, district names, or the ID of the 1 km^2 grid cells. To include these attributes, use the **Intersection** tool to intersect the 200 m^2 clipped layer with the 34-feature grid layer (Figure 8.59).

Figure 8.59: Intersecting a Clipped Layer with a 34-Feature Grid Layer

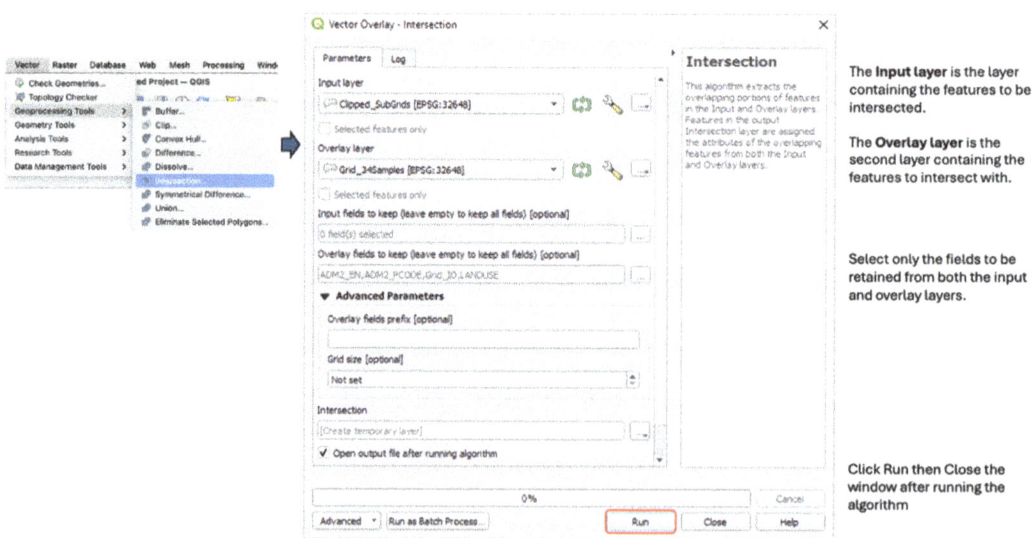

Note: The clipped layer represents 200 square meters (m^2).
Source: Screenshot generated by the Asian Development Bank using QGIS 3.34.14. 2025.

Note: At this stage, multiple layers with similar names (e.g., with "grid," "clipped," and "joined") may be present in the Layers panel. It is recommended that the user rename these layers to clearly distinguish them from one another.

(iii) Identify the 200 m^2 sub-grids and assign each a number that links it to its parent **Grid_ID**. Then create a new ID field named "**SubGrid_ID**." Figure 8.60 shows the numbering system used by QGIS for assigning grid and sub-grid IDs.

Figure 8.60: Numbering of Sub-Grids in QGIS

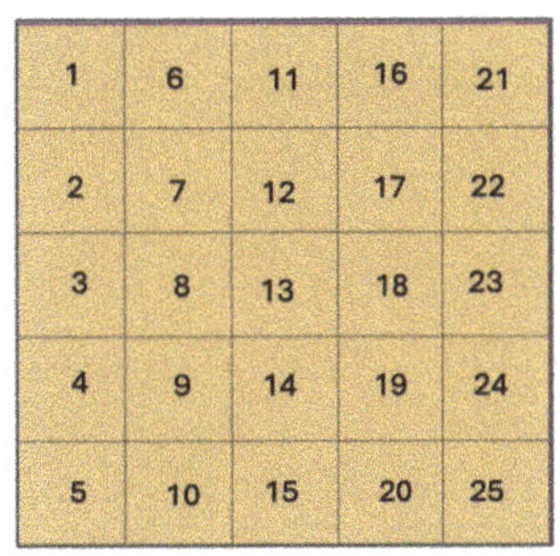

Source: Screenshot generated by the Asian Development Bank using QGIS 3.34.14. 2025.

To proceed, right-click on the **Sub-Grid (Intersection) layer**, open the **attribute table**, and click on the **Field Calculator** icon, . In the dialog box, check **Create a new field**, name the output field **SubGrid_ID**, and set the output field type to **Whole Number (integer)**; also, indicate **10** for the output field length. In the **Expression box**, enter **array_find (array_agg ($id, "Grid_ID"), $id) + 1** (Figure 8.61).

Figure 8.61: Assigning Hierarchical Identifiers Using the Field Calculator

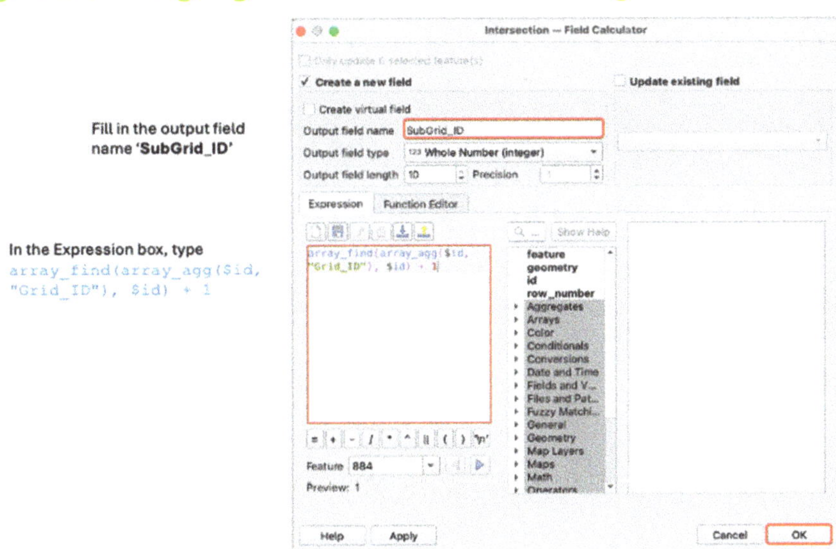

Source: Screenshot generated by the Asian Development Bank using QGIS 3.34.14. 2025.

The terms used in the **Expression box** have the following functions:

- **(array_agg ($id, "Grid_ID"):** This part of the expression groups all feature IDs ($id) into arrays or lists of values, one for each unique Grid_ID. It essentially collects the IDs of all the sub-grids that belong to the same parent grid.
- **array_find:** This function searches for the current feature's $id within each array created by array_agg. It returns its index, or the position number of that ID within the array.
- **+ 1:** Since arrays in QGIS start counting from zero, adding 1 adjusts the index so that it starts from 1 (e.g., the first sub-grid will be numbered 1, instead of 0).

Figure 8.62 displays the attribute table showing sub-grids 1–25 linked to Grid 1951, with sub-grids 1 to 5 selected, as reflected in the map view.

Figure 8.62: Grid Identification in the Map View

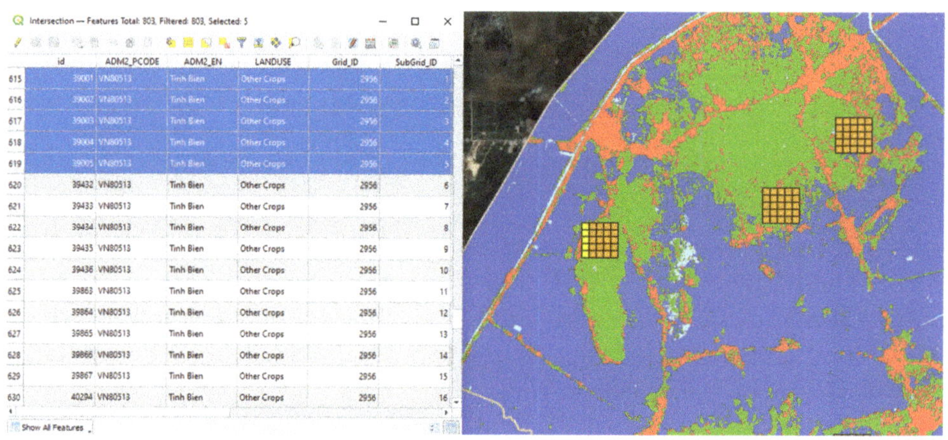

Source: Screenshot generated by the Asian Development Bank using QGIS 3.34.14. 2025.

(iv) Select four 200 m² sub-grids. Figure 8.63 illustrates the spatial distribution of the sub-grids to be selected, ensuring that they are not too close to one another geographically.

Figure 8.63: Locations of the Sub-Grids to Be Selected

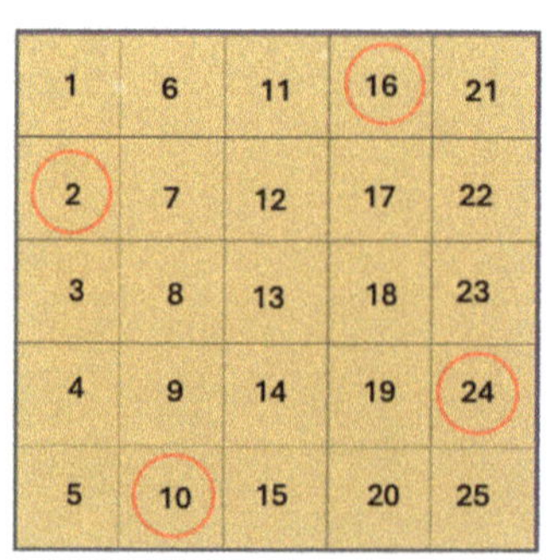

Source: Screenshot generated by the Asian Development Bank using QGIS 3.34.14. 2025.

To proceed, open the **attribute table** of the 200 m^2 sample layer. Click on the icon for the **Select by Expression tool**, , which you will use to identify and select the target sub-grids (Figure 8.64).

Figure 8.64: Identifying Sub-Grids Using the Select by Expression Tool

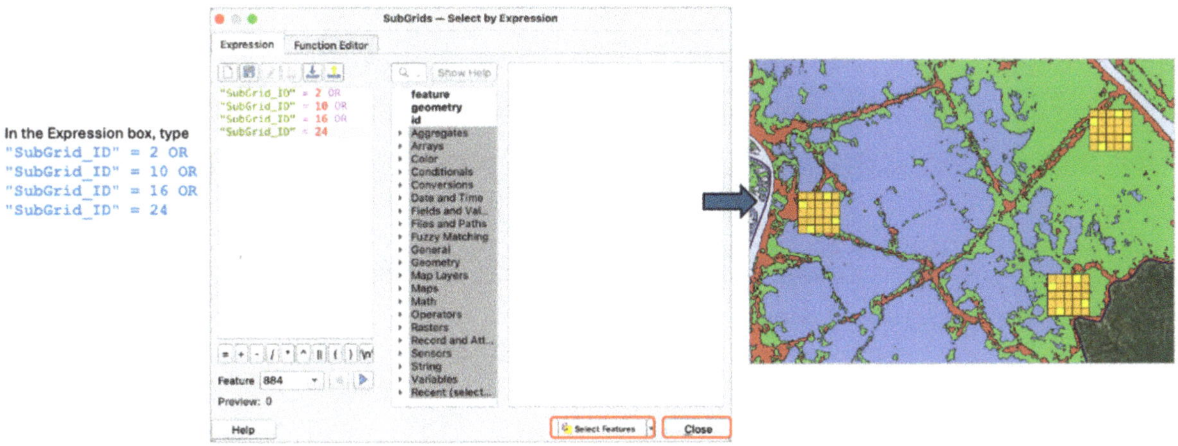

Source: Screenshot generated by the Asian Development Bank using QGIS 3.34.14. 2025.

Export the selected sub-grids and save them as a new shapefile.

Note that grids located partially within the area or along its boundaries may result in a different total grid count. Consequently, the numbering of sub-grids may also vary. These variations can affect both grid numbering and the subsequent sampling or sample locations. It is recommended that users consider alternative approaches for handling such scenarios.

7. Sample Selection—Third Stage

(i) Overlay a 50 m^2 grid onto the selected sample of sub-grids. Use the extent of the **AnGiang_District** layer, ensuring that its spatial coverage aligns with the overall extents of both the 1 km^2 grid and the 200 m^2 sub-grid layers. **Clip** the 50 m^2 grid layer to the sample sub-grid layer to retain only the areas within the selected sample sub-grids. This 50 m^2 grid will be referred to as the "**Segment**" (Figure 8.65).

Figure 8.65: Clipping a Segment into 200-Square-Meter Sample Sub-Grids

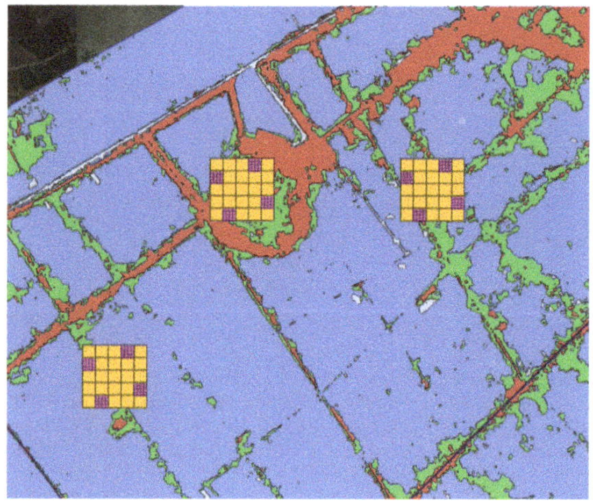

Note: The segment represents 50 square meters (m^2).
Source: Screenshot generated by the Asian Development Bank using QGIS 3.34.14. 2025.

(ii) The resulting clipped layer does not include attributes such as land use, district ID, district names, the ID of the 1 km^2 grid cell, or the numbering of sub-grids from the previous sampling. To add these attributes, use the **Intersection tool** to intersect the **50 m^2 clipped layer** with the **200 m^2 sample sub-grid layer** (Figure 8.66).

Figure 8.66: Intersecting a Clipped Layer with a 200-Square-Meter Sub-Grid Layer

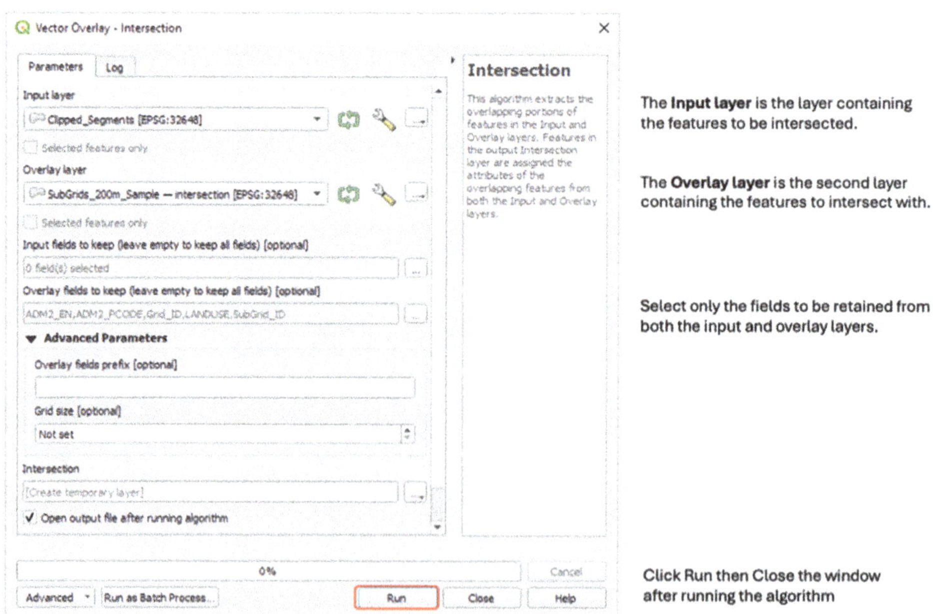

Note: The clipped layer represents 50 square meters (m^2).
Source: Screenshot generated by the Asian Development Bank using QGIS 3.34.14. 2025.

(iii) Identify the 50 m^2 segments and assign each a number that links it to its parent grid and sub-grid. Figure 8.67 shows the numbering system used by QGIS for assigning segment IDs.

Figure 8.67: Numbering of Segments in QGIS

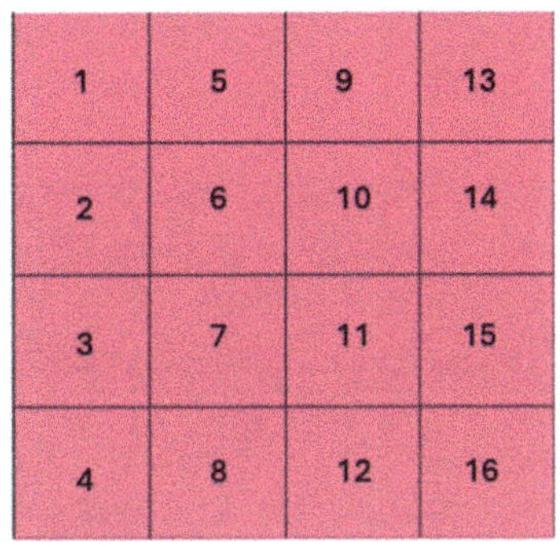

Source: Visualization by the Asian Development Bank. 2025.

Create a new ID field name: **Segment_ID**. Right-click on the **Segments (Intersection) layer**, open the **attribute table**, and click on the icon to access the **Field Calculator**. In the dialog box, check **Create a new field**, name the output field **Segment_ID**, and set the output field type to **Whole Number (integer)**; also indicate **10** for the output field length. In the Expression box, enter **array_find (array_agg ($id, concat ("Grid_ID", "SubGrid_ID")), $id) + 1** (Figure 8.68).

Figure 8.68. Assigning Nested Grid Identifiers Using the Field Calculator

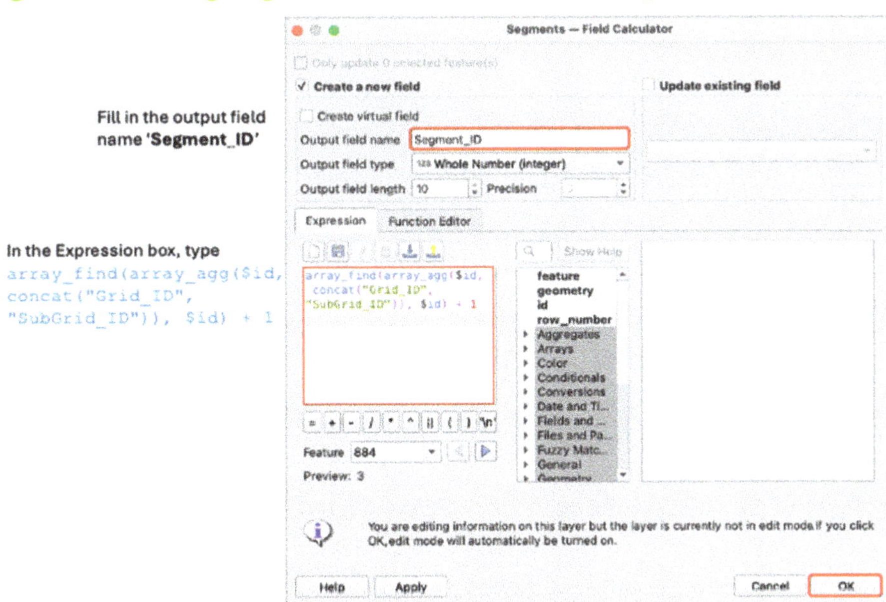

Source: Screenshot generated by the Asian Development Bank using QGIS 3.34.14. 2025.

The terms used in the **Expression box** have the following functions:

- **concat ("GridID", "SubGrid_ID")):** This function combines the parent-grid ID, sub-grid ID, and segment numbers into a single string, separated by underscores.
- **array_agg ($id, concat ("GridID", "SubGrid_ID")):** This function groups all feature IDs through the combination of Grid_ID and SubGrid_ID.
- **array_find (..., $id) + 1:** This finds the position of the current feature within its group and adjusts it to a 1-based index.

Figure 8.69 displays the attribute table showing segments 1–16 linked to Grid 1277, Sub-Grid 2, as reflected in the map view.

Figure 8.69: Segment Identification in the Map View

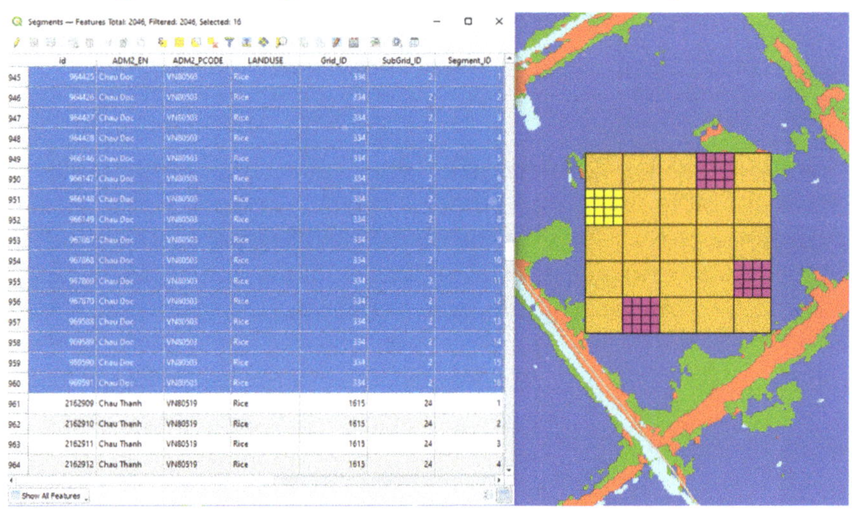

Source: Screenshot generated by the Asian Development Bank using QGIS 3.34.14. 2025.

(iv) Select two 50 m^2 segments. Figure 8.70 illustrates the spatial distribution of these segments, ensuring that they are not too close to one another geographically.

Figure 8.70: Locations of the Segments to Be Selected

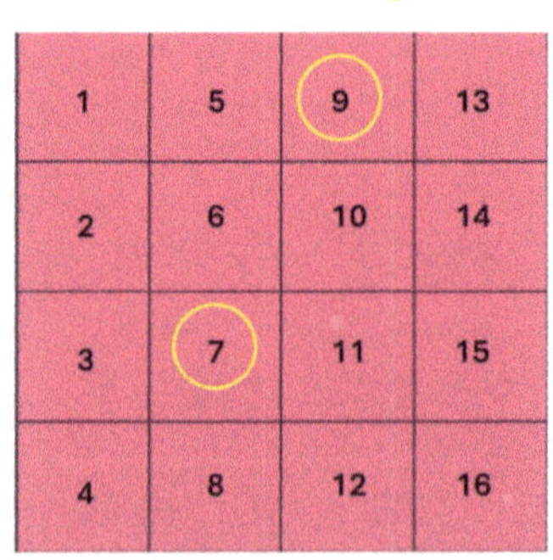

Source: Visualization by the Asian Development Bank. 2025.

To proceed, open the **attribute table** of the **50 m² sample layer**. Use the **Select by Expression tool** to identify and select the target segments (Figure 8.71).

Figure 8.71: Identifying Segments Using the Select by Expression Tool

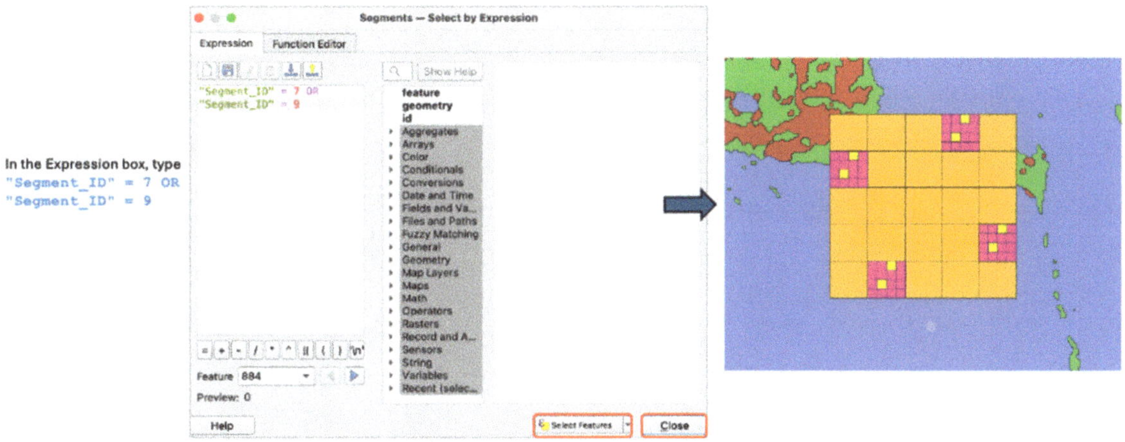

Source: Screenshot generated by the Asian Development Bank using QGIS 3.34.14. 2025.

Export the selected features and save them as a new shapefile.

8. Adding Centroids to the Sampling Units

(i) Add centroids to each selected sample segment. Click **Vector | Geometry Tools | Centroids**. Alternatively, open the **Processing Toolbox**, then search for **Centroids**, and select the appropriate tool. Set the sample segment layer as the input layer. Choose an output location or set it as temporary, then click **Run** (Figure 8.72).

Figure 8.72: Adding Centroids to Selected Sample Segments

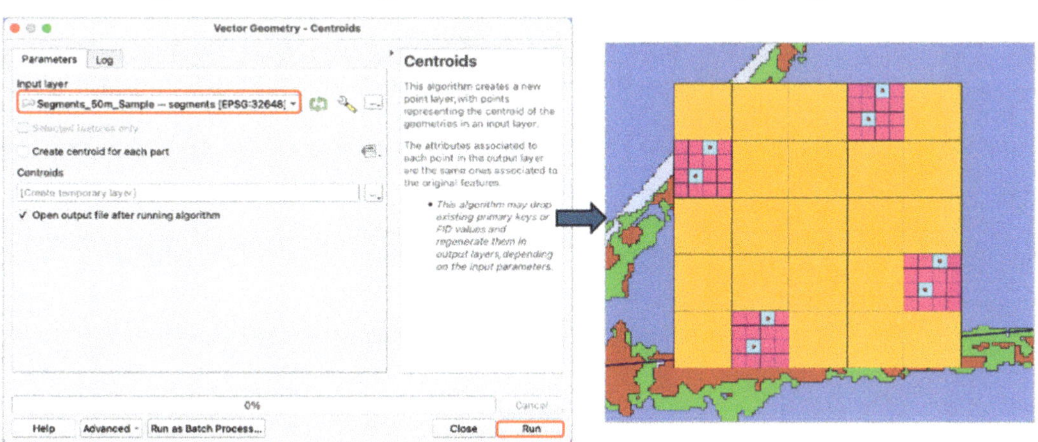

Source: Screenshot generated by the Asian Development Bank using QGIS 3.34.14. 2025.

(ii) To assign coordinates to each centroid, go to **Processing | Toolbox** and search for the tool **Add X/Y fields to layer**. Select the **centroid layer** as the input, then choose whether to save the output as a temporary or permanent layer. Click **Run** to start the process.

The resulting layer will include the original attributes along with the newly added coordinate fields, as shown in Figure 8.73.

Figure 8.73: Output Attribute Table with Added X/Y Fields for Centroids

Added fields — Features Total: 257, Filtered: 257, Selected: 0

	Grid_ID	SubGrid_ID	Segment_ID	ADM2_EN	ADM2_PCODE	LANDUSE	x	y
1	2956	10	9	Tinh Bien	VN80513	Other Crops	104.9453007288	10.5744469535
2	2157	10	9	Cho Moi	VN80517	Water	105.5667120642	10.4835041650
3	446	24	9	Chau Phu	VN80511	Rice	105.1336289600	10.6304975346
4	558	16	7	Tinh Bien	VN80513	Rice	105.0947216520	10.4631929023
5	2795	10	7	Thoai Son	VN80521	Rice	105.3466932998	10.2568102239
6	558	2	9	Tinh Bien	VN80513	Rice	105.0896956642	10.4622899254
7	2945	2	7	Tri Ton	VN80515	Other Crops	104.9338966674	10.4704355698
8	203	24	9	An Phu	VN80505	Rice	105.1245677634	10.8294724422
9	643	24	7	Chau Phu	VN80511	Rice	105.1422972948	10.5934124782
10	1455	24	9	Thoai Son	VN80521	Rice	105.2065895807	10.3591306747
11	952	2	9	Phu Tan	VN80509	Rice	105.2360705196	10.6883196766
12	952	16	7	Phu Tan	VN80509	Rice	105.2411005884	10.6892202458
13	2956	10	7	Tinh Bien	VN80513	Other Crops	104.9448438418	10.5735424489
14	1904	10	7	Long Xuyen	VN80501	Water	105.4473314588	10.3923492804
15	1904	24	7	Long Xuyen	VN80501	Water	105.4528152737	10.3941504056
16	334	24	9	Chau Doc	VN80503	Rice	105.0422064437	10.6576557710

Show All Features

Source: Screenshot generated by the Asian Development Bank using QGIS 3.34.14. 2025.

(iii) To export the **attribute table** of sampling units that include coordinate fields, right-click on the layer containing the added X and Y fields in the **Layers panel**. From the menu, select **Export**, then choose **Save Features As**. In the export dialog, set the format to **Comma Separated Value (CSV)** and specify the desired file name and destination folder. To automatically add the exported file to the current QGIS project, enable the option **Add saved file to map**. Click **OK** to complete the export process.

Figure 8.74 shows a sample 1 km^2 grid map that indicates the locations of 200 m^2 sub-grids, 50 m^2 segments, and the centroids within each grid cell for ground validation.

Figure 8.74: Sample Grid for Ground Validation

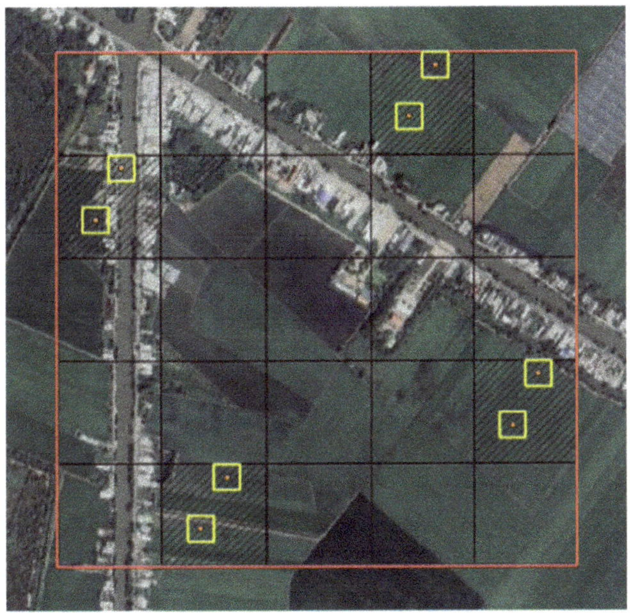

Source: Screenshot generated by the Asian Development Bank using QGIS 3.34.14. 2025.

The resulting CSV file will contain all the relevant attributes, including segment identifiers and centroid coordinates, making it suitable for ground validation and further spatial analysis.

References

ADB (Asian Development Bank). 2024. *Plotting from Above: Enhancing Agricultural Mapping in Asia and the Pacific.* https://dx.doi.org/10.22617/TCS240326-2

Buckley, A. 2010. Using and Citing Esri Data. *Esri: ArcGIS Blog.* 3 December. https://www.esri.com/arcgis-blog/products/product/mapping/using-and-citing-esri-data.

Burgard, A. et al. 2025. *Mapping Rice in Viet Nam: Estimating Rice Area Using Remote Sensing and Area Frame Survey Methods in An Giang Province.* ADB. https://www.adb.org/publications/mapping-rice-viet-nam.

Campbell, J., and M. Shin. 2011. *Essentials of Geographic Information Systems.* Saylor Foundation. https://open.umn.edu/opentextbooks/textbooks/67.

Chatterjee, S. 2018. Getting the Most out of Zonal Statistics. *Esri: ArcGIS Blog.* 2 July. https://www.esri.com/arcgis-blog/products/spatial-analyst/analytics/getting-the-most-out-of-zonal-statistics.

De la Rúa, L., P. Bright, and G. Le Gall-Queguineur. 2020. *QGIS for Census and Survey Mapping: Training Manual.* Secretariat of the Pacific Community. https://sdd.spc.int/digital_library/qgis-census-and-survey-mapping-training-manual.

Dempsey, C. 2024. Types of GIS Data Explored: Vector and Raster. *Geography Realm.* 12 August. https://www.geographyrealm.com/geodatabases-explored-vector-and-raster-data/.

Esri (Environmental Systems Research Institute, Inc.). 1998. *ESRI Shapefile Technical Description.* An ESRI White Paper. https://www.esri.com/content/dam/esrisites/sitecore-archive/Files/Pdfs/library/whitepapers/pdfs/shapefile.pdf.

Esri. 2012. ArcGIS Desktop 9.3 Help: Using the ArcGIS Spatial Analyst Toolbar to Create a Zonal Histogram. https://webhelp.esri.com/arcgisdesktop/9.3/index.cfm?TopicName=Using_the_ArcGIS_Spatial_Analyst_toolbar_to_create_a_Zonal_Histogram.

Esri. n.d.-a Georeference CAD Data. https://pro.arcgis.com/en/pro-app/latest/help/data/cad/georeferencing-cad-data.htm.

Esri. n.d.-b GIS Dictionary: GPS. https://support.esri.com/en-us/gis-dictionary/gps.

Esri. n.d.-c. What is GIS? https://www.esri.com/en-us/what-is-gis/overview.

Esri. n.d.-d. XYZ Tiles Satellite Map. https://server.arcgisonline.com/ArcGIS/rest/services/World_Imagery/MapServer/tile/{z}/{y}/{x}.

European Space Agency (ESA). WorldCover V2 2021: Layers. https://viewer.esa-worldcover.org/worldcover/ (accessed 16 September 2025).

FAA (Government of the United States, Department of Transportation, Federal Aviation Administration). n.d. Satellite Navigation - Global Positioning System (GPS). https://www.faa.gov/about/office_org/headquarters_offices/ato/service_units/techops/navservices/gnss/gps.

Food and Agriculture Organization (FAO) of the United Nations. 2017. World Programme for the Census of Agriculture 2020: Volume 1; Programme, Concepts and Definitions. *FAO Statistical Development Series*. No. 15. FAO. https://www.fao.org/3/i4913e/i4913e.pdf.

Food and Agriculture Organization (FAO) of the United Nations. 2018. World Programme for the Census of Agriculture 2020: Volume 2; Operational Guidelines. *FAO Statistical Development Series*. No. 16. FAO. https://www.fao.org/3/CA1963EN/ca1963en.pdf.

Food and Agriculture Organization (FAO) of the United Nations, Global Strategy to Improve Agricultural and Rural Statistics (GSARS). 2017. *Handbook on Remote Sensing for Agricultural Statistics.* GSARS. https://www.fao.org/3/ca6394en/ca6394en.pdf.

Food and Agriculture Organization (FAO) of the United Nations. The Survey Design. https://www.fao.org/economic/the-statistics-division-ess/world-census-of-agriculture/conducting-of-agricultural-censuses-and-surveys/chapter-6-the-survey-design/en/.

Food and Agriculture Organization (FAO) of the United Nations. n.d. Geospatial Information for Sustainable Food Systems. https://www.fao.org/geospatial/en/

Food and Agriculture Organization (FAO) of the United Nations. n.d. Geospatial Data | FAO's Data Lab. https://www.fao.org/datalab/geospatial-data/en

Free Software Foundation. GNU Operating System. https://www.gnu.org.

Gandhi, U. 2024. QGIS Tutorials and Tips: Working with Attributes (QGIS3). https://www.qgistutorials.com/en/docs/3/working_with_attributes.html.

Garmin Ltd. 2020. *ETREX® 10/20/20X/30/30X: Owner's Manual.* https://www8.garmin.com/manuals/webhelp/eTrex_10_20x_30x/EN-US/eTrex_10_20_20x_30_30x_OM_EN-US.pdf.

Garmin Ltd. n.d.-a. About Us. https://www.garmin.com/en-US/company/about-garmin/.

Garmin Ltd. n.d.-b. eTrex® 20x. https://www.garmin.com/en-US/p/518046/.

Garmin Ltd. n.d.-c. What are Tracks, Routes, Waypoints, Activities, and Courses on Garmin Outdoor Handhelds? https://support.garmin.com/en-US/?faq=uJZh4J9rF20RVCRNfCtXV9.

GISDATA.io Docs. Raster Data. https://docs.gisdata.io/raster-data.html.

Government of Japan, Japan Aerospace Exploration Agency (JAXA). n.d. High-Resolution Land Use and Land Cover Map of Vietnam for 2020 (Released in September 2023 / Version 23.09). https://www.eorc.jaxa.jp/ALOS/en/dataset/lulc/lulc_vnm_v2309_e.htm (accessed December 2024).

Government of Viet Nam, General Statistics Office. 2024. *Statistical Yearbook of Viet Nam 2023.* https://www.nso.gov.vn/wp-content/uploads/2024/06/NIEN-GIAM-THONG-KE-2023_Ban-quyen.pdf.

Humanitarian Data Exchange (HDX). https://data.humdata.org (accessed 11 October 2024).

Humanitarian Data Exchange (HDX). Viet Nam Buildings (OpenStreetMap Export). https://data.humdata.org/dataset/hotosm_vnm_buildings (accessed 4 February 2025).

Humanitarian Data Exchange (HDX). Viet Nam Points of Interest (OpenStreetMap Export). https://data.humdata.org/dataset/hotosm_vnm_points_of_interest (accessed 4 February 2025).

Humanitarian Data Exchange (HDX). Viet Nam Roads (OpenStreetMap Export). https://data.humdata.org/dataset/hotosm_vnm_roads (accessed 4 February 2025).

Humanitarian Data Exchange (HDX). Viet Nam–Subnational Administrative Boundaries. https://data.humdata.org/dataset/cod-ab-vnm (accessed 11 October 2024).

Kraak, M.J. et al. 2020. *Mapping for a Sustainable World.* United Nations / International Cartographic Association. https://digitallibrary.un.org/record/3898826?ln=en.

Lee, S. 2025. The Ultimate Guide To Raster Data in GIS: Unlocking the Power of Raster Data for Spatial Analysis and Mapping. *Number Analytics.* 27 May. https://www.numberanalytics.com/blog/ultimate-guide-raster-data-gis.

Lim, H. 2016. Raster Data. In S. Shekhar, H. Xiong, and X. Zhou, eds. *Encyclopedia of GIS.* Springer. pp. 1–7. https://doi.org/10.1007/978-3-319-23519-6_1080-2.

Malki, M. al-. 2011. *Combining Geospatial and Statistical Data for Analysis & Dissemination (with Special Reference to Qatar Census 2020)* . Presentation at the United Nations Committee of Experts on Global Geospatial Information Management (UN-GGIM) thematic seminar on Integrating Statistical and Geospatial Information: Issues and Challenges. 27 October. https://ggim.un.org/meetings/2011-Seminar_ISGI-Korea/documents/2.%20Qatar.pdf.

Partnership in Statistics for Development in the 21st Century (PARIS21) and Statistics Sweden. 2021. *Guide on Geospatial Data Integration in Official Statistics.* https://www.paris21.org/geospatial.

Pebesma, E. 2017. Tidying Feature Geometries with sf. *r-spatial.org.* 19 March. https://r-spatial.org/r/2017/03/19/invalid.html.

Qader, S. et al. 2019. A Novel Approach to the Automatic Designation of Predefined Census Enumeration Areas and Population Sampling Frames: A Case Study in Somalia. *Policy Research Working Papers.* No. 8972. World Bank. https://documents1.worldbank.org/curated/en/810771565268760366/

pdf/A-Novel-Approach-to-the-Automatic-Designation-of-Predefined-Census-Enumeration-Areas-and-Population-Sampling-Frames-A-Case-Study-in-Somalia.pdf.

QGIS. http://www.qgis.org.

QGIS Documentation. 2024. A Gentle Introduction to GIS: 8. Coordinate Reference Systems. https://docs.qgis.org/3.34/en/docs/gentle_gis_introduction/coordinate_reference_systems.html.

QGIS Project. 2025a. A Gentle Introduction to GIS. QGIS Documentation. https://docs.qgis.org/3.40/en/docs/gentle_gis_introduction/index.html.

QGIS Project. 2025b. A Gentle Introduction to GIS: 7. Topology. QGIS Documentation. https://docs.qgis.org/latest/en/docs/gentle_gis_introduction/topology.html.

QGIS Project. 2025c. QGIS Training Manual: 4.1 Lesson; Using Print Layout. QGIS Documentation. https://docs.qgis.org/3.40/en/docs/training_manual/map_composer/map_composer.html#follow-along-the-layout-manager.

QGIS Project. 2025d. QGIS User Guide: 9.2. List of Functions. QGIS Documentation. https://docs.qgis.org/latest/en/docs/user_manual/expressions/functions_list.html.

QGIS Project. 2025e. QGIS User Guide: 12. Working with Vector Data. QGIS Documentation. https://docs.qgis.org/3.40/en/docs/user_manual/working_with_vector/index.html.

QGIS Project. 2025f. QGIS User Guide: 12.2. Working with the Attribute Table. QGIS Documentation. https://docs.qgis.org/3.40/en/docs/user_manual/working_with_vector/attribute_table.html.

QGIS Project. 2025f. QGIS User Guide: 25.2.2. Geometry Checker Plugin. QGIS Documentation. https://docs.qgis.org/latest/en/docs/user_manual/plugins/core_plugins/plugins_geometry_checker.html.

UN DESA (United Nations Department of Economic and Social Affairs). 2009. *Handbook on Geospatial Infrastructure in Support of Census Activities*. https://unstats.un.org/unsd/demographic/standmeth/handbooks/Series_F103en.pdf.

UN DESA (United Nations Department of Economic and Social Affairs). 2019. *The Global Statistical Geospatial Framework*. http://ggim.un.org/meetings/GGIM-committee/9th-Session/documents/The_GSGF.pdf.

UN Sustainable Development Goals Knowledge Platform. Geospatial Information for Sustainable Development. https://sustainabledevelopment.un.org/topics/informationforintegrateddecision-making/geospatialinformation.

United Nations Human Settlements Programme (UN-Habitat). 2013. *GIS Handbook for Municipalities*. https://unhabitat.org/gis-handbook-for-municipalities.

www.ingramcontent.com/pod-product-compliance
Lightning Source LLC
Chambersburg PA
CBHW041753220926
51847CB00014B/186
9789292776152